A Textbook of Applied Aquatic Biology

Dr. B.B. Hosetti (b. 12-1-1957) M.Sc; Ph.D. Reader and Chairman, Department of Applied Zoology, Kuvempu University, B.R. Project, Shimoga District, Karnataka, possesses many years of experience as an outstanding teacher and researcher. He was a recipient of all the meritorial fellowship awards of CSIR, New Delhi, India (JRF, SRF and RA).

He did his Ph.D. on the topic related to biological treatment of wastewater using stabilization pond process from Karnatak University, Dharwad in 1987. He has received Young Scientist Award from the Society of Biosciences, India during 1989-90 for the work on wastewater enzyme assay. Worked as visiting faculty member at the Environmental Resources Department, University of Salford, U.K. under Leverhulme Fellowship Programme during 1990-91.

He has written two books viz., *Concepts in Wildlife Management* and *Environmental Impact Assessment and Management* published from the Daya Publishing House, New Delhi. He has written more than 100 research papers and reviews in national and international journals. Recently A Review: Aquatic plants for toxicity assessment published in the Environmental Research, Section A81-259-75 (1999) by the Academic Press. He has been nominated as Research Board of Advisors for the ABI Institute for Who's Who in the World, U.S.A. He has also successfully guided a few candidates for Ph.D. and M.Phil. degree. Presently he is researching on the Ecotoxicology of Pentatomids bugs, Cyclopelta cissifolia and Ecobiology of Weaver birds of Western Ghats of Karnataka State. His special interests include: Environmental Pollution Abatement and Wildlife Management

Dr. Arvind Kumar (b. 23-3-1953) Professor and Head, Department of Zoology, S.K. University, Dumka, Bihar has 23 years of experience as an outstanding teacher and researcher. He has also a recipient of meritorial fellowship of CSIR through his College and University years. Dr. Kumar did M.Sc. in Zoology in the year 1976 and obtained Ph.D. in 1984 from Bhagalpur University respectively. His field of specialization include limnology, aquatic insect ecology and biodiversity.

Dr. Kumar has guided many candidates for Ph.D. and published 130 research papers in national and international journals. He is also on the Editorial Board of many journals and books. He is also the Editor of the Journal titled *Indian Journal of Environment and Ecoplanning* and he is also serving as Chairman of Board of Environmental Impact Assessment and Ecoplanning, Bihar.

A Textbook of Applied Aquatic Biology

B.B. Hosetti
M.Sc; Ph.D., Reader
Department of Applied Zoology
Kuvempu University, Karnataka

Arvind Kumar
Professor and Head
Department of Zoology
S.K. University, Bihar.

2016
Daya Publishing House®
A Division of
Astral International Pvt. Ltd.
New Delhi - 110 002

First Published, 2002
Reprinted, 2016

ISBN: 978-93-5130-988-8 (International Edition)

Published by : **Daya Publishing House®**
A Division of
Astral International Pvt. Ltd.
– ISO 9001:2008 Certified Company –
4760-61/23, Ansari Road, Darya Ganj
New Delhi-110 002
Ph. 011-43549197, 23278134
E-mail: info@astralint.com
Website: www.astralint.com

Laser Typesetting : **Classic Computer Services**
Delhi - 110 035

Printed at : **Sanat Printers**

**DEDICATED
TO**

Educationist, Academician
and a great Scholar

DR. M.I. SAVADATTI
Former Vice Chancellor
Mangalore University
and U.G.C. Member

Acknowledgments

We owe our thanks to the following persons, academicians, professionals, editors, researchers and journals.

1. Professor S.P. Hiremath, Vice-Chancellor, Kuvempu University, Shankarghatta, Shimoga, Karnataka.
2. Dr. S.K. Konar, Editor, Environment and Ecology, Kalyani, West Bengal.
3. Dr. T.C. Khatri, Scientist, Zoological Society of India, Chennai.
4. Dr. J.S. Munshi, Professor of Botany, Bhagalpur University, Bhagalpur and Dr. J.S. Datta Munshi Emeritus Professor of Zoology, Bhagalpur University, Bhagalpur.
5. Dr. Mahadev B. Nadoni, Head Department of Zoology, Sri Siddartha First Grade College, Tumkur.
6. Professor H.S. Patil, Department of Zoology, Karnataka University, Dharwad.
7. Prof V. Vasudev, Department of Applied Zoology, Kuvempu University, Shankarghatta, Shimoga, Karnataka.
8. Dr. Venkateshwaralu, Department of Applied Zoology, Kuvempu University, Shankarghatta, Shimoga, Karnataka.
9. Dr. Baban Ingole, Senior Scientist, National Institute of Oceanography, Goa.
10. Professor Rajendra Chenni, Department of English, Kuvempu University, Shankarghatta, Shimoga, Karnataka.
11. Professor S. Siddappa, Vice-Chancellor, Bangalore University, Bangalore.
12. Professor Uday Pratap Singh, Vice-Chancellor, S.K. University, Dumka.
13. Dr. K.R. Sridhar, Department of Bioscience, Mangalore University, Mangalore.
14. World Health Organisation, Geneva.

The first author would like to thank his students, Mr. Mohan, B.S., Ms. Tharavathy, Mr Lokesh, K.L., Mr. Naveed, A. and Mr. Shasikumar, wife Mrs. Bhaghya Hosetti and sons, Kumar Sachin and Chetan Hosetti.

The second author would like to thank his counterpart, Prof. Kumari Bimla Devi and sons Kumar Pallav and Prasan.

B.B. Hosetti
Arvind Kumar

Foreword

I am pleased to note that Dr. B.B. Hosetti, Department of Applied Zoology, Kuvempu University, in collaboration with Dr. Arvind Kumar, Professor of Zoology, S.K. University Dumka are bringing out a book, ***A Textbook of Applied Aquatic Biology***. Both are eminent teachers and good researchers in the field of aquatic biology and have played a key role in the development of the subject on Environmental Biology in India. This book is the outcome of several years of their teaching and research experience in various aspects of aquatic biology and aquatic toxicology. The book is nicely illustrated. It has a broad coverage on applied aspects of aquatic biology. I hope that all students, teachers and researchers appreciate and find this as a very useful publication.

I look forward to see more books from these authors.

Dr. S.P. Hiremath
Vice Chancellor,
Kuvempu University,
Shankarghatta - 577 457,
Karnataka State
India

Preface

A Textbook of Applied Aquatic Biology has been written covering different topics related to hydrobiology in order to cater to the needs of students and teachers of aquatic biology in Indian Universities and Colleges. A textbook of this kind will be of immense use to all those who study or teach a special course in aquatic biology or as part of their curriculum.

The water sources are exploited by man for his various needs such as domestic use, irrigation, in swimming pools, navigation, fisheries, aquaculture, pearl culture and recreation. In all his activities and due to disposal of sewage and industrial wastes into aquatic systems, no part of hydrosphere is left free from human impact. The pollution problem has jeopardised not only the general fabric of human life but it also affected the flora, fauna and the quality of life.

We having been involved in teaching and research on various aspects of aquatic biology and pollution during the past two decades felt the need to write this book. We thought that we should begin by trying to explain the relationship of hydrosphere with other components of the ecosphere and hence a chapter on biosphere is incorporated as Chapter 1.

Various physico-chemical properties of water and their interaction with the environmental factors and the responses of organisms to their physical environment are narrated in Chapters 2 and 3 respectively. During the 1990s there was a mad race to convert the paddy fields and mangrove swamps into aquaculture ponds to grow the tiger prawn and earn money along the both coasts of India, specially in Andhra Pradesh, Tamil Nadu and Karnataka States. The Coastal Zone Regulation Act, 1992, posed stringent measures and the EIA was extended to aquaculture also. Now all the aquaculture ponds with more than 20 ha area are to be scrutinised by EIA before establishment. The importance of ecology and status of mangrove ecosystems of India is included in Chapter 4.

Another important aspect of aquatic biology is study of interstitial and intertidal organisms. There is very meagre information available on these topics. With a view to discuss this, the topics are dealt in Chapters 5 and 6. Another most fascinating avenue yielding field is pearl culture. In India aquaculture scientists are now producing indigenous pearls using modern techniques. Details about the pearl culture practised and its status is described in Chapter 7.

Chapter 8 embodies information of drift animals and phytoplankters. Pollution problem is a global concern. According to the report of NIO, Goa and NEERI, Nagpur most of the Indian rivers are polluted. These become sewers at least near the cities and towns during summer months when the water volume in the river is low and sewage pollution is more. The water pollutants, their impact, drinking water purification and water pollution abatement methods adopted in India and elsewhere are illustrated in Chapter 9.

We have carried out research on waste stabilization ponds for waste treatment during the past 20 years. The stabilization pond systems are found to be the most suitable cheap techniques to a tropical country such as India. Ecobiology of these ponds is covered in Chapter 10. The last Chapter of the book deals with fresh and marine biotoxins which is another rare information being made available for the readers.

We have consulted numerous scientists, journals, reviews reports and research papers and books for the information, all over the world. Finally we hope that the present title will be useful to all the readers and teachers of aquatic and environmental biology.

B.B. Hosetti
Arvind Kumar

Contents

Chapter 1
Biosphere

Introduction

The term biosphere was used by an Australian geologist, Edward Suess in 1875. The concept was further developed by a Russian mineralogist, Vernadsky in 1926. According to Hutchinsón (1967) it is defined as "That part of the earth where life exists or that place which envelops a pocket of life". The largest biological system in which several ecosystems can operate together is known as biosphere or ecosphere. The three sub-divisions of the biosphere include, hydrosphere, lithosphere and atmosphere.

Hydrosphere means the sum of total of all liquid components, the water in the oceans, lakes, rivers, and on land. Thus, hydrosphere includes two major biocycles the fresh water and the marine water. Lithosphere comprises the solid mass that is rocky substances of the continents. The soil components of the lithosphere often also called as pedosphere (Odum, 1969). The atmosphere is the gaseous mantle which covers the hydrosphere and lithosphere as well. There are no characteristics inhabitants of the atmosphere. However, air is traversed by many kinds of animals, plant propagules and micro-organisms. Microbes and spores will be present in the air always.

The plants and animals require inorganic metabolites which are obtained from various compartments of the ecosphere. The sub-divisions of the ecosphere or biosphere viz. hydrosphere supplies water, lithosphere supplies all necessary minerals like, Phosphorous, Sulphur, Iodine, etc. and the atmosphere supplies

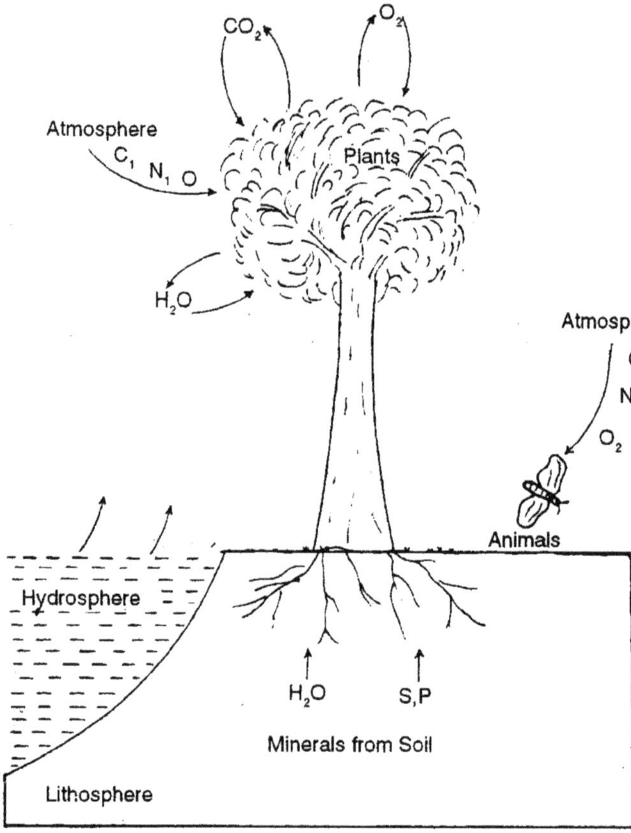

Fig. 1.1: The material contributions from the three sub–divisions of ecosphere to maintain life

oxygen, nitrogen and carbon dioxide. Together these inorganic materials provide all the chemicals elements needed in the building and maintenance of living organic matter. Fig 1.1 explains the material flow from the three sub–divisions of the ecosphere in the living world.

Hydrological Cycle

Water is the most abundant mineral of the earth. It covers about 73 per cent of the earth's surface. It is the major constituent of the protoplasm constitutes about 80-90 per cent. It is also the major component of the lithosphere. For the metabolic processes water is the only source of hydrogen and one among the several sources of oxygen. Hydrological or water cycle is a balanced continuous cycle of evaporation, transpiration, precipitation of the surface run-off and groundwater movements.

Every year about 507 T m^{-3} of water is evaporated and the same amount is precipitated over the whole surface of the earth including the land and the ocean. The amount of water which flows from land to sea through rivers and streams is about 44.5 T m^{-3} which is supposed to be available for man. The solar heat evaporates water from soil, surface of vegetation, surface of oceans to atmosphere. Subsequently the water vapour forms into clouds, condenses at high altitudes and precipitates as rain or snow thus returns to the hydrosphere.

About nine per cent of the natural evaporation from sea moves over the land surface and precipitates as rain water and balances the hydrological cycle and hence provides additional water to mankind. Thus, water as rain, hail or snow is precipitated over the land and water surface. The water on the land eventually percolates into soil as groundwater and reaches the water-table. The soil below the water table is always saturated and sustains water percolation further by its underlying impervious rock layers.

Groundwater don't remain static but move in various directions. Hence, water at low altitudes emerges out as streams and rivers help the man to provide natural water. In some regions on the earth water percolates through porous rocks and forms underground reservoirs. The water present underground blocked in porous rocks and the lower strata will be impervious. Such water body is called aquifer. Water can be taken out from the aquifers through holes or borewells to the surface.

All the water precipitated on the land will not percolate into soil, but flows through surface run-off join to streams, rivers and eventually into oceans. Some of the groundwater that absorbed by the plants passout the leaf surface as vapour by transpiration. Thus, through natural process of hydrobiological cycle water is exchanged between the hydrosphere, land area, sea surface, all living organisms and industrial plants.

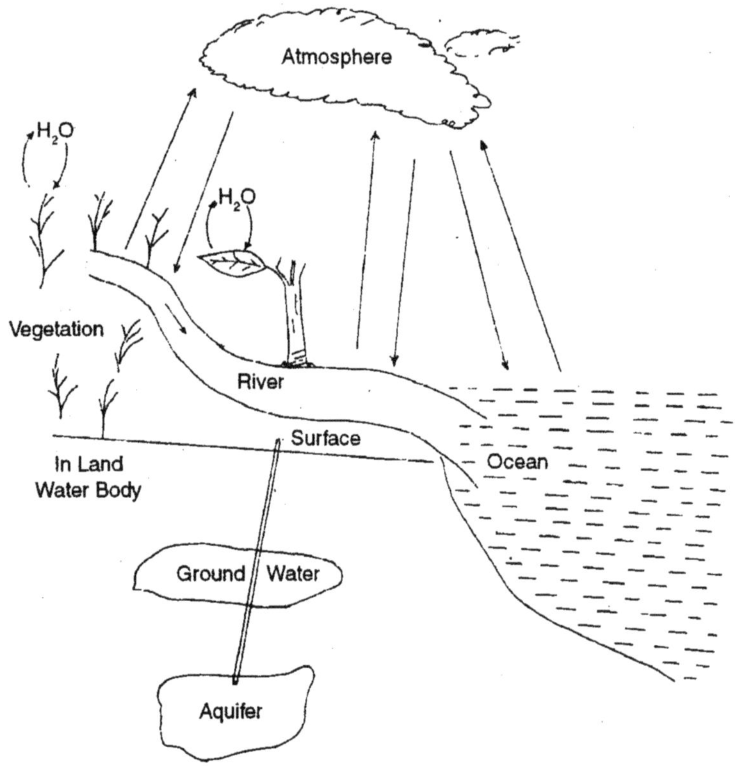

Fig. 1.2: Hydrological Cycle

Lithosphere

It plays two important roles in the metabolism of organisms:

1. It acts as source of most of the mineral metabolites of all organisms.
2. It forms a bulk component (Soil) or pedosphere required for terrestrial plants.

Minerals and their Cycle

Similar to hydrobiological cycle rock substances of the earth surface is a gigantic cycle but the circulation is dead slow, measured in terms of millions of years. One part of this movement is diastrophism. The movement of land mass of India towards China at a speed of about 5 cm/year is an example of diastrophism.

The vertical uprising of parts of the earth crust is slow process. Major parts of the continents or indeed whole continents may undergo diastrophic movement. They occur when land mass moved up from below or is subjected to great lateral pressure generated by adjacent portions of the earth's crust, uplifting that follows. This sort of changes are very slow. An important example of diastrophism is mountain building. At present the youngest and highest mountain ranges are Himalayas, Rockies, the Andes and the Alps. All these were thrown up about 75 million years ago. The formation of mountains has many effects on the earth's climate and hence on the metabolism. These mountains interfere with the air circulation. That is the moisture loaded winds from ocean will not cross the mountains and continuous rain will fall nearby the areas making that region more fertile.

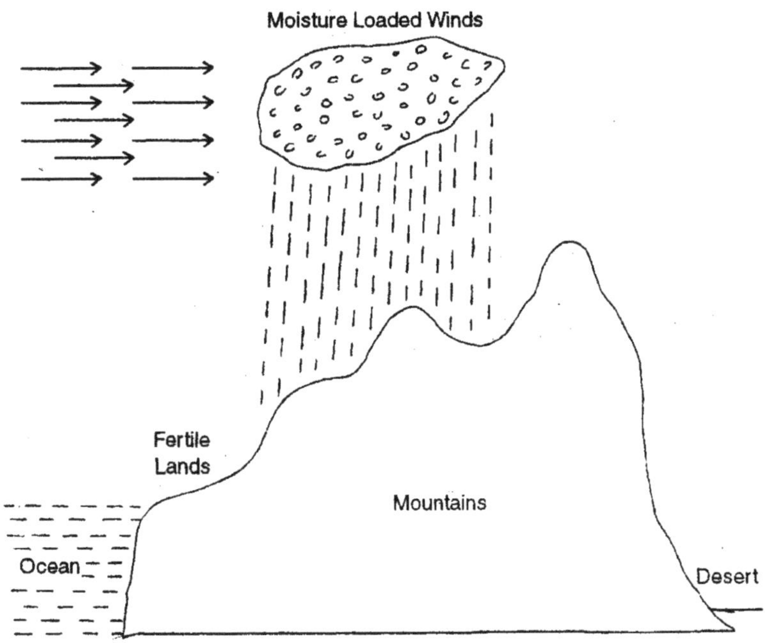

Fig. 1.3: Effects of mountain on climate

The second segment of lithosphere cycle involves lowering and levelling of mountains is brought about by actual geological sinking and also by the actions of hydrological and atmospheric actions, which include, erosion and dissolution of rocks. The lithosphere also plays an important role in the formation of soil. The soil provides anchorage to plants and it also serve as a source of minerals and water for plants and animals.

Soil: Soil is a part of the lithosphere plays an important role in the metabolism of land plants. Soil serves two important functions: 1. provides mechanical anchorage and 2. it holds water and minerals.

Before the ancestral plants invaded the earth or terrestrial environment the land was without soil. But the process of soil formation was underway. The weathering action of winds, powdering of rocks, milling action by rivers, the sheer force of ice are all important in the process of soil formation. Silt was deposited in the flood planes along the rivers and coastal areas. Winds have spread the dust far inland. Small rock grains of rough texture are formed. Sand and clay become the hard mineral fraction of the soil. Other major component of the soil was contributed by the living organisms. Early terrestrial organisms which don't require soil they shed excretory products into sand and upon their death the dead bodies also added to it. Decay ensured the formation of soil that is organic component of the soil called humus. Acids produced during decay also speeded the fragmentation of rocks and thus helped in the formation of sand. Humus mixed with sand and clay constituted the soil.

On the land soil is the cheaper means of supplying minerals to plants. The chemical values of soil depends upon its usable water and mineral contents. The rock particles are covered by a thin layer of water which is bound electrochemically. Some minerals also dissolved in it. The rotation crop pattern and adding of mineral fertilizers help to maintain the soil fertility and quality.

The Atmosphere

Our atmosphere is mixture of gases completely surrounded the earth. It is usually divisible into four zones. Those are: 1. Troposphere, 2. Stratosphere, 3. Mesosphere and 4. Thermosphere. The troposphere is in contact with earth's surface and is about 8 km thick at poles and 16 km at the equator. In an unpolluted condition, it consists the gases in the following order:

	Nitrogen	78.08 per cent
	Oxygen	20.95 per cent
Trace Gases	Argon	0.93 per cent
	Carbon dioxide	0.0325 per cent
	Neon	0.0018 per cent
	Helium	0.0005 per cent
	Krypton	0.0001 per cent
	Ozone	0.000002 per cent

The troposphere also contains various amounts of water vapour (1-4%) by volume. This water may be present in the form of gas or in condensed form as clouds (fog, mist or solids as ice or snow). The second layer is the stratosphere present above the troposphere. It extends to an average height of about 12-50 km above earth surface. It differs from troposphere in that it contains very little water and no clouds. Though its gaseous composition is similar but it contains more ozone. The third layer called mesosphere extends from about 50 to 80 km above the earth surface. It consists less gases, no water and contain more ozone. The fourth layer called the thermosphere or heterosphere, starts at an altitude of about 80 km onwards. Here gases are present in atomic state and exhibit a sort of layering from 80 to 120 km and oxygen and nitrogen are abundant. But above 500 km helium and hydrogen gaseous ions are the main gases.

Beyond 500 km onwards, the thermospehre becomes rare field and extends to the extreme outer limits of the atmosphere to about 40,000 km. This heterosphere contains no water vapour and ozone. The outer most layer where gases are present in ionic form is called as ionosphere. These ions are formed by the photochemical action of UV rays upon oxygen and nitrogen molecules.

Role Atmospheric Ozone and Evolution of Life

The ozone (O_3) is an isotope of oxygen exists in a so-called ozone layer in the middle and upper stratosphere and lower mesosphere at a height of 15-60 km. By natural chemical process sunlight creates ozone layer when it strikes the stratosphere. It is now generally believed that when life began on the earth more than three billion years ago, the atmosphere contained nitrogen, ammonia, hydrogen, carbon monoxide, methane and water vapour. But no free oxygen.

Because of lack of oxygen there was no ozone layer to check suns deadly ultraviolet radiation. Such radiation would kill any exposed organisms but it is not strange to say that it was this radiation that is thought to have created a chemical evolution leading to complex organic molecules such as aminoacids, which become the building blocks for primitive life. The very small amount of non-biological oxygen produced by UV radiation on water may have provided enough ozone to form a shield against lethal action of UV radiation. Yet as long as the atmosphere oxygen and ozone remains scarce, the life could only develop under the protective cover of water.

The first living organisms were naturally aquatic, yeast like forms that obtained energy for respiration by the process of fermentation. As the level of oxygen in the atmosphere was increased, the ozone layer formed in the atmosphere, thickened and covered the earth surface, become shielded and life could then move up to the surface of the sea. It is thought that the first multicellular animal appeared when oxygen level reached three per cent of its present level of 20 per cent.

Still it is not clear that why the ozone whole is depleting every year. But according to many scientists chlorine molecules from chloroflurocarbons used to pressurize aerosol spray might be responsible. Prof. Augustein (1987) said that depletion of ozone layer in winter is mainly due to lack of sunshine. If the trend of depletion in ozone continues, the earth's inhabitants would be exposed to more UV rays, which can cause, cancer, cataracts and related diseases.

Chapter 2

Physico-chemical Properties of Water

A. Properties of Water

For the first time life has been originated in aquatic medium. Water makes up 80 per cent of the body. In simpler forms of organisms water is taken in by direct contact with the environmental water vapour. The complex higher forms of plants and animals may have outer impervious coatings, but their cells must remain in contact with the aqueous environment of the internal medium.

When the cell looses its water, life activity is suspended or it may be extinguished. Activity reappears in desiccated dormant cells, when water again becomes available. The cells of some dried mosses, lichens, rotifers, tardigrades and nematodes may survive under variable conditions of desiccation for years. They resume activity quickly when moisture and other conditions become normal. Many protozoa encyst and some bacteria and fungi sporulate when deprived of water. Such dried resistant stages are especially characteristics of fresh water and terrestrial organisms which are subjected to seasonal drying.

Many of the special properties of water make it suitable not only as the main constituent of the living cell, but also as the main factor in the environment. The following factors are responsible to make the water as a temperature stabilized medium. Those properties include, high specific heat, high heat of vapourization and high heat of fusion.

Specific Heat

If a dry landmass made up of rock and soil with low specific heat were to absorb the huge amount (10 cal/km) of incoming radiant energy from sun, it would rise to a lethal temperature by day and fall below the freezing point by night, as it apparently happens on the moon. But the water on the earth and water vapour in the atmosphere absorb some of the incoming radiation of sunlight, thus, preventing over heating in day time. Some of the sunlight which gets through the blanket of atmospheric water vapour is absorbed and heats up the land masses. As the heat is re-radiated from the land by night, the atmospheric water vapour absorbs the outgoing infra-red rays, preventing the earth from cooling at night. Thus, the so-called "Green house effect" of the atmospheric water vapour, moderates the temperature of the land mass. Because of the high specific heat of water, the temperature of the great bodies of water covering the earth's crust is much less than that of the land. The oceans act as thermostats of the earth's climate, resisting temperature change.

Heat of Vapourization

Because of the high heat of vapourization of water, a great amount of the solar heat is absorbed during evaporation from the vast surface of the oceans. The vapour holds the heat until the warm air is chilled in a cooler part of the earth. Then the vapour condenses and liberates the heat, thereby counteracting the coolness of the region in which precipitation occurs. The water loaded air may move far from its original source before it condenses and gives off its heat. The heat absorbed at the surface of the ocean thus moderates the climate of the earth, the heat absorbed in the tropics is carried to the Northern polar region.

Heat of Fusion

The high heat of fusion of water is another factor leading to radical changes in temperature, because ice is formed from water only after a considerable loss of heat, and it melts only when this amount of heat is replaced.

$$\text{(Water) } 0^\circ\text{C} \underset{\text{Heat}}{\overset{\text{Heat}}{\rightleftarrows}} \text{(Ice) } 0^\circ\text{C}$$

The system remains at 0° C, during the withdrawal of heat, accompanied by ice formation or the addition of heat while melting the ice.

Specific Gravity

An additional thermostabilizing property of water is its high specific gravity (weight per unit volume) at 4° C. Because of this property, pure water cooled below 0° C rises and freezes only at the surface. Otherwise, such bodies of water like lakes might freeze to ice and become uninhabitable.

Heat Conductor

Water is a very good heat conductor than any other common liquid. Consequently the heat absorbed is readily distributed throughout its volume. Because of its usefulness as coolant, water has been used extensively in industry, specially for cooling nuclear and thermal power plants.

Water as a Solvent

Water is the best solvent as it dissolves more substance in it than any other solvent. Electrolytes dissolved in water readily ionize because water molecules are polar solvent of high dielectric constant (78.5). Ionization and ionic reactions are of great importance to life. Water is remarkably inert and most substances are unchanged by being dissolved in it. Because of its excellence as a universal solvent, water has been used extensively in industry.

Water is Needed for Hydrolysis and Biological Oxidations

Cell nutrients are generally stored in cells in insoluble form, mostly as polymers. Before such nutrients can be used within the cell they must be hydrolyzed, when the molecule is split by the addition of water.

Oxidation is a process by which the cell liberates energy from nutrients for performing work of the cell and it occurs only in the presence of water. A mixture of the appropriate enzymes and substrates do not react in the dry state. Water is thus necessary for hydrolysis and for biological oxidations, the two chemical reactions most fundamental to the living cell.

Special Properties of Water

Water has several other properties which make it particularly fit as part of the external and internal environment. It is transparent to light, enabling the specialized photosynthetic organelles, the chloroplasts, inside the plant cells to absorb sunlight, a process upon which depends all life on earth. The transparency of aquatic bodies enables the submerged plants to grow below the surface of water.

Physical Characteristics

The physical characteristics of water such as temperature, pressure, density, currents etc. in one way or another, have significant influence on aquatic life. It should be understood that each physical factor is described not as an isolated physical phenomenon but as a contributor to the sum total of all aquatic environment where the organisms exist, grow and reproduce.

Buoyancy and Density

Buoyancy varies directly with the density of water. The law of Archimedes states that a body is buoyed up by a force equal to the weight of the water displaced. In denser water the floating body will rise up more.

Wave Action and Current

In streams and lakes the distribution of vital gases, salts and small organisms depend on movement of water. The principal forms of water movement are waves, currents and seiches.

Waves

Waves are mainly produced by winds and occur in all types of water bodies. The forms and magnitudes of waves depend upon various local conditions, such as an area of open water, direction, duration and velocity of winds, shape of shore-line, and relative amount of deep and shallow water. If the wind blows over a large area of water, the height, length and velocity of wave action will be more.

Wave of Oscillation

When the water particles follow a vertical circular path and returns to the position from which it started without horizontal movements of water, it will be known as wave of oscillation.

Wave of Translation

When a wave from the open water approaches the shallow water of the shore, the wave height is increased and breaks into foam water between the shore and outermost breaker forming surf. Waves in which there is a definite forward movement of water itself are knows as waves of translation

Currents

A relatively rapid (about 1-7 km/hr) movement of mass of water is known as current. The current in freshwater bodies may be of three types, vertical, horizontal and returning (sometimes called under low currents) type:

(i) Vertical currents are generally present in large water bodies such as great Lakes. These occur due to unusual thermal, morphological or hydrostatic circumstances. Upwelling of water from the deep source is an example.

(ii) Horizontal currents are common in lakes. They are usually produced by wind and often modified by the shape of the shore line and form of the basin.

(iii) Returning currents are formed when water is piled up on an exposed shore as a result of an onshore (from the shore) wind. Such action raises the water level at that position and then the excess water returns along the bottom. When the velocity and duration of the onshore wind becomes high the return currents may extend to the opposite side. In summer season when a lake is thermally stratified and the upper portion of the water (epilimnion) is approximately of the same temperature and hence of same density, the returning currents if produced will encounter the colder and denser water of thermocline. As such the returning current will be diverted in a horizontal direction and will flow towards the opposite shore, maintaining a level above the thermocline. In such a circulation, the upper stratum of epilimnion flows towards one shore; the lower stratum (the return current) flows towards the opposite shore, and the stratum between the two, which is known as the shearing plane is practically without motion.

Seiches

In lakes and along the sea coast, oscillations of the water level occur under certain circumstances, which are called seiches.

Thus, a seiche consists of a local, periodic rise and fall of water level. Any influence which produces a temporary local depression or elevation of water level may produce a seiche. Those influences which most commonly produce seiches in lakes are (a) winds, temporarily strong, which pile up water on the exposed margin of the lake; (b) sudden change in barometric pressure over a portion of the lake areas, (c) earthquakes, (d) landslides and (e) sudden very heavy rainfall at one end of the lake.

Temperature

Water has several unique thermal properties which combine to minimize temperature changes. So the range of variation of water temperature is smaller and changes occur more slowly in water than in air. The most important of these properties are:

(i) **High Specific Heat:** A relatively large amount of heat is involved in changing the temperature of water. One gram calorie (gcal) of heat is required to raise one milliliter (or one gram) of water to one degree centigrade between 15°C and 16°C. Only ammonia and few other substances have values higher than 1.

(ii) **High Latest Heat of Fusion:** Eighty calories are required to change 1 gram of ice into water with no change in temperature and vice-versa.

(iii) **Water has its greatest density at 4°C:** Water becomes lighter above and below 4°C and prevents water bodies from freezing. Though the temperature is less variable in water than air, it is a major limiting factor as aquatic organisms often have narrow tolerances (stenothermal). Even moderate thermal pollution by man can have widespread effects. Temperature changes produce characteristic patterns of circulation and stratification in lakes which greatly influence aquatic life. Large bodies of water greatly modify the climate of adjacent areas of land.

Transparency

Penetration of light is often checked by suspended particles and when the aquatic bodies have appreciable depth, this may restrict the photosynthetic zone. When turbidity is caused by clay and salt particles, this becomes an important limiting factor. Conversely, when turbidity is due to living organisms measurements of transparency become indices of productivity. Transparency can be measured with a very simple instrument called Secchi disk (after A. Secchi an Italian Scientist who introduced it in 1865), which consists of a white disk about 20 cm in diameter that is lowered from the surface until it just disappears from view. The depth of visual disappearance becomes the Secchi disk transparency, which will range from a few centimetres in very turbid bodies of water to 40 metres in very clear, unproductive high altitude lakes.

In general the minimum intensity of sub-surface light that permits photosynthesis has been set up at about one per cent of incident surface light. Thus the region from surface to the depth at which 99 per cent of the surface light has disappeared is called the euphotic zone, or simply the photic zone. The energy flux in this level is well within the photosynthetically active spectral band of 380 to 720 mm. Below this, primary productivity is usually considered as zero. However, shade adapted algal communities carry photosynthesis beneath the snow or ice of Alaskan and Swedish lakes where light intensities are well below one per cent of ambient light.

Intensity of Light at the Surface

The intensity of light at the surface of water varies under a number of circumstances, such as magnitude of cloudiness or clearness of sky, presence of fog, smoke, dust, time of day and season of the year. On a clear day, the light reaching the surface of the water is the sum total of light coming directly from the sun and from the hemisphere of the sky (reflected and refracted).

Difference in Latitude

The vertical penetration of the incident light at the surface of a water body is maximum at equator. The more remote a water body is from the equatorial region the greater is the departure of the sun rays from the vertex.

Diurnal Difference

The daily effects of sun rise in the east and sunset in the west results in the lighting of the sky from the eastern horizon to the meridian and then to the west. The angle of contact of light rays increases from 0° to 90° and then declines to 0 in the evening. Thus, the angle of contact of light with the water surface changes during the day, reaching its nearest approach at mid-day.

Dissolved Materials

Chlorides of calcium and magnesium and similar salts diminish light absorption. Traces of ammonia, proteins and nitrate in solution in water greatly reduce its transparency to ultraviolet light.

Colour

Light coming from the water surface give an apparent colour that is the result of many factors. Of these the materials in solution and particulate matter, both living and non-living are important. Some reddish ponds are due to suspended inorganic particles eroded from the surrounding soil. Organisms also account for other types of red waters. As for example the purple sulfur bacteria, the brine shrimp Artemia, blooms of *Oscillatoria rubescens* (in eutrophic lake), surface scums of the flagellate *Euglena sanguinea* impart red colour to water. Some ponds are bright green because of a superficial layer of some other species of Euglena. Phytoplankton populations consisting of diatoms or of the common dinoflagellate *Ceratium* give a dark yellow tint, and heavy blooms of blue greens produce a green colour.

B. Light as an Abiotic Factor

For satisfactory understanding of an organism the knowledge of the environment in which the organism live which is directly or indirectly influenced by the environmental factors should be studied properly. Leaving the morphology and systematics, marine biological problems are concerned with basic process producing organic matter. Life in the sea is dependent upon the synthesis of organic food. Chemosynthesis is very less when compared to photosynthesis.

The physical, chemical and biological processes which influence the life activities of an organism, collection of these factors and their inter-relationship should be studied. These factors are classified into biotic and abiotic factors.

Biotic factors are living components of the environment and the abiotic factors or abiotic components include physical factors like light, temperature, salinity, pressure, waves, tides and currents etc. The chemical factors include dissolved salts, dissolved gases, and chemical properties of water.

Light is regarded as the radiant energy roughly covering the range of 380-780 nm which is capable of stimulating human eyes to produce sensation of vision. The natural sources of vision are the sun, the moon and the stars. Of these sun is the most important source since all the energy needed for the organisms is derived from it. The life activities in the ocean are fueled by the suns visible light acting on green plants. Of the every millions of photons of sun's light reaching the earth's surface some 90 per cent enter the net production of basic food. Main energy source of photosynthesis in plants is light. Without light organic food production in plants will be disastrous and their will be no life.

Every animal directly or indirectly depends on plants, which inturn need energy for their existence. The intensity and wavelength of light are important factors. The optimum wavelength for photosynthesis varies for different species of plants. Thus, it enables the aquatic plants and animals to take up different positions in water thus leading to sort of zonation. The light entering the water undergoes qualitative and quantitative changes. Light is reduced in intensity both by absorption and scattering as it penetrates into the water medium. Absorption is a selective performance in which certain wavelength are absorbed more quickly than others. It has been observed that very small quantities of violet, blue and green rays penetrate to a depth of at least 1800 m in clear water. Light is measured by two methods:

1. Limit of visibility test and
2. Light measurements.

Based on light availability in aquatic environment we notice three zones: a. Euphotic zone b. Dysphotic zone and c. Aphotic zone. Euphotic zone extends from surface to 80 m or more. This zone is abundantly supplied with light sufficient for photosynthetic process. The dysphotic zone is dimly lighted extends from 80 m downwards or it starts from the closure of euphotic zone. Further the aphotic zone is a dark zone lies below dysphotic zone. In this only carnivores and detrivores can live and this zone varies depending upon turbidity, season and extension of the season.

Light and Photosynthesis

Light plays an important role in the aquatic system during photosynthesis by all chlorophyll bearing plants. These plants which directly or indirectly become the source of food for the entire ecosystem of aquatic world. The illumination diminishes rapidly with the depth even in clear water. UV rays are of little consequence in water. Experiments showed that photosynthesis is affected by different wavelengths. It is being greatest in the red light and least in blue violet rays. Photosynthesis is the first step in the ecological cycles in the sea. It is carried out by microscopic plankton and other algae. The effects of light on photosynthesis is studied experimentally by productivity tests at different depths which decreases as we go down. At one stage the oxygen production equals with the rate of respiration. This is known as compensation intensity and the depth is called as compensation depth. With the decline in light intensity the process of photosynthesis is also declined. However, the respiration remains the same provided the temperature and other factors remain unchanged.

Pigment Production and Chromatin Adaptations

Critical comprehensive reports are lacking on the light requirements by marine algae, however, the marine flagellate *Dunaliella tertiolecta* rapidly increases its chlorophyll contents upon transferring to complete darkness just similar to most fresh water algae examined.

Light intensity plays a vital role in determining the chlorophyll content of algae cells. Unicellular algae grown under low light intensities (shade cells) have much higher chlorophyll content than cells grown under high light conditions (sun cells). When cultures of *Ekeltonema costatum, Dunaliella tertiolecta, Amphidinum ertheri, Manachrysis lutheri* were transferred from low light to high light intensity the rate chl-a synthesis decreased. If the reverse is done then chl-a synthesis, cell division and total biomass are increased. In the *Euglena* absence of light reduced the amount of carotenoids but didn't entirely inhibit their production.

Major pigments of algae are chlorophylls, carotenoids and phycobilins. Each algal group contain specific pigments. Thus, Rhodophyta contain phycobillin, a-pheoerythin and xanthophyll.

Careful studies on AAS and photosynthetic properties of algal cells reveal that photosynthesis in algae is associated with the presence of chl-a although chl b, c, d, and e are present. In Rhodophyta phycobillins are the major light absorbents. This transfers energy to active form and increase the photosynthetic efficiency.

Light and Colouration

Light has a profound influence on colouration in plants and animals in the aquatic biosystems. The physiological colour change is a rapid event where alteration occur in the degree of dispersion of pigments in the chromatophores after appropriate stimulation.

Morphological colour change is a long-term event and may be due to the changes in the quantity of the pigment within the chromatophores possessed by the animals or by the combination of both. Various shades of grey and black are common in the littoral flat fishes. Many animals show colouration which give protection from their enemies. Many littoral crustaceans show same colouration as their background.

The isopod, *Pentidotea reseeta* upon which it lives is almost indistinguishable from the kelp grass and eel grasses. The prawn *Hippolyta varianus* is a striking example of colour adaptation. It corresponds not only to give background but it also changes readily with the changes in the colour. Flat fishes have been shown to assume shoals and patterns appropriate for making them blend with their background. The mechanism that brings about these changes in the colour of shades are associated with the chromatophores or pigment cells. Thus, contents of these pigments cells are activated directly by the intensity of incident light or mediated through eyes and CNS in response to albedo of background radiation. Dorsoventral differentiation of body colouration is present in pelagic fishes. This indicates that stronger illumination has received from above and with the result the animal blends with its background.

The uniform dark pigmentation of fishes in the abyssal layers as contrasted with the ventrodorsal colour differentiation of those inhabiting euphotic zone indicates that the direction of light is significant in the distribution of pigments.

The blue and green colour are indeed striking characteristic of the surface and open sea life. In many of the smaller forms copepods and other crustaceans pale red is very common. Hyaline or colourless forms are characteristic of open sea life. Colour change serve as protective way from high illumination and helps in thermoregulation or as associated with the breeding as in certain lizards, fishes and squids.

Light and Reproduction

Some detailed studies have been made on the environmental stimuli which trigger the various functions of reproduction. High light intensity and long day length may be important for the reproduction of marine diatoms. Under short day conditions sexual resting spores are formed, while sexual stages are induced under long day conditions. In many marine algae endogenous lunar or daily rhythm seems to be involved in triggering the discharge of reproductive cells. In general release of spores or gametes takes place around daybreak indicating the direct importance of light transition. Light probably also influences the formation or reproductive cells. Photic stimulation also influence the extrinsic factors controlling reproduction in marine invertebrates. The manner in which control takes place ranges from the direct effects of illumination on tissues to the psychological impacts of specific visual impression. The photoperiod is an impertinent component of the seasonal calendar by which many organisms seems to regulate their reproductive activities.

In culture tanks or ponds light was shown to be necessary for the spawning action of *Clupea harengus* both in the act of choice of substratum and depositing their egg mass.

Light and Growth

Light is a important ecological factor which controls plant growth. Growth rates are usually expressed in terms of increase in cell number or biomass. Photoautotrophic plants depend upon their photosynthetic capacity to supply the necessary organic substances and energy for growth. Nutrient conditions and temperature may influence the relation between light intensity and growth in marine algae. In general higher light intensities are required for maximum growth rate at optimal conditions of nutrition and temperature.

Some algae show better growth under continuous light whereas some require day and night cycles. Growth of marine diatom, *Biddulphia sienensis* is strongly inhibited by continuous light. Two diatom species *Melosira moniliformis* and *Biddula aurita* show little or no dependence on the day length. From the limited information available it appears that mature sea weeds may have higher light intensities requirements for growth than younger stages. Growth of marine decapod crustaceans appear to be a discontinuous event as they increase in size only when animal sheds its shell.

Vertical Migration

Light is an important factor in the local or restricted distribution or organisms. One of the unsolved puzzles of pelagic natural history is vertical distribution. The correlation of animal movement with the light is best studied in the pelagic division of the ocean. Numerous investigations on the vertical distribution of plankton of the open sea have shown that many of the pelagic forms make daily journey from deeper water to the surface at the approach of darkness and begin returning to deeper layer before the break of dawn. This rhythmic movement is called as diurnal migration.

These organism spend a lot of energy daily in migratory activities (e.g. Protozoa, Coelenterates, Medusae, Siphonophores, Ctenophores, Polychaetes, Salps). Most young fishes undertake shorter or longer migrations. The changing light intensity seem to be one of the causes for this migration. It is thought that most plankton react negatively to both gravity and light. In day time bright light forced them away from the surface but when its influence faded away at the evening negative geotrophism made the animal to swim up against the gravity.

According to Eyden (1923) vertical migration is due to a change in the specific gravity of the animal correlated with feeding. She worked on Daphnia (Water Flea) and explained that the animal come to feed upon phytoplankton near the surface have good meal, become heavier and sink and then when the food is digested and unwanted food is excreted out they become active and are able to swim to the surface again because they are lighter. This was supported by Clarke (1932) that Daphnia well fed tended to show negatively phototrophic and the starved are positively phototrophic.

Bioluminescence

Light of biological origin is known as bioluminescence. It is the only source of light in the deep sea. Emission of light from living beings is a wide spread phenomenan found in most bacteria, fungi, protozoa, sponges, hydroids, jelly fishes, cetenophores and fishes. Luminescence or emission of light in animals is the result of chemiluminscent reactions in which a substance is oxidized. In many organisms the light producing reactants are expelled exterior where actual reactions in the production of light occurs. This is known as extracellular luminescence. The biochemical activities involved in the bioluminescence are partly understood.

Prof. Newton Harvey states that light is produced as a result of oxidation of a substance, probably a simple protein known as luciferin. It is produced by living cells. Before it is produced a enzyme system is also necessary to oxidise and produce the light. That enzyme is called luciferase.

$$\text{Luciferin + luciferase + oxygen} \longrightarrow \text{Oxyluciferin + luciferase + water + light.}$$

The most striking property of animal light is that it is cold unaccompanied by heat of any kind and more efficient than any form of artificial light. Greater the concentration of luciferin longer is the duration of luminescence. Different organisms use different chemical compounds for light production. If luciferin of a bacterium is an aldehyde complex of reduced riboflavin-5-phosphate and bacterial luciferase is flavin. On oxidation FNH_2 is reduced to FMN and a flash of light is produced. In crustacea definite light organs are present. Some of which are simple and others have great complexity. Luminescence has many roles to play. But definite explanation is attributed in smaller organisms. In Chetopterus light serve to attract towards the burrow and feed on them. The bright light reveals that the whole animal seems to advertise its presence to other free swimmers. A sudden flash will serve to threaten the predators.

Periodicity

It is an important factor of the environment. The variation in the light intensity is the most important factor behind the behaviour. Two types of rhythm are recognized:

a. *The exogenous rhythm* - This occurs as an immediate response to environmental changes, but do not occur in constant conditions.

b. *The Endogenous rhythm:* This persists even under constant conditions.

Diurnal Periodicity

The behavioural response of an organism changes to the environmental responses in 24 hours. The most fundamental diurnal rhythm was that of photosynthesis. Photoperiodism was first discovered in relation to plants that flowers only when the days are longer and the night are correspondingly short. Short day plants naturally flower under short day conditions. However, reproductive cycle of some species is not affected by day length. Such plants are called as indifferent plants. The examples for long day plants include radish, red clover and spinach.

Short day plants are mostly perennials in which flower buds are set in the previous autumn. Photoperiodism plays an important role in the life cycle of many animals. Wing production in Aphids, metamorphogenesis in mosquitoes and other life processes are influence by day light. The fresh water pulmonate snail *Lymnaea palustris* will not lay eggs on the day with less than 11 hours of light. It lays eggs only on 13 hours day. The opening and closing of flowers and folding of leaves is regulated by day length.

Light and Structural Adaptations

The structural adaptations to light are presented on three lines: 1. Tactile structures. 2. Food producing ways and 3. Light production:

1. Development of tactile devices is an important feature in numerous instances and is well illustrated in *Macropharynx longicaudatus* and various other deep water fishes.
2. Food production devices of deep sea fishes consists an extraordinary immense mouth in proportion to body size. The stomach and body wall are so elastic that it is possible for some of these fishes to swallow other fishes three times their own size and many time their weight.

The possession of these unusual devices is correlated with the scarcity of food in the deep sea, pelagic habitat and difficulty in procuring it. Some fishes posses very large eyes. The enlarged lenses are enable to receive better and dim rays of light. The loss of cones in these leads to telescopic eyes.

Temperature as an Abiotic Factor

In order to understand the effects of temperature on the diverse types of ecological processes and behavioural aspects of living organisms, we must first clearly define what we mean by "Temperature", and distinguish this concept from the related phenomenon of "heat". A concise way to distinguish between these two concepts is to refer to temperature as the intensity of the heat energy present in the system. Temperature is the measure of the intensity of the kinetic energy of the constituents of system. Temperature is expressed in units of degrees Celsius (^{0}C) or degrees absolute (^{0}K). In contrast to the temperature heat is an expression about the total energy in a given system and is expressed in Calories or Joules (1 Joule = 4.18 calories). One can of course relate the temperature to the substance to the heat present in the system. Thus, for water, 2.6 Kcal/mole of thermal energy is present at 0^0 (273°K) and 3.8 Kcal/mole of thermal energy is present at 60^{0}C (333°K).

Temperature is the most important ecological factor exerts its influence in several ways. The body temperature of many animals corresponds to the environment. They are referred as cold blooded or poikilotherms or ectothermic animals. On other hand warm blooded animals or endothermic animals or homothermic animals such as birds and mammals regulate their body temperature. Some animals have limited power of temperature regulation they are called as heterotherms. Examples include, monotremes and marsupials.

All metabolic process of organisms are regulated to the controlling effect of temperature. An exponential law of controlling or governing the metabolism or temperature relationship has been suggested by Krough. Temperature regulates all chemical processes of metabolism and many physical process. It controls reproduction, feeding, respiration, growth, survival and geographical distribution of pants and animals.

Temperature is the most familiar environmental factor with multisided effects on plants and animals. The amount of solar energy per unit time on a surface or the mean distance from the earth to the sun is estimated at 2.0 cal/cm^2/min. The value is called as solar constant. Out of this one cal/cm^2 reaches the earth surface. Temperature regulates all chemical process of metabolism and many physical process.

Variations in Temperature

The variations in temperature of fresh water, marine and inland waters will affect the living organisms. The minimum temperature recorded in ponds is never below 0.8° and in ocean never below –2.5°C and minimum in terrestrial environment is –70°C (in Siberia 1947).

The maximum in the oceans goes up to 36°C in fresh water lakes and ponds. It may go up to 40°C where as in the deserts the maximum temperature may go up to 60°C. Similarly the difference in temperature during day and night (diurnal) are more marked on land than in water. Diurnal fluctuations in the surface waters are around 1°C whereas in oceans it is about 4°C. The thar desert covers an area of 10,000 sq. km starting from South of Rajasthan in India to Sindh in Pakistan, shows diurnal variations in the range of 20-30 or some time up to 50°C.

Thermal Stratifications

It is an outstanding phenomena in aquatic conditions. The freshwater has maximum density at 4°C and hence any warmer or colder water will float on the top of heavy water (4°C). In summer atmospheric temperatures may be 27°C and this will rise the temperature of upper layers of water to 22-23°C while the water at the bottom may be at 5°C only. Between these two ranges (5-22°C) we find different zones. That is thermocline or metalimnion, epilimnion and hypolimnion are noticed. The thermocline is an intermediate zone with rapid vertical temperature changes. This is also called a transitional zone. Layers of warmer water above the themocline are called epilimnion and the cold water below thermocline are called hypolimnion. In winter the temperature of surface waters becomes zero and the lakes are covered with ice. However, the temperature of these zones vary in different lakes of temperate areas. Depending upon the temperature relationship Welch (1938) classified lakes of the world into three categories as polar lakes, temperate lakes and tropical lakes. In polar lakes the surface temperature will be never above 0°C. In temperate lakes temperature varies above and below 4°C whereas in tropical lakes surface temperature is always above 4°C (Fig. 2.1).

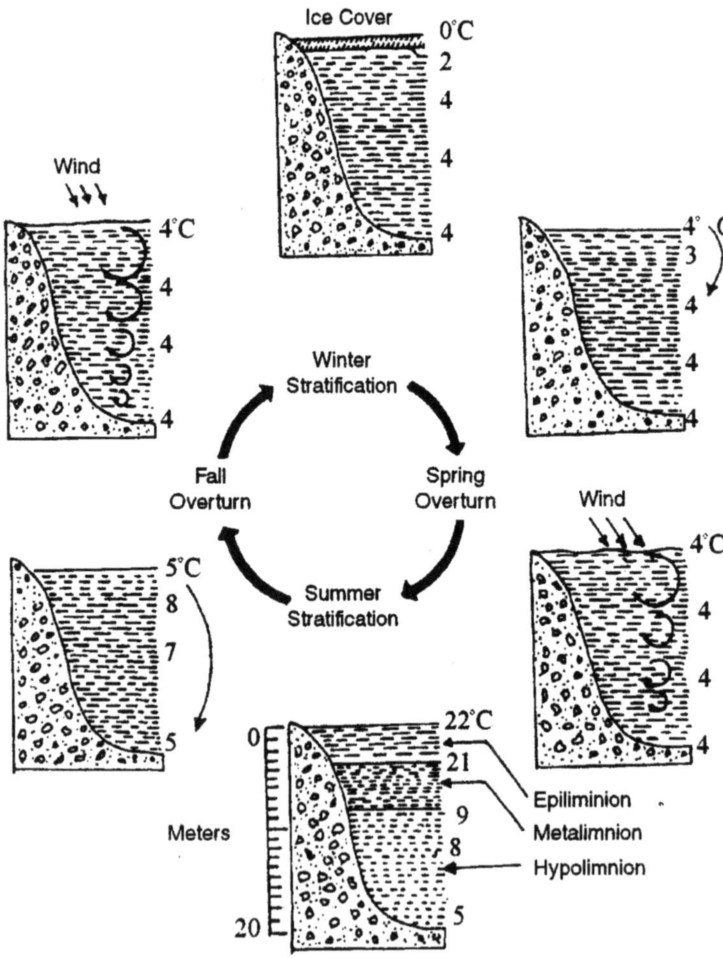

Fig. 2.1: Diagrammatic vertical section of a deep temperate lake showing thermal stratification.

Temperature Tolerance

Plants and animals which can tolerate very large fluctuations in temperature are called as eurythermal. Examples include, Cyclops, wall lizard, snake and toad. Whereas the others which can tolerate only small range of temperature are called as stenothermal. Those are fish, snail, coral reef etc. All the metabolic activities function in a optimum range of temperature. The change in temperature outside the optimal range retards or inhibits or stops metabolic processes. This is of great ecological importance and this limits the distribution of plants and animals.

Minimum Temperature

The lowest temperature at which organism can live actively is called as minimum effective temperature. If the animal enters below the minimum effective temperature, it becomes inactive, a condition called chill coma. The activities of all plants and animals in chill coma can be restored by warming them to minimum effective temperature (Fig. 2.2).

The lowest temperature at which survival is possible is called as minimum survival temperature. This depends on the time factor or exposure. For example in an experiment egg and larvae of fruit fly, *Ceratitis capitata* were killed after 7 weeks when kept at 7°, after 3 weeks at 4°C and after 2 week at 1°C. Because at 0.0°C the water freezes and produce mechanical harm as in the form of rupture of cell wall and stoppage of circulation. Some animals and plants have adaptations whereby they can tolerate temperature even below 0.0°C. For example the green alga, *Chlorella sp* can withstand temperature below 0.0°C up to –182°C for an hour without any harm.

Optimum Temperature

The temperature above minimum temperature increases physiological activities of animals and plants. The range of temperature at which the physiological activities of plants and animals are maximum and the reproduction is vigorous is called as optimum temperature. This optimum temperature also varies in different species at different time. Animals have optimal ranges of activity through which temperature is favourable. Beyond these they may become greatly excited if the temperature becomes too extreme, they finally die as observed in the case of *Musca domestica* (Table 2.1)).

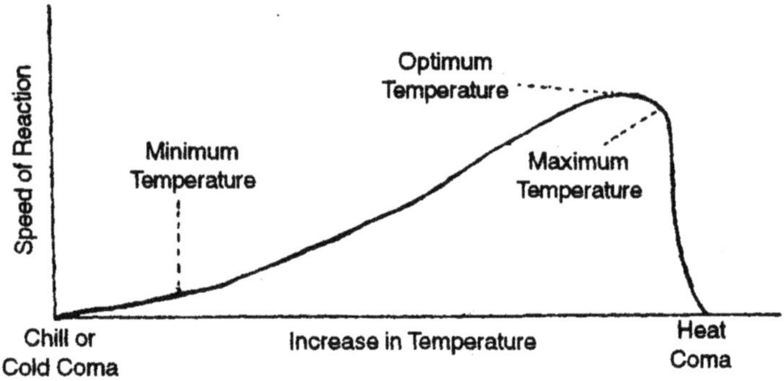

Fig. 2.2: Graph showing effect of temperature on the physiological activity of animals (From Allee *et al.* 1968).

Table 2.1: Effects of temperature on the activity of *Musca domestica*

Feebly active or live	3.5 to 12.0
Decreased activity	7.5 to 14.0
Normal activity	15.0 to 22.5
Increased activity	24.0 to 34.5
Restlessness	43.5 to 45.0
Dies	45.5 to 47.5

The intensity of heat shows daily and seasonal variations in different regions of the earth. On land temperature shows a wide range of about –70°C to 60°C, whereas in the sea the range is much less being –2.5 to 36°C, while in the fresh water it ranges from 0.0 to 90°C in hot springs. Temperature influences the activities and distribution of plants and animals and acts as a limiting factor in the growth of organisms. One of the important evidence of influence of temperature on the organisms is the classification of organisms into tropical, temperate and polar biomes.

Minimum, Optimum and Maximum Temperature

No animal can live in active condition if temperature fails below the freezing point of its body fluids. Upper limit of temperature for many vital activities lies between 30.0°C and 50.0°C. Above this there is a suspension of the activity of protoplasm causing eventual death of organisms. So the life functions only in a narrow range of temperature. This is the reason why very hot summer or an unusually cold winter causes death of a large number of animals owing to the adverse effects of very high and very low temperature. Different species of animals and plants are adopted to different range of temperature and each species has an optimum temperature in which it survive best. There are maximum and minimum temperatures as well. The minimum, optimum and maximum temperatures are vary from one species to another. Effects of temperature on the physiological activities of animal is depicted in Fig. 2.2.

Eurythermal and Stenothermal

Animal that can tolerate a wide range of temperature are said to be eurythermal. Such as some species of Limnaea, the sperm whale and the tiger. Animal with narrow range of temperature tolerance are referred as stenothermal forms. Such animals include coral reefs, termites, carps and reptiles. There are two categories of stenothermal organisms noticed which include, cold limited called as psychropiles and other are warm limited called thermopiles. Some species though capable of tolerating extremes of temperature need a optimum for normal reproductive processes.

Homotherms and Poikilotherms

Based on the body temperatures animals are classified into two groups. Birds and mammals have constant body temperature referred as homotherms, while animals those body temperature vary according to their surroundings are called as poikilotherms, such as invertebrates, fishes, amphibians and the reptiles.

Homotherms can maintain an almost constant body temperature by temperature regulation and remain active both in hot and cold climates. For example in man the normal body temperature is 37°C with a diurnal range of 1°C, in mouse it is 36.5°C, in Horse 37.7°C, in Elephant 36.2°C and in birds it may range from 40.0°C to 43°C. It will be seen that body temperature from species to species does not show great difference and a tropical monkey, a polar bear and a temperate Walrus have all the same body temperature of 38°C. In poikilotherms on the other hand the body temperature is nearly the same as its surroundings. Thus, the fish has the temperature of water where it swims. It is the fact that homotherms can flourish in a wider range of temperatures than the poikilotherms. Changes occurring instantly or rapidly will not be tolerated by many animals, particularly those that live in the almost constant conditions of the sea. Many may perish when the temperature suddenly fluctuate.

Many physical processes such as viscosity of the medium and the solubility of the oxygen in the water are dependent on the temperature and these in turn have influence on the animal living in the medium. Viscosity of water rises with the decrease in temperature so also the solubility of oxygen in water. Effectiveness of temperature difference as barrier for distribution is evident from the pattern of distribution of bipolar species or very closely related ones seen accepting zones of similar temperature in middle or high altitude in both northern and southern hemispheres. They are absent in the intervening warm water area.

Some among several examples of this bipolarity are *Balanus balanoides, Botryllus schlosseri,* and *Didemnum albidum* the genus *Engraulis* and the entire order Lycernariidae. The fauna around north and south poles have many genera in common and in the warm area between the poles they are known to occur only in deep waters where they have the desired type of temperature. Thus, in some cases for example *Eukrohnia hamata, Parathemisto abyssorum* and *Dimophys arctica* the apparent bipolarity is in fact a case of the continuous distribution through deeper colder layers of water beneath the warm surface layers of the tropics. This continuity in the superficial distribution of species is due to tropical submergence.

Biochemical Adaptations to Temperature

Physiologists have devoted a great deal of study to the effects of temperature on metabolic rates and other critical processes.

Benefits and Costs of Endothermy

In view of the pervasive influences of temperature changes on the structures and functions of organisms many benefits accrue from maintaining their body temperature independent of variations in ambient temperature. Organisms with the ability to regulate their body temperature via heat generated within their cells (endotherms) can maintain the rates of their vital processes ot stable levels even in face of large changes in ambient temperature. The stability of neural function may be an important advantage to the endothermic homeotherm. A stable brain temperature may preclude periods of torpor as observed in most reptiles faced with low temperatures. The ecological consequences of stable body temperature in mammals include the ability to remain active at night, when ambient temperatures are generally much lower than day time temperatures which allows them to conduct important activities of life. Perhaps partly for this reason, most mammals are nocturnal in habit.

A related advantage of enodhtermic control of body temperature is the ability to select its habitat other than thermal criterion. Dramatic example include rich availability of krill population for certain whales and seals in Antarctica. This function is mainly enabled them through the endothermy. Because of the endothermy the mammals could function actively in an environment where water temperature is below freezing point of their blood. Similar example is offered by warm blooded fishes such as bluefin tuna *Thunnus thynnus*. Because of its high metabolic rate and large body mass, the fish has a substantial capacity to hold its body temperature above the ambient levels. The benefits of endothermy do not arise without substantial costs, however, even for well insulated endotherms like many birds and mammals, the basal metabolic rate necessary for supplying heat needed for body temperature control is roughly ten times the basal metabolic rate of similar sized ectotherm (Bennett and Ruben, 1979). The endotherm must acquire and process larger amounts of food than an ectotherm of the similar size. Consequently the population densities of endotherms cannot be as large as those of ectotherms. The important tissue which help the homotherms in the endothermy are brown adipose tissue (BAT) and a single protein called mitochondrion from the other tissues (Brooks *et al.*, 1980).

Basic Considerations of Ectotherms

In ectotherms who's body temperature is determined by the ambient temperature, these lack any physiological or biochemical defences for close tracking of any ambient temperature by the body temperature. However, many ectotherms possess behavioural thermoregulatory capacities and seek appropriate micro environments for holding body temperature within a preferred range. In case of seasonal variations in addition to behavioural regulation modifications in their biochemical systems e.g., enzymes and membranes are required. These changes involve phenotypic regulation of gene expression and thereby changes in quantities and types of proteins in the cells. Genetic changes over many generations which allow the evolution of new macromolecular variants that adapted to the new thermal regime may occur.

Temperature Variations

Temperature varies during day and night on land and the range is more than in water. Extreme variations of temperature are experienced under special habitats like the tidal zone. Winds and similar movements are caused as a result of unequal heating of air horizontally and vertically. The greatest of temperature is near the equator and at higher latitudes takes partly due to air currents.

Temperature shows variations in vertical directions also. In the great depth of oceans and deep lakes there are no temperature variations. Variations in temperature are found only in the upper strata. Warm water will float over the cold water. In the vertical column of the water there is a transitional area where temperature shows maximum with the depth. This zone is called thermocline. In the oceans boundaries of this layer starts from the division of surface water the intermediate and the deep water below.

The addition or loss of heat at the surface couples with mixing processes which transmit the temperature change to the depths determines the existence of thermocline. The deepest thermocline noticeable in the sea is referred as the permanent thermocline because of its great stability than other seasonal thermoclines which related to seasons.

The smallest of detectable thermoclines are the diurnal thermoclines which develop very near the surface during the day and disappear during night. There is thermal stratification on land also. This can be measured in forest canopy with the depth in soil and also in the air.

In the ocean and lakes the temperature is too high nor too low for the active life of animals and changes in temperature are much less and more gradual than in the terrestrial environment. Animals can avoid high or low temperature by moving to appropriate places.

Structural Modifications Induced by Temperatures

Temperature affects the absolute size of many animals and plants species and their parts. Marine animals present suitable examples of correlations of structure with the temperature. Some arctic fish show an increase in length owing to an increase in the number of vertebrae. Marine animals of temperate and arctic regions are generally larger in size than their tropical counterparts, perhaps because of delayed sexual maturity and the resultant longer period of growth.

Among birds and mammals the same species attains greater individual size in colder regions than in warm regions. Among closely related species the larger ones inhabit colder water. This type generalization is called as Bergmans Principle. Nevertheless exception have also been noticed in regard to this generalization. For examples in case of diatoms of warmer water are on the whole large in size than those of colder waters.

Similarly the sea urchin, *Echinus esculentus* grows to bigger size in warmer waters than in colder water. In marine organisms it has, however, been shown that the maximum growth rate is attainable in any fixed temperature or in particular fluctuation temperature and it may be essential for the completion of life cycle. According to Allens rule tail, ears and legs of mammals are shorter in colder climate, that is there is a shortening of extremities.

A phenomenon called cyclomorphosis is observed in certain aquatic organisms such as Cladocerans. Here the body form changes with the seasons of the year. During warm months as an adaptation to lower density of water there is an increase in body surface by the lengthening of the head and tail spines. This is an instance of a relationship between body form and seasonal changes in temperature.

Temperature of the environment can control egg type and sex ratio in certain animals. Thus, cladocerans under normal conditions produce parthenogenetic eggs which hatch into females. On the onset of extreme temperature they produce winter eggs which can withstand desiccation or cold.

Thermal Migration

Journey undertaken by animals to escape discomfort of extreme temperature are called as thermal migrations. These trips may be short such as moving into shade, changing to wet places or into water. Soil inhabiting forms burrow deeper into the earth, mountain animals descend to sheltered valley and fishes and plankton move up or down to get desired temperature. These migrations may be long as seen among bisons, caribou, insects and birds.

Influences of Temperature on Metabolic Process

An increase in temperature hastens the chemical reactions on which the release of energy depends. Many animals react favourably to the increase of temperature to a certain point. The development of many animals is accelerated by rise of temperature. Reproduction of protozoa and development of the eggs of certain rotifers, sea urchin, fishes and the pupal stages of insects all shown increased rate with the rise in temperature. Thus, clam and haddock eggs are known to develop more rapidly at higher temperature than at lower temperatures.

In the edible oyster, *Ostrea edulis* the free swimming life of the larva is one week at 27°C and 17 days at 20°C. It has, however, been shown that in oyster there is an optimum temperature above or below which the activity decreases. In certain instances temperature may have little or no influence on the rate of biological activities. This is seen from the fact that the hens egg has no temperature control and it will hatch only if kept between 40-41°C for a incubation period of 21 days. The incubation period of reptiles is also not influenced by changes in temperature of the surroundings. Experimental evidence proved that different body process such as action of enzymes, respiration, muscular and nerve activity, growth and maturity of the gonads all show different temperature optima.

Vont Hoff's Law

A fish consumes three times as much food at 20°C as it does at 10°C. The speed of animal increases with the rise in temperature, so also the heart beat, rate of respiration, oxygen consumption and fission rates. Rise in temperature usually causes two or three times increase in the activity by 10°C increase in temperature. The process is said to have a Q_{10} (increased rate of metabolism) of 2 or 3 based on the Vont Hoffs law. Q_{10} is calculated as follows:

$$Q_{10} = \frac{K_2}{K_1} \times \frac{10}{T_2 - T_1}$$

whereas Q_{10} = increased rate of metabolism at 10°C rise in temperature while K_1 and K_2 represent the rate of reaction ot temperature T_1 and T_2.

Example: The temperature T_1 and T_2 are the rate of boring of a shipworm *Toredo megotara* is about 1.01 gram for five days at 5°C and 3.5 gram for same period at 15°. If we substitute these values in the formula it will give:

$$Q_{10} = \frac{3.5}{1.01} \times \frac{10}{15 - 10} = 3.5 \text{ approximately}$$

So the metabolic activity increased 3.5 times at a rise of 10°C. Similarly relationship between temperature and rate of respiration have been demonstrated in many marine animals such as mussel *Mytilus* and the copepod *Calanus calanus*. Vont Hoffs law was enunciated in 1884 and later it was variously modified. This is clearly seen among poikilotherms and not so well in warm blooded animals. Since they carry their own body temperature around with them. Thus, the rate of ciliary beat in molluscans and the frequency of chirping in crickets show a direct relationship with the rise in temperature. But increase in temperature beyond the critical levels may cause a reduction in the rate of activity and general metabolism.

Influence on Reproduction Process

Cyclic metabolic process like reproductive activity and growth are often influenced by the change in temperatures. This is evident from the fact that tropical animal show a rapid rate of growth and attain sexual maturity at a comparatively early age. Further as a result of the fairly uniform high temperatures all the year round some animals breed almost continuously and several generations of animals are produced in rapid succession within a single year. The speeding up of life histories favours acceleration of evolutionary processes.

This is the reason why tropical zone is rich in genera and species but has a smaller number of individuals per species than are found else where. It has also been shown that length of the life cycle tends to be inversely proportional to temperature with certain limits. That is if the diversity is more population will be less.

Animals living in shallow waters experience seasonal variations of temperature much more than those in deeper waters of oceans and thus are therefore subjected to diurnal variations in temperature as well. Shallow water forms breed earlier than those in deeper water owing to the fact that shallow water forms get warmed up much earlier than deeper water forms.

The other effects include, temperature along with moisture, light and other factors affect coloration of animals such as insects, birds and mammals. These are melanic with deep pigmentation at high temperatures. In the terrestrial environs temperature and humidity operate together as co-factors because temperature affects humidity and the rate of evaporation. Temperature is affected by condensation. If the atmosphere becomes very dry, it becomes unfavourable for many animals. A certain amount of humidity is essential for the normal activity of many organisms. Certain behavioural patterns are influenced by the heat factor. Some snakes like pit viper can locate and pursue their mammalian prey through their ability to detect the heat of the body temperature of the prey.

Effects of Low Temperature

Adjustments to temperature require modifications both in structure and function and in behaviour. Usually no animal can tolerate a low temperature less than freezing point of its body fluids. This is much lower than 0°C owing to the presence of salts. Freezing produces mechanical harm to the cells. Ice formation in side the cells causes a drainage of cellular water leading to total drying. Further effects of freezing are cessation of circulation and lowering of the rate of respiration. Lower temperature slows down chemical reactions in the body cells leading to slow down of vital activities which may cause death. Further when temperature is very low death may occur due to direct mechanical injury or as a result of loss of water to precipitation of proteins.

The term minimum effective temperature is given for the least temperature at which organisms can live in an active state. Any lowering of temperature below this leads to death. In cold winter actual freezing may takes place and this can lead to mass mortality of organisms.

Adaptations to Low Temperature

Animals and plants subjected to frequent low temperatures show adaptations to withstand its bad effects. The most common adjustment shown by animals to the cold of the winter is the development of special seasonal insulations such as fat, fur or feathers on their bodies. Some caterpillars spin cocoons and pass the winter in a pupal stage, other like black fish of Alaska recover even if it is frozen for periods. It has been found that in certain plants and animals resistance to cold is achieved by the removal of water from tissues. By removal of water from tissue the osmotic pressure of the body fluids is increased. This lowers the freezing point of these organisms, so at extremely low temperature dehydration of the tissues to a certain extant is essential for survival. Removal of water prevents the formation of ice crystals in the protoplasm which would cause rupture of the cells. Thus, some insects are known to have the ability to be super cooled to as low temperature as – 40°C to – 50.0°C. Along with freezing a great reduction in the rate of respiration has also been reported. Another method adopted by the animals to pass unfavourable conditions is inactivity or dormancy. The dormancy during winter is called hibernation while that of summer is called as aestivation.

Hibernation

It is a unique physiological state between sleep and death. It is the method by which nature protects her helpless offsprings which otherwise would die of starvation and cold. Hibernation is different from sleep even though it is some time referred to as winter/summer sleep. Sleep is temporary and brief period of complete relaxation. It is natural phenomena but its exact physiological mechanism are not understood. It differs from cold necrosis and from sleep in that it is a total body phenomena. The respiratory organs are adopted to hibernation but in cold necrosis death usually results from respiratory failure. In sleep there is no lowering of metabolism although there is a reduction in nervous activity and slowing of heart beat and lung ventilation rate. But it is known that sleep is associated with the hypothalamus and it is on account of this fact that sleep and hibernation share common factors. Hibernation is a lengthy and prolonged process by which animal are able to tide over unfavourable conditions of winter when temperature is low and food is scarce. In this state all movements are suspended and general metabolism is reduced to a minimum. The body undergoes many physiological changes. Many warm blooded mammals, reptiles, and amphibians undergo this process of hibernation. Mass hibernation is seen among bats, snakes, and alligators. The winter sleep of bears and suspended animation of insects, moluscans and nematodes are however not hibernation. Species well known for hibernation belong to order rodentia. These are generally mammals which are relatively poor regulators and small enough to have to maintain high rate of metabolism per unit body weight. The wood duck or ground hog, the ground squirrel, the dormice and the hamster are true hibernators.

Similarly among insectivora and hedgehog, the fenrice of Madagasker, the hedge hog of Ethopia also hibernate. Among the bats the microchiroptera for example *Rhinabatus, Myotis, Macrotus* and *Plecotus* of northern hemisphere are all true hibernators. None of the Microchiroptera of tropics move to hibernation.

Animal preparing for hibernation feed actively during summer and autumn when food is plenty. This helps them to store great quantities of fat and then the blood system of these animals take charge of the process. The blood changes in its characters. This is to prevent any putrefaction during hibernation. After this WBC return back to blood stream. The creatures become dull, drowsy and curls up and enters into a trance like existence.

Histological studies showed that the thyroid, the adrenal cortex and adenohypophysis all involute. The adrenal medulla, however, don't undergo any apparent change but the parathyroids are hyperactive so also the brown fat which is in fact an endocrine gland. During this period of hibernation the creature slowly consumes the fat stored in their body. The metabolism is reduced some 20-100 times below the normal level. Very shallow breathing and a feeble circulation are maintained. The body temperature drops sharply, heart beat slow down and pulse rate reduced and all apparent movements cease. This state of torpor is reached not all at once but in gradual stages. Previously it was believed that hibernation was achieved by animals drugging themselves by constantly breathing their own carbon dioxide exhalations but now it seems that the process starts in the blood stream.

During waking up process the pulse rate leaps up, the remaining fat reserves are utilised quickly and the body temperature return to the normal state. If a hibernating animals is artificially awakened before its proper time it may prove fatal. Certain animal like ground hog may be induced to hibernate by injecting insulin into their blood.

Effects of Very High Temperature

The greatest intensity of heat at which a species can live actively is its maximum effective temperature. At very high temperature the dissolved albuminoids probably undergo destructive chemical changes. Death may be due to asphyxiation or as a result of failure of certain enzymes. There will be suspension of protoplasmic activity leading to a state of coma and even death. Thus, high temperatures are intolerable for many animals and such conditions can act as an effective limiting factor in the distribution of organisms. Instances are on record which indicate that rise of temperature from normal range can lead to mass mortality. A curious effect has been noticed in respect of tropisms of certain animals increased temperature causing a change from positivity to negatively phototropism.

Adaptations to High Temperature

Some adaptations of terrestrial animals for high temperature include devices for evaporation from skin or lungs and in certain social hymenoptera form the nests. Some animals migrate when the place become too hot, some burrow into the soil, while others may become nocturnal in their habits. Many animals and plants withstand high temperature by passing the adverse effects as highly resistant spores, seeds, eggs, pupae or cysts. In some special parts of the body very hard and resistant to thermal extremes as in the case of snail, *Ariophanta, Pila,* the slug, *Vaginulus,* the red hairy caterpillar of south India avoid high temperature by entering into a state of dormancy until the onset of next normal conditions. Larvae of sedentary animals like barnacles, tube worms avoid high temperature through their behavioural patterns. The free swimming states show a tendency to settle in the neighborhood of previously settled organisms of their own species thereby exhibiting a sort of gregariousness. The majority of aquatic animals live at temperature below 40°C. Animals living in small ponds are adopted to withstand high temperature than those live in streams and lakes.

Salinity and Osmoregulation

Salinity refers to the salt concentration of water. The inland waters exhibit a wide range of salinity where as open sea waters maintain fairly constant levels of salinity, uniform through out the open ocean. Average salinity of the ocean is about 35 per cent. Variations is salinity of surface waters in the ocean are small ranging between 30 to 36 per cent. In general salinity is minimum near the equator and maximum at the latitudes of tropic of cancer and tropic of capricorn. From the regions of the surface salinity shows declining trend towards the respective poles. A remarkable feature of the sea water is that although the total amount of dissolved materials vary from place to place, there is no variations in the relative amounts of various substances dissolved.

Their ratio of dissolved salts in the sea water shows great consistency. This enables the calculations of the total salinity of a sample of sea water by measuring only one of the constituents and then multiplying by a known factor to get the total dissolved salts. That is chlorine content will help in calculating the complete salinity of the water.

In the areas of tropics and in the red sea salinity will be as high as 40 per cent or more, where as in the Baltic sea it is reduced only 8 per cent owing to dilution by rain, snow or river waters. In the polar areas melting of ice causes dilution of sea water during summer. The salinity of sea water around India is subjected to great seasonal fluctuations due to monsoon rains. The effect of this rainfall and the consequent changes in salinity vary from region to region. Six large rivers namely, Ganga, Krishna, Godavari, Kaveri, Narmada, Mahanadi and Brahamputra and a large number of small rivers open into Bay of Bengal. On the eastern side Irrawadi, Sittang, Salween, and Tenasserim open into the Andamans sea. All these greatly influence the salinity and other properties of water in the Bay of Bengal. Dilution to the tune of 18 per cent has been reported.

The influence of river water on the salinity is evident from the fact that the Arabian sea which is on the same latitude as the bay of Bengal has a salinity of 35 to 37 per cent almost throughout the year. This high salinity is due to intense evaporation and influx of high salinity water from the Red sea and the Persian Gulf.

Only three major rivers namely, Indus, Narmada and Tapti empty into it. In the back water along Indian coast also salinity shows great fluctuations. In the estuaries and coastal waters receiving large amounts of run-off salinity may vary over the whole range from nearly freshwater to full sea water. Many animals can tolerate decrease in salinity while others like plankton of open sea perish and die and are sensitive to such changes. Marked dilution occurs along the coasts during the monsoon season due to the result of influx of river water.

Variations in the salinity of the medium will affect the organisms chiefly through specific gravity control and variations in osmotic pressure. Since specific gravity of the soft tissue is equal to the sea water, it is advantageous. Some forms such as Sargassum and Siphonophores counteract in their excess weight through the devices like afloats filled with gas.

A study on the role of variations in the salinity level on the intertidal meiofauna of tropical waters was conducted along the west coast of India (Ingole and Parulekar, 1998). It reveals that distinct invertebrate communities were associated with different salinity regimes. Since the occurrence and abundance of dominant species and dilution of oceanic waters is influenced greatly by monsoonal rain, the presence or absence of known species during a particular season could be predicted on the basis of field salinity.

The specific gravity of benthic forms and sedentary organisms is generally high. Since both Sp. Gr. and salinity ranges are small in the open sea, those do not seem to have much influence on the distribution of organisms. However, some inhabitants of sea seem to be sensitive to difference in Sp. Gr. The salt contents of marine cells is greatly similar to salinity of sea water and the changes in the salinity are compensated by suitable osmoregulatory mechanisms which can be seen in all groups from protozoa to fishes.

It is probable that sea water contains traces of every chemical elements known to mankind. The difficulty in detecting them is that they are present only in very minute traces and our methods are not sensitive enough to detect their presence. Scientists have already detected the presence of nearly 50 elements in sea water. Thirty nine elements that are present in detectable quantities in sea water are given in Table 2.2 in the order of the amount of each dissolved in the sea water.

Table 2.2: List of Elements

Chlorine	Cobalt	Tin
Sodium	Lithium	Cerium
Magnesium	Phosphorus	Gold
Calcium	Barium	Silver
Potassium	Iodine	Vanadium
Bromine	Arsenic	Lanthanum
Strontium	Iron	Yetrium
Boron	Manganese	Nickel
Silicon	Calcium	Scandium
Fluoride	Zinc	Mercury
Nitrogen	Lead	Radium
Aluminum	Selenium	Cadmium
Radon	Cesium	Chromium

Some of these elements are very essential for the growth of plants and animals. For example calcium is essential for molluscans and corals for making their shells and skeletons respectively. These animals remove huge amounts of calcium from water and sea water. Shells of the dead animals settle to the bottom of the sea to become limestone. Fresh supplies of calcium are brought in by the rivers. Similarly silica is required by the diatoms for making their glassy shells and large quantities of silicon is brought in by the rivers into the sea.

It is interesting to note that many marine animals have the capability to extract various elements and retain them with their bodies. Some of which are found only in small amounts in the sea water. For example sulfur bacteria can absorb sulfur, radiolarians absorb silica and srontium. The jelly fish absorbs zinc and lead. The crustaceans and molluscans contain copper as haemocynin and the squirts contain vanadium. Certain elements such as cadmium, bismuth are present only in mall concentrations in sea water so that their occurrence is detected only from the ash of marine organisms. The greater part of the salts in the water consists of chloride of sodium, magnesium and potassium of which sodium chloride is most common. Sulfate of magnesium, Calcium and potassium as well as calcium carbonate and magnesium borates are also present.

Methods of Meeting the Osmotic Problem

Fresh water organisms have a body fluid, which is far richer in salts than the medium which surrounds them. Such animals are hypertonic to their surrounding medium. If a cell is hypertonic to the medium and is surrounded by a differentially permeable membrane, water will enter the cell with a certain force. This is equal to the osmotic pressure. The osmotic pressure of a solution is proportional to the number of dissolved particles or ions in it. The higher the concentration of solution, the greater is the osmotic pressure. The problem of fresh water animals is to retain the salts and remove excess water which inevitably penetrates through the membranes of the surface. The problem of a marine fish is the just opposite to gain water and secrete the excess salts. To maintain a favourable salt balance most aquatic animals have mechanisms for osmotic regulations, through kidneys or have special secretary cells.

Water which is isotonic with protoplasm i.e. water containing salts in such concentrations that osmosis between it and the organism is balanced and is the most favourable medium for animal life. This is the condition of sea life and it is by virtue of this fact that the sea has been the ideal dwelling place for most of the vertebrates. Here animals experience little difficulty in maintaining water balance. In the case of fresh water protozoans the protoplasm has a much higher concentration of salts than the medium. So there will be a flow of water into their bodies which they remove by the action of their contractile vacuoles. Thus, a paramecium expels several time its own volume of water in an hour.

The salt content of marine animal cells is usually close to that of the surrounding sea water and change in salt content of the medium is attended with osmoregulatory mechanisms to maintain balance. Many methods have been adopted by animals for osmoregulation such as external protection from the ambient medium. Protection of cell membranes is attained through the excretory mechanisms for the removal of excess water from body. The body fluids of Coelenterates, Echinoderms and Molluscans have the same osmotic pressure as the outside medium.

Several invertebrates in the sea remain reconciled to the outside medium keeping their body isotonic with the medium. Thus, the sipunculid, *Phascolosoma* and the polychaete, *Arenicola marina* and *Neris pelagica* can with stand change in salinity. Such organisms are osmolabile or poikilosmotic in distinction to forms like *Nereis diversicolor* the flat worm and Gunda, the crab which have the ability to regulate the concentrations of their internal environment.

Crustacean blood may have higher or lower osmotic concentration than the environment. Many teleosts are hypertonic to sea water. So in this case they lose water from their tissues to the surroundings and the effect of it is like entering into a very dry zone. One way of gaining water is through drinking is an adaptation of teleosts. An eel (teleost fish) can drink up to 200 ml of sea water daily and about 60-80 per cent of it may be extracted as water.

Another method of correcting this imbalance is to secrete chlorides, through chloride cells found at the base of gills. In fresh water, teleosts are hypertonic to their medium. So water diffusion through differentially permeable membranes, diluting the body fluids and swelling up the body which may interfere with the normal functioning of the protoplasm would occur. The excess amount of water that thus enters the body is eliminated by kidney as very dilute urine. The salt lost from the body is balanced by absorption from the medium by gills. This method involves a constant expense of energy.

Elasmobranchs keep a higher osmotic pressure in sea water by retaining the nitrogenous excretory products and trimethylamine oxide in the blood instead of excreting them. Fish such as eels and salmon migrate from fresh to salt water or vice versa in relation to their breeding cycles, usually they spend several days in estuarine areas before entering into the fresh waters. During this period they get acclimatized to the altered environment and are able to maintain their water balance in the new surroundings.

Stenohaline and Euryhaline

Organisms with a relatively low tolerance for difference in salinity are known as stenohaline forms. Such animals are characteristic of oceans where salinity variations are very narrow. These forms usually die if carried into coastal or estuarine areas with low salinity. Plant and animals that are able to tolerate a wide range of salinity are termed as euryhaline. Such forms are characteristic of coastal waters and estuaries where great fluctuation of salinity occur. Euryhaline forms are able to withstand these wide changes of the medium by adjusting the permeability of the skin and by secreting a mucous covering for the body fluids. Euryhaline forms have enzyme systems which can tolerate a great range of salinity and have efficient power of osmoregulation.

The term euryhaline and stenohaline are, however, entirely relative without any sharp dividing line. Some are known to acclimate gradually and withstand wide range of salinity. These may not tolerate rapid change in salinity. It has been shown that ability to tolerate low salinity is greater at high temperature. This is natural and more research should be carried out on osmoregulation and for this higher temperature levels are conducive. The optimum salinity may also decrease with high temperature.

Adult animals seem to have greater tolerance ranges for salinities than young ones. Egg and larvae cannot tolerate low range but adult can do it. However, exceptions to this have shown in case of the crab, *Carcinus* sp. in which the young ones are more tolerant to high salinities than adults. Most stenohaline animals are isotonic with the sea water, therefore, there is no waste of energy for osmoregulation. Consequently they have not developed any osmoregulation and are unable to penetrate into brackish waters.

Salinity can limit the distribution of many plants and animals. This is seen in case of salt lakes and estuaries where organisms are distributed from high tolerant to least tolerant areas. Variations in salinity of the medium affect the organisms chiefly through specific gravity control and variations in osmotic pressure. Since the specific gravity of soft tissues is generally close to normal sea water, it is advantageous for both swimming and floating forms.

In many planktonic organisms oil droplets are present and some seem to have the ability to regulate these oil droplet as required. Some algae like Sargassum and Siphonophores increase their buoyancy through special devices such as floats filled with air. The specific gravity of benthic and fixed forms is generally high. Since both salinity and specific gravity are in small range in open sea, this do not seem to have much influence on the distribution of organisms. However, inhabitants of deep sea are sensitive to difference in specific gravity.

C. Pressure

Pressure influences the organisms in two ways. It may be mechanically or physiologically through osmotic phenomena. The mechanical impact is related to gravity. Organisms living at the sea level are subjected to an average total pressure of one atmosphere and this is used as a standard of measurement. This pressure can support a column of mercury 760m height. The pressure at any given point in the sea level is function of water density and depth. Therefore, the pressure at 10,000 m depth is equal to 1000 atmosphere. Pressure increases with the depth in water and decreases with the height of the altitude. It get reduced to half atmosphere at a height of 5,150 m and one fourth at 10,300 m height on land.

There seems to be a correlation between animal activity and changes in the barometric pressure. It is difficult to isolate the effect of pressure in nature from those of other environmental factors. For example at high altitudes there is decrease in the partial pressure of atmospheric gases particularly the gases like oxygen and carbon dioxide. Further there is a lowering of temperature which in turn affects the growth of plants and, therefore, the food of animals. Birds are unable to fly owing to the thinness of air and reduction in the amount of oxygen.

The effects of pressure increased with the increasing depth of water. It is more significant in water than that which decreases with the increase altitudes on land. It has been said that there is an increase of one atmosphere of pressure for every 10 metres increase in depth. Since average depth of oceans is about 4000 m, the pressure there will be about 400 times as great as that at sea level. In the Marina trench (10,860 m) the pressure is 1086 atm. This pressure alteration with the depth constitutes an important environmental variable in the ecology of marine organisms.

An important effect of pressure on biological system is as follows: (1) Protein molecules are compressed and altered in structure and chemical activity and denatured at high pressure. (2) The viscosity of sea water and protoplasm varies with the pressure and colloid gels take up more water at high pressure. (3) It is also known that the action potential and rate of propagation of nerves can decrease their threshold of stimulation. Owing to the fact that gases are much more soluble in water under high pressure, carbon dioxide at the deep layers goes into rapid solution making the water acidic which renders calcium carbonate easily soluble. This makes the deposition of calcium difficult in deep sea animals and accounts for the comparatively week skeletal structures. Our information is incomplete regarding the influence of high hydrostatic pressure on the physiology of marine organisms. (4) Bacteria are known to occur at great depths and those which grow best at high pressure are called barophilic organisms. Recent studies on bacteria indicate that there exist a special adaptation of bacteria to high hydrostatic pressure and if it is probably multicellular organism which will demonstrate even more strictly pressure dependent physiology. It has also been found that the action of some beteriophage is retarded by exposure to 1000 atmosphere pressure for 45 minutes. Some yeast fail to carry out fermentation at a pressure of 600 atmosphere.

The putrefaction of variety of organic substances is inhibited by prolonged application of 700 atmosphere pressure. Increased pressure beyond 250 atmosphere causes cessation of movement in amoeba. Many invertebrates are inactivated by the pressure above 400 atmosphere. Experiments have demonstrated a variation in the response of cells of different species at different pressure at room temperature. The pressure required to block the cell division or to cause the rounding of certain cells is 400 atmosphere in amoeba, 330-400 atmosphere in echinoids, more than 800 atm in ascaris, between 200-270 atm in some molluscans and annelids.

Among certain fish eggs cleavage is retarded and abnormal embryos develop when subjected to high pressure. Animal inhabiting shallow waters are usually killed by exposure to a pressure of about 400 atm. The bivalve, *Mytilus edulis*, the *Balanus*, *Amphitrite*, the barnacle attached to rocks are reported to have survival at a dive from the surface to 2000-3000 m. Other effects of increased pressure reported are the retardation of the heart beat of fish, an increase in rate of ciliary beat of molluscans and stoppage of muscle contraction and blocking nerve conduction in frog and an increase in the tension of isometric contraction of turtle heart.

Deep sea animals are usually dead when brought to surface as result of release of great pressure to which they subjected in their habitats. It may be generally stated that forms like bacteria and protozoa are less sensitive than other aquatic invertebrates and the sensitivity increases with the increasing complexity of the organization. As stated earlier our information is meager on the effects of pressure on organisms. Nevertheless there is no doubt of the fact that marine invertebrates too will show special physiological adaptations to high pressure. It is known that even small pressure changes can elicit behavioural responses in certain marine invertebrates but the anatomical basis for this still remain unknown.

Dissolved Gases

We are usually more interested to associate with atmospheric gases than the gases dissolved in aquatic systems. It is understood that at least 5-6 important gases are found to be dissolved in lakes, streams, ponds and seas. They all have biological and physico-chemical functions. But they differ from one another in their bahaviour and origin. Surface water is in contact with mixture of gases and water vapour called air, absorb some of its components. Among them the nutrients like C, N and O have special importance because they are essential as biogenic salts. Among the gases CO_2 and O_2 are the two important gases exchanged between organisms and their environment in a reciprocal way. These two gases are important in the vital processes like photosynthesis and respiration.

$$6CO_2 + 6H_2O \xrightarrow{\text{in presence of light}} C_6H_{12}O_6 + 6O_2$$

Dissolved Oxygen

Oxygen is needed by all organisms to make the energy available in organic food materials. Majority of plants and animals use free oxygen from air or water for oxidation of organic matter.

The anaerobic forms obtain their energy by partial decomposition of organic matter without free oxygen. The abundance of oxygen in the environment may become critically low for aerobes, but it never becomes harmfully high for these forms. However, natural concentration of oxygen may be become seriously detrimental to some anaerobes.

Oxygen occurs as a physical mixture in the air and dissolve in simple solution in water. Dissolved oxygen does not combine with water but it does that with iron and other inorganic materials in water. Oxygen is taken in and released by organisms, hence it is referred as a modifiable factor under such condition of the environment (air). But it is a non-modifiable factor as such.

Availability of Oxygen

It constitutes about 21 per cent in the air. Plants and animals continuously remove oxygen from their environment and plants add to it. These life activities will not alter the oxygen level due the great volume and mobility of the air medium. Most part of the terrestrial environment is provided with uniform and adequate supply of oxygen. Two places in the terrestrial environment i.e. at high altitudes and in subsurface of soil, shortage of oxygen is felt. The amount of reduction in oxygen with altitude is proportional to the reduction in atmospheric pressure. Thus, the oxygen level at an altitude of 5500 m oxygen is only half as abundant as it is at sea level. The level which was 21 per cent in the air is reduced to 10 per cent in loamy (aerated) soils and still low in the soil below water table. The air circulation is locked in very fine textured soils.

Aquatic Environment

The saturation level of oxygen in water varies with temperature, salinity and pressure. Some saturation levels are showed in Table 2.3

We know that the 21 per cent oxygen in atmosphere is equivalent to 210 cc per litre in aquatic environment. The world's reservoir of oxygen is atmosphere. The chief source of oxygen in water is the one which can be absorbed from air and the second source is the oxygen released during photosynthesis by aquatic plants. Water may loose oxygen by diffusion from its surface into atmosphere. Oxygen is also used up by aquatic plants and animals

Table 2.3: Dissolved oxygen levels in fresh and marine waters

Temperature °C	*Fresh water mg/L*	*Sea water cc/L*	*Salinity 36 per cent cc/L*
0	14.7	10.3	8.0
15	10.3	7.2	5.8
30	8.3	5.6	4.5

during respiration and decomposition of organic matter in the water. Living organisms seem to influence both the supply and depletion of oxygen in natural water bodies. Oxygen is a modifiable factor in aquatic environment but a unmodifiable factor in the air. The actual amount of oxygen at a given time in water may be the result of a balance between the process of supply and depletion.

The factors considered important in the study of oxygen in the environment are as under:

1. The surface layer.
2. The subsurface zone.
3. Deep layers.

The water at surface is found to be in equilibrium with the air above and so that the oxygen value near the surface may be further influenced by the composition of surface air and salinity of the medium. In the subsurface zone a great variation in oxygen level may be due to that the water does not have easy exchange with the atmosphere. The respiration and decomposition activities or organisms cause depletion of oxygen in all layers. Photosynthesis occur up to the compensation depth i.e. up to the termination level of the euphotic zone. Under polluted conditions free oxygen may not be available or exhausted rapidly. If the rate of photosynthesis exceeds the rate of respiration, then the oxygen level as high as 180 per cent in the sea and 300 per cent in the inland waters will be noticed.

Cyclic Changes

Often cyclic fluctuations in oxygen levels have been observed in natural water bodies. In some cases the diurnal fluctuations were observed which are known as oxygen pulse. For example, oxygen pulse or diet cycles have been studied in fresh water ponds also. The oxygen pulse is studied in waste stabilization ponds (Hosetti 1987). In these ponds the depth will be in the range of 4-6 ft. The upper layers are always aerobic and bottom layers are towards anaerobic, such facultative systems show a wide range of diurnal cycles in the oxygen levels.

Table 2.4: Variations in oxygen levels around the clock in waste stabilization ponds at Karnataka University Dharwad

Time	*Oxygen(mg/l)*
6.0 am	2.0
9.0 am	5.0
12.0 noon	12.0
3.0 p.m.	16.0
6.0 p.m.	9.0
9.0 p.m.	6.0
12.0 midnight	4.0
3.0 am	2.0

During the day time large amounts of oxygen is produced by the photosynthetic action of green algal cells. Majority of it is used by the bacteria in the process of degradation and a part of it was diffused to the air. Since there is no oxygen production in the night. Hence, oxygen gradually increases in the day time as the day progresses and declines with the consumption by bacteria in the night hours and thus, the diet cycle continues (Fig. 2.3).

Oxygen also reaches the bottom of oceans and lakes slowly through water by direct diffusion or by circulation of water itself. The oxygen rich surface water is carried downward by processes like wind stirring, eddy conduction and by mass sinking. Wind circulation is feasible in oceans and large lakes and reservoirs. Wind circulation is stopped when water become ice or trapped due to lowered temperature and salinity and density. In the open sea where highest winds and largest waves occur, little influence of wave actions felt below 100 m.

In lakes

Oxygen may be carried to the depth beyond the limit of effective wave action by eddy induction. The exchange will bring

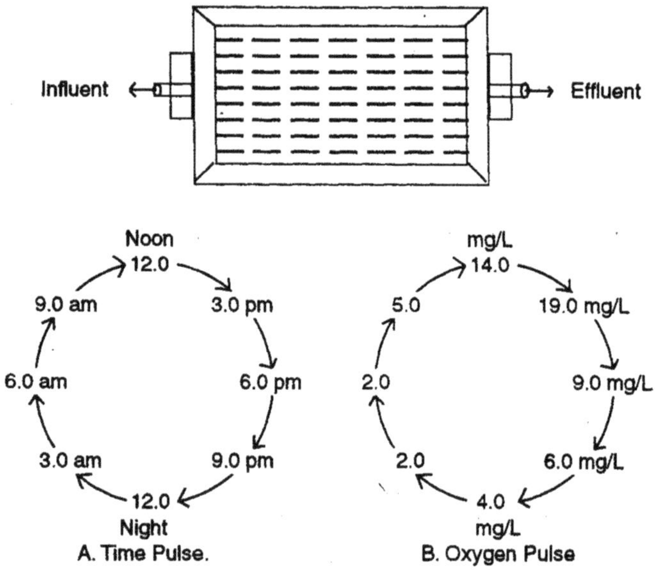

Fig. 2.3: Schematic diagram showing waste stabilization pond with diel cycle.

about a transfer of any characteristics of the water that differ in the adjacent water masses. The rate of transfer depend upon the following factors:

a. The rate of mixing of water units.

b. The difference in concentration of each characteristic.

The rate of mixing or constant coefficient is the property of the water under existing conditions of turbulence and it affects all characteristics that are being exchanged between water masses. During eddy's action oxygen and heat may be transferred to the lower layers and phosphate may be moved upwards. Eddy conduction is a process by which materials are transferred at right angles to the direction of current. The transfer may take down vertically or horizontally or in some other direction. The downward transfer of oxygen by eddy transfer is slow but it is commonly thousand times greater than the diffusion rate. The eddy conduction helps to furnish the oxygen to deep layers in ocean and lakes. In case of abyssal animals oxygen reaches them by mass sinking or deep circulation of water. In a typical deep lake of temperature region the water masses merged with oxygen at surface sink to the bottom during spring and the fall overturns. As a result of extensive circulation the water becomes muddy and turbid.

Sea water in tropical area near the surface contain about 4 cc/L. But as it becomes colder at high latitudes oxygen is absorbed from the atmosphere and produced by photosynthesis until a concentration of 8 cc/L is attained. When the water mass sinks to the bottom, depletion begins due to respiration and decomposition but oxygen of 5 cc/L remains at the abyssal layers and this process is very slow.

The concentration of oxygen in deep layers at the end point depends upon these factors:

1. Amount of oxygen at the surface.
2. The rate of which oxygen used in respiration and decomposition.
3. The time lapse.

Usually in atlantic areas minimum is 3 cc/L but in certain cases it may be nil between the depths 300 to 1300 metres. Complete lack or very low oxygen level found only in special cases in which water circulation is cut off and decomposer organisms are accumulated. In case of black sea where oxygen is absent at depths 150 to 2000 m. Oxygen availability in water is also influenced by redox potential or oxidation-reduction potential. The redox potential is measured in volts. It is an expression of electropositivity or elctronegetivity of water.

Effects of Oxygen

Now let us know ecological effects of oxygen in various habitats. Since oxygen exists in abundance in the lower layer of atmosphere, it has no limiting action in most terrestrial situations. Its shortage in high elevation limits the distribution of animals and plants. No mammal can live permanently at latitudes at which the partial presence of oxygen is less than 45 per cent of its value at sea level. Oxygen shortage and thinness of air at high altitudes imposes a restriction on the latitudinal organisms having high oxygen requirement.

The lack of oxygen in poorly aerated soils becomes seriously harmful to the roots of most plants at concentration below 10 per cent. If the oxygen level falls below 3 per cent then even root penetration in the soils was stopped. Under flooding conditions most of plants are drowned and killed. Roots of some plants, however, posses special adaptations enabling them to grow in soil layers devoid of oxygen. In some cases like herbaceous hydrophytes, air space are present within the stem and major root allows the oxygen to move from atmosphere and delivered to the lower root system. The mangrove found along the coasts of Florida and on both the coasts of India produces root branches called as pneumatophores which grow vertically upward through the mud and water until they project out into air and absorb oxygen. Some plants are capable of withstanding flood conditions by anaerobic existences.

Most cryptogamic plants and soil animals depend upon aerobic respiration and are not reported in the soil which lacks oxygen. A few higher forms like earthworms may survive for short period in the absence oxygen, when soil is soaked. In order to overcome oxygen shortage most soil animals migrate to surface layers. This behaviour can be observed while irrigating the field most of the soil insects and larvae come to surface and egrets prey upon them in the tropical areas. Many organism undergo dormant stage, when oxygen become too scarce. Many facultative microbes take up anaerobic life. Special nitrogen fixing bacteria *Rhizobium* and *Clostridium tetani* have special concern to man. It is obvious that the variability in oxygen in soil in turn limits the distribution and ecology of both aerobic and anaerobic organisms of soil environment.

Aquatic Environment

The range of oxygen in aquatic environment is between surpersaturation to complete exhaustion. The oxygen is always high in atmosphere than it is in the water. Due to this reason many aquatic insects and fish come to surface, engulf air and go back into sublayers. For example the water bug, *Notonecta*, Larvae of Odonata (damsel and dragon fly), the water scorpion (*Ranatra*) and mosquito larvae possess breathing tubes that extend to surface. Some snails refill their lungs with air.

Most animals get along with available meager amount of oxygen in water. The inhabitants of water environment display a wide variety of adaptations that serve to obtain sufficient supply of oxygen from the medium. Sessile animals cause water to flow through their respiratory chambers. The echiuroid worm keep a current of water flowing through the burrows in the sand. Higher mechanisms for pumping water continuously over its gills is noticed in *Mackerel*.

Some organisms posses special respiratory pigments or physiological adaptation that helps in absorption of oxygen at low partial pressure. The chironomus larvae in the muddy bottoms of ponds are provided with a type of hemoglobin that becomes 95 per cent saturated under a partial pressure of oxygen of only 10 mm/g where as mammalian hemoglobin is less than 1 per cent saturated at this pressure. Such insects are also capable of simple anaerobic metabolism whereas lactic acid formed from glycogen breakdown is excreted and in turn eliminate this compound rather than oxidising it as most animals do. It is difficult to determine the minimum concentration of oxygen required by various animals. The value is influenced by temperature, pH and other modifiable factors and also the rate of accumulation to availability of oxygen. Some fishes can live at oxygen concentration 0.1 to 0.3 cc per litre. Other muddy fishes like mackerel requires a minimum of 3.6 cc/l at 70 mm Hg at 24°C.

Under oxygen depleted conditions the animals migrate to other regions. The copepods are known to leave hypoliminion of lakes during winter and summer stagnation periods. Many plants and animal die due to exhaustion of oxygen. Sudden fish kills are reported in polluted rivers of India. Every years regular fish kills have been reported from the River Thungabhadra due to pulp and paper industry effluents mixing in water at the downstream of Harihar in Karnataka (Hosetti and Patil, 1988). Most organic pollutant cause oxygen sag in the water bodies. In permanently anaerobic zones of water no aerobic organisms can sustain. Under such situations like deep lakes and black sea and heavy polluted waters only anaerobic forms of bacteria and anaerobic protozoa can live. Slime worms, tubicidae members, pea clams, and insect larvae *Corythra, Chironomus* etc., can live with minimum oxygen in the deep water bottom of lakes.

Carbon Dioxide

Carbon dioxide is one of the essential component required for photosynthesis. Carbon dioxide provides carbon source required for the growth of all green plants and indirectly for all tropic level organisms. Carbon dioxide also influences other features of the environment. Quite in contrast, to the oxygen, carbon dioxide combines with water and forms carbonic acid. Though this reaction affects pH, but it forms ecological important compounds. Carbon dioxide affects respiration of animals and in combination with calcium takes part in their bone and shell formations.

During the respiration of organisms carbon dioxide is produced. This process is the reverse of photosynthesis. Carbon dioxide also released when dead organic matter is decomposed by oxidising or fermenting activities of micro-organisms. On other hand carbon dioxide is withdrawn from environment by the growth of green plants and by lime secretion. Since the abundance of carbon dioxide affected by living organisms, we may certainly say that is a modifiable factor of the environment. Because of the reciprocal relationship between oxygen and carbon dioxide whenever one is depleted the other will increase.

Carbon Dioxide in Terrestrial Environment

It is present in earth's atmosphere in extremely small amounts. It constitutes only 0.03 per cent of the air medium. Hence carbon dioxide is only 1/10th of oxygen abundance in the air. Because of great mobility of air carbon dioxide is well distributed and its low concentration is sufficient for a moderate amount of photosynthesis. Probably reduction in carbon dioxide at high altitudes curtails the process of photosynthesis of plants. But low temperature and pool soil conditions are usually more important in limiting plant growth in mountains.

In nature some instances of local increase in atmospheric carbon dioxide are recorded. In some volcanic areas carbon dioxide concentration is more which is harmful to organisms living around. Such places are referred to as death valleys where all the life are excluded. Such conditions exist in the volcanic places in Java. The carbon dioxide level in soil environment is increased due to decomposition and respiration in underground surface where poor aeration exists. The air just above cultivated lands found to show 10 fold increase carbon dioxide level.

Our atmosphere receives carbon dioxide from geological and industrial sources. The supply increased by carbon dioxide released from the oceans and from the respiration and decay of organic materials. Carbon dioxide is withdrawn from atmosphere during photosynthesis by vegetation. The existing balance between supply and consumption of carbon dioxide is low in the air. If this minute gas is absent in the air then all the life on land would come to a stop. The lives of terrestrial organisms thus depends on a slender thread of carbon dioxide balance in the nature. At present carbon dioxide is a general limiting factor.

Carbon Dioxide in the Aquatic Environment

The occurrence of carbon dioxide in water is more complex than on the land. Carbon dioxide is readily soluble in water but is very less in air (0.3 cc/l). Water is in equilibrium with it to hold it only about 0.5 cc/l of free carbon dioxide at 24°C. Actually carbon dioxide level in most natural waters is considerably larger due to the presence of bicarbonate and carbonates dissolved in it. Sea water at a salinity of 35 per cent normally contains a total of about 47 cc of CO_2/l. The ocean is a great reservoir of world's carbon dioxide and this marine source help in the regulation of carbon dioxide in the air just reverse of oxygen in water.

$$CO_2 + H_2O \longrightarrow H + HCO_3 \longrightarrow H + CO_3$$

The amount of carbon dioxide in simple solution plus that in the form of H_2CO_3 is called as the free CO_2. The carbon dioxide in bicarbonate ions (HCO_3 and in carbonate ions (CO_3) is called combined CO_2. When HCO_3 dissociates it releases hydrogen (hydronium) ions (H_3O) these affect pH. At high pH values more carbon dioxide is present as CO_3 ions and at low pH value it is present in free condition. Addition or removal of carbon dioxide will affect pH conditions of the water medium.

The bicarbonate, carbonate and other anions of week acids form a buffer system, which tends to resist changes in pH. The buffer capacity of water is determined by the abundance of these anions and it is directly related to alkalinity. Sea water and hard fresh water are relatively highly buffered media.

Ecological Effects of CO_2

Important effects of carbon dioxide in the aquatic environment is its role on photosynthesis of green plants just as higher plants do on land. In sea water with a pH of 8.3 contain less than 1 per cent carbon dioxide in free form. However both in sea and hard water contain a large supply of carbon dioxide as carbonate and bicarbonates. In soft water lakes the total supply of carbon dioxide is small. Extensive photosynthesis may reduce the level of carbon dioxide to the extent that further growth of plants is stopped.

The withdrawal of carbon dioxide from water causes a rise in pH leads to hindrance of phytoplankton growth. The abundance of carbon dioxide also exerts certain specific effects on animals of aquatic environment. The increased carbon dioxide level brings stress and it enhances respiratory activity in molluscans and arthropods and in others it has no effect. Special adaptation have been observed in fish to high carbon dioxide levels in littoral seas and inland waters in blood. Changes in carbon dioxide also tend to modify the pH and alkalinity of blood of fish under migration and also affected by increased carbon dioxide levels.

Hydrogen Ion Concentrations

For a proper understanding of the term hydrogen ion concentration, it is essential to know the definitions of some terms. A base is a substance which gives off the negatively charged hydroxyl ion (OH) for example potassium hydroxide dissociates to form potassium and hydroxyl ions.

$$KOH \longrightarrow K^+ + OH^-$$

An acid is a substance which gives off the positively charged hydrogen ions H+

$$\text{e.g. } HCl \longrightarrow H^+ + Cl^-$$

Salts: These are substances which provide on ionization a positive ion other than hydrogen and a negative ion other than hydroxyl ion.

Acidity: A solution where the hydrogen ion concentration exceeds to any degree of hydroxyl ion concentration.

Alkalinity: A solution is alkaline when the hydroxyl ion concentration exceeds to any degree of hydrogen ion concentrations.

Because of mathematical relation of OH ions to the hydrogen ion in the same solutions the alkaline intensity can also be expressed in terms of hydrogen ion concentrations.

Neutrality: In chemically pure water the hydrogen and hydroxyl ion concentration are equal. This is neutral condition. Among the molecules of water, about one out of ten million dissociates into hydrogen (H+) and hydroxyl OH-ions which have positive and negative electrical charges respectively. As stated above the positive hydrogen ions are acidic and the negative hydroxyl ions are basic. If these are equal then the water is neutral.

Hydrogen ion concentration is expressed by the symbol pH which is explained as the power of hydrogen or potential of hydrogeni. In a scale for the measurement of hydrogen ion concentration the pH value zero represents the pH of solution normal in hydrogen ions. pH 14 represents a solution normal in hydroxyl ions. All values between 0-7 represent acidic and those between 7-14 represent alkaline. Each unit increase in pH represents a ten fold decline in hydrogen concentration and a ten fold increase in OH ions. The pH scale, therefore, is a logarithmic scale. pH is a measure of the degree in which the water is acidic or alkaline.

pH Levels in Natural Waters

Natural water show pH range of at least 3.2 to 10.5. It may be beyond this range in some special situations. In case of lakes and rivers the range will be much less being about 6-9, but in case of open sea pH don't show much variations. While in smaller ponds variations are more and irregular. These small bodies of water show changes on different days and at different times on a single day. The magnitude of change depending on the abundance of aquatic plants, the air photosynthetic activity and the reproductive activities of both plants and animals. In the depths pH variation are due to changes in the nature of bottom substances. Based on the circumstances of the season natural waters may change to acidic or alkaline or vice versa. Sea water is considerably buffered and these changes in pH are not much significant. Such minute changes will have very little effect on organisms.

pH of soil and mud in the terrestrial habitat also shows variations depending on several factors such as nature of the parent material and the activities of organisms and the process of decomposition. pH value records for soils vary between 2.2 to 9.6. Values below 4.5 and above 8.5 are unusual. Volcanic ash is neutral. In the tropics the soils found to be acidic and dry areas are usually alkaline. But deviation to generalization are uncommon. Fungi can grow well in acidic soils and bring about decay. Earthworms and certain bacteria are sensitive to acidic soils and so are found in neutral or alkaline soils. Activities of plants can bring about changes in pH of the substances.

The distribution of organisms is not related to the pH of the medium. In contrast to this, some organism prefer separate pH ranges. Some species of Euglena grown well at pH 1.8 rotifers grow well at pH 10.0. The rotifer, *Monostyla bulla* tolerates pH range 3.7 - 9.1. Eggs of sea urchin, *Arbacia sp.* can develops in a medium of pH range 5.2-9.4 (cell sap). The protozoan, *Stentor coeruleus* live within a pH range 7.7 to 8.0. Isolated vertebrate and crustacean hearts can tolerate a wide range of pH. Changes in pH can bring about significant changes in physiological reactions to tissues, enzyme and other reactions. Marine algae can tolerate an acidic medium. Diatom can grow well in pH range between 6.5-9.0 with an optimum of 8.2. The cell sap of many algae is acidic. Experiments proved that in animals temporary changes is pH are not of serious consequences.

Electrolytes

Electrolytes are acids, bases or salts in a state of solution in water which can conduct electricity. These acids, bases and salts have different degree of solubility in water and based on this they have different degree of conductivity as electricity. So a week electrolyte will naturally have low solubility and low conductivity. On the other hand a strong electrolyte has high solubility and high electrical conductivity. By measuring electrical conductivity it is possible to find out the total electrolytes in natural waters. It is a fact that bodies of water with high productivity are rich in electrolytes. Due to various chemical and biological process electrolytes show horizontal stratification in aquatic systems.

Redox Potential of Eh

Oxidation is a process of addition of electrons and removal of electrons called reduction. It is possible to make quantitative investigations of oxidation or reduction depending on its relative position on oxidation reduction potential or redox potential or Eh (ratio). This is expressed as electrical potential. Redox potential is, therefore, an expression of the electropositivity or electronegetivity of the substance in solution as referred to a hydrogen standard. It is measured electrometrically in volts. A positive redox potential indirectly the reverse condition tending to bring about reduction. The measurement of Eh in aquatic environment and muds has been recently utilized as a possible means of studying the dynamics of these environments. The capacity to withstand changes in Eh is called poising capacity. This capacity is a function of weekly dissociated materials, that can act by dissociation, to overcome changes in oxidising or reducing substances.

Chapter 3
Freshwater Communities

A. Freshwater Ecosystems

The study of freshwater systems with respect to physical, chemical, geological and biological parameters is termed as Limnology. Freshwater biology emphasizes the ecology of organisms in relation to their fresh water habitat. Freshwater habitats occupy a relatively small portion of the earth surface as compared to marine and land habitats, but their importance to man is far greater than other areas because, they are the most convenient and cheapest source of water for domestic and all other activities of man. They also provide the important sites for waste disposal systems.

Distribution of Species

Freshwater habitats are isolated from each other by land and sea and the organisms with little dispersal mechanism fails to get established even in favourable places. The niches in different rivers and streams which are separated from each other will be occupied by different species. But small organisms like algae, crustacea, protozoa and bacteria having significant dispersal power become cosmopolitan in distribution, but not large animals like crocodiles and dolphins. A small animal like *Daphnia* is found universally in freshwater ponds in the USA, England, India and elsewhere.

As the concentration of salts is greater in the internal fluids of the body cells than in fresh water environment (hypertonic), water tends to enter the body by osmosis and salts must be excreted. Thus, freshwater animals must have efficient means of excreting water (contractile vacuole, nephridia or well developed kidney) otherwise the body will be bloated up with water. Because of difficulties in osmoregulation the marine animals are unable to invade the freshwater environment.

Examples for lotic system include all forms of inland waters in which the entire body of water moves continuously in a definite direction. They are in the form of brooks, rivulets, channels and rivers. Whereas the lentic environments include all forms of inland waters, lakes, ponds, swamps and their various intergrades in which the water does not flow continuously in a definite direction. Essentially, the water remains static, though a certain amount of water movement may occur, such as wave action, internal currents or flow of water in the vicinity of inlets and outlets. The sequence of genesis of these units is lake-pond-swamp in series.

There are no sharp boundaries between the two series as described above. In course of time the lakes tend to fill up while streams tend to reduce its bed to base level. When base level is reached, sedimentation occurs and a base-level meandering river results. This represents a more or less "climax" stage of the lotic system. If the effect of allogenic processes (geochemical forces) exceeds that of autogenic ones the small lakes or other forms of lentic systems which are transitory in geological sense may become extinct and get transformed into land area.

Difference Between Lentic and Lotic Environments

As a rule, the depths of all running water units are comparatively small. The water is confined to a relatively narrow channel and flows in one direction. All conditions, physical, chemical and biological gradually change with distance along the main channel in a definite direction in a lotic system. Extension of channel takes place as the stream increases its length and width and decrease their depth to base level with increasing age. This is in distinct contrast to the shrinking processes characteristic of all standing water units, like lakes and ponds.

At any position along the course of a running water unit, materials eroded at that point and all materials momentarily suspended or dissolved at that level are transported to the down stream with no chance of return. In lentic systems such materials commonly remain within the same basin. Constant flow of the whole water body of the lotic system eliminates prolonged summer stagnation of the bottom waters unlike lakes. Unlike lentic system the producers in the main channel of the lotic system are not able to produce sufficient food material because of their high water motions. In a stream the animals have to depend for their food produced in the surrounding land masses (catchment area) and in their tributaries.

Ecological Classification

Ecological classification of organisms are different from their taxonomic hierarchy. This is based primarily on their mode of nutrition or position in the trophic level and their life forms. Based on nutrition, the organisms are grouped into autotrophs or producers and heterotrophs. The heterotrophs are further divided into phagotrophs and saphrotrophs. These are green plants and chemosynthetic bacteria which can manufacture their own food by themselves and do not depend on other organisms for nutrition. Thus, these form the T_1 (Trophic level one) in the food chain.

Heterotrophs or Consumers

These organisms live on organic matter produced by the autotrophs and occupy the trophic level next to T_1. Heterotrophs are further subclassified into two categories.

Saprotrophs (Microconsumers or Decomposers)

These organisms mostly depend on dead organic matter for their nourishment and decompose it into simpler forms. They are generally kept in a separate position in the trophic level considering their various links with different trophic levels. Bacteria and fungi are examples of microconsumers. In the deep water aquatic systems some microconsumers like zooplankton also play an important role in the process of decomposition.

Phagotrophs or Macroconsumer

Mostly these are animals which seek their food by themselves and engulf and digest them with the help of well developed sensory neuromotor and digestive system. Aquatic phagotrophs are crustacea, insect larvae, fish and some other zooplankton. Based on the life forms these are further classified.

a. ***Benthos***

These organisms live in the bottom sediments. Mostly these are animals. The alga Chara, however, can grow well in deep waters of lakes. The benthic animals according to their feeding habit are subdivided into filter-feeders and deposit-feeders. The filter feeders like Unio, Clam, and other bivalves in the absence of their mouth depend on the water flow carrying small organisms which can pass through the gillsieve system. The deposit-feeders like Snails (gastropod) with the help of their mouth consume deposited organic matter from the bottom.

b. ***Periphyton***

These are plants or animals which remain attached to stems or leaves of rooted plants or other substratum projecting above the bottom in lentic or lotic waters. Examples are filamentous algae, protozoa and other attached forms.

c. ***Plankton***

These are microscopic floating plants and animals whose movements are more or less dependent on currents. Plankton as a whole can not move against strong currents, though some zooplankters have active swimming movements which help to keep them in vertical position. Plankton are mostly composed of algae, protozoa, copepods, ostracodes, cladocera, rotifers, crustacea and nauplii larvae.

d. ***Nekton***

They are fish, amphibia and insects which can navigate according to their will power.

e. ***Neuston***

Organisms which rest or swim on the surface of water are referred as neuston. Examples include, Duck weeds *Lemna, Spirodella, Pistia* and large insects. These mostly spend their time in the joining places between water surface and the air.

The Freshwater Communities

Among the existing diversified flora and fauna of aquatic habitats some are well represented in Indian waters. Algae are the most important aquatic producers and then comes the aquatic spermatophytes (seed bearing plant) ranking second. In animal kingdom from the biomass point of view the mollusks, aquatic insects, crustacea and fish are the dominant forms of consumers. The annelids and zooplankton like rotifers, protozoa rank second, but this group of consumers seem to be of more aquatic importance in the "economy" of the aquatic system.

1. *Lentic Communities*

The lentic systems like lakes and ponds with their various zones have characteristic organisms. In general vertical profile of ponds especially those of abandoned ones show the characteristic lentic communities.

Littoral Zone

Producers

The littoral zone include rooted or benthic plants belonging mostly to spermatophytes (seed bearing plants) and musk grass *Chara* and *Nitella* as producers. With these plants, phytoplankton and filamentous algae grow profusely in this zone. Sometimes neuston, like duck weeds, *Lemna* and *Azolla, Salvinia* during summer months cover up the whole surface of the aquatic body with their luxuriant growth. These plants reduce the penetration of light and wipe out the other submerged green plants (Fig. 3.1).

Important flora of this zone comprise: Cattail (*Typha*), Bulrush (*Scirpus)*, waterlily (*Nymphea*), pond weeds (*Hydrilla, Potomageton*) and *Vallinsneria* belonging to Spermatophytes, musk grass, (*Chara, Nitella*) of Charophyta (algae), green algae like *Spirogyra, Zygnema, Oedogonium, Scenedesmus, Coelastrum, Closterium* (desmids), diatoms like *Navicula, Nitschia, Fragellaria, Synedra, Gomphanema*, blue green algae like *Oscillatoria, Anabaena, Nostoc*, and *Mmicrocystis*.

Consumers

More diverse varieties of animals are found in littoral zone in comparison to other zones of lentic system. Large and weak

Fig. 3.1: A freshwater pond near Harige area of Shimoga City, Karnataka State India (Note: The pond is infested by various hydrophytes).

Place	:	Harige
Area	:	6 acres
Depth	:	8 ft.
Type	:	Eutrophic
Vegetation	:	Nelumbo nucifera, Nymphea stillata, Ipomea aquatica Ottelia alismoides, Hydrilla verticillata Jussiea repens, Azolla, Nitella
Purpose	:	Irrigation
Source	:	Channel and rainfed

swimming animals which can rest or cling to the aquatic vegetation predominate in this zone. Zooplankton are herbivorous (Primary consumers) of which crustacea such as *Daphnia, Alona, and Simacephalus* of Cladocera (water fleas), *Neodiaptomus*, Cyclops of Copepods, *Cypris* of Ostracods and some species of rotifers from the main bulk.

The periphyton animals (primary consumers) exhibit zonation which are parallel to that of rooted plants. Important periphyton animals on leaves and stems of rooted plants consist mainly of testaceous protozoa (*Areclla, Difflugia, Centropyxis*), ciliate protozoa (*Vorticella*) and sessile rotifers like *Rotatoria, Philodina* and developing embryo and young ones one snails of gastropods and eggs of frogs and toads.

Littoral neuston mostly are composed of aquatic beetle like Whirligig beetle of Gyrinidae (*Deneutes*), predatory diving beetle (*Dytiscus*), water scorpion (*Ranatra*) and damsel fly and dragon fly. Macrobenthos of littoral zone consists of snails (viviparous Pila amphibious habit) and mussels (like Unio) act as primary consumers. Vertebrates of this zone are adult frogs, toads, water snakes, turtles which act as secondary consumers. Important littoral fishes are air-breathing fish like Channa (Garai), Anabas (Kawai) etc.

2. *Limnetic Communities*

Producer

Phytoplankton are the important producers of this zone though in some seasonal blooms of macrophytes like *Azolla, Spirodella, Lemna* etc., also contribute to the total production. In general, members of Chlorophyceae, Bacillariophyceae and Myxophyceae form the main bulk of phytoplankton (Table 3.1). Along with this community species of Xanthophyceae (Txribonema) are also found. Dinophyceae (*Ceratium*) and Euglenoids (*Euglena, Phacus*) are very common. Sometimes blooms

Table 3.1: List of phytoplankton commonly found in the freshwater ponds of Karnatka.

Class : Chlorophyceae	5. *Asterionella*
Unbranched filamentous algae	6. *Pinularia*
1. *Spirogyra sp.*	7. *Stauroneis*
2. *Zygenema sp.*	8. *Gomphanema*
3. *Ulothrix.*	9. *Fragilaria*
Desmids	
1. *Closterium sp.*	**Class : Myxophyceae : Blue green algae**
2. *Volvox* : free swimming green algae	1. *Oscillatoria*
3. *Ankistrodesmus sp.*	2. *Anabaena*
4. *Chlorella sp.*	3. *Anacystis*
Class : Bacillariophyceae (*diatoms*)	
1. *Diatoma*	**Class : Xanthophyceae**
2. *Cymbella*	1. *Synura*
3. *Synedra*	2. *Peridinium*
4. *Nitzschia*	3. *Certium*

or phytoplankton are marked in the limnetic zone due to changes in the season or in the abiotic variables of the lentic system.

Consumers

The limentic zooplankton are made up of copepods, cladocera and rotifers, the species of which are different form that of littoral zone. Though copepods like *Diaptomus* are very common, cyclops, *Diaphanosoma* and *Bosmina* sp. are also noticed. Characteristic genera of rotifers are of the zone are *Branchionus, Polyarthra* and *Keratella* etc.

Many of the Crustacean zooplankton graze on water bacteria, phytoplankton and organic particles by means of combs of setae on their thoracic appendages. The zooplankton are primary consumers, which exhibit pulses just after the phytoplankton pulse as mostly they depend on the latter. Zooplankton show diel migration in lakes. With the sunrise occur they move to the deeper layer of lakes and come up at the dawn. The limnetic nekton consists entirely of fish, such as *Labeo rohita* (Rahu). *Catla catla* (Katla) and trash fishes like *Lepidocephalithys guntea.*

3. *Profundal Communities*

As there is no light, the inhabitants of this zone depend on the limnetic and littoral zones fot food and are mainly the consumers of different trophic levels. Flora of this zone is represented only by bacteria and fungi which are abundant especially in the watermud interphase where organic matters get deposited. Important profundal Carp fishes are Kalbasu (*Labeo calbasu*), Mirgal (*Cirrihinus mirgala*) and Siluroid fish *Bagarius bagarius.* Benthos animals like chironomid larvae or blood-worm, Tendipes, burrowing red annelids, Tubifex and bivalve mollusks, Lamellidens and a few fresh fishes are common.

Lotic Communities

Current is the major limiting factor in a lotic system. Most of the lotic communities show adaptations for maintaining position against the current of the streams. Producers of lotic communities have permanent attachment to a firm substrate such as stone log, or leaf. Periphyton as such contributes a large portion of the total plant population in a stream. Animals of streams develop such devices as hooks and suckers, sticky undersurfaces, streamlined or flattened bodies. The body of stonefly and mayfly nymphs of a stream is much flatter than the body of nymphs of related species living in lentic systems. The stream animals generally show positive rheotaxis (rheo = current; taxis, arrangement) which help them to move against the current. Others show positive thigmotaxis (thigmo = touch/contact), having inherent behaviour to cling close to a surface. Thus, when a group of streams stonefly nymphs are placed in a dish, in the absence of other surfaces to cling, they even cling to each other. In streams structural changes including physico-chemical characters are more pronounced in the upper part of streams because the gradient, volume of flow and chemical composition of water. When a narrow water stream is changed into a broad-channel river, we get more plankton, periphyton, nekton and benthos. The following vertical zonations of a river become more prominent as given below.

Marginal Water Zone

The chief producers of this zone are the algae in the form of periphyton and plankton of which Chlorophyceae ranks first followed by Myxophyceae and Bacillariophyceae. The common free floating algae are *Cosmarium auriculatum, Closterium tumidum, Pediastrum simplex, Senedesmus dimorphus, Actinastrum hantzchii, Hydrodictyon reticulatum, Melosira ambiguns, Synedra ulna, Fragillaria* sp., *Lyngbya magnifica* etc. The dominant periphyton constitue *Cladophora glomerata, Oedogonium crassum, Spirogyra borgeana, Ulothrix zonata, Cymbella turgida, Oscillatoria rubbrevis, Phormidium purpurens* and many diatoms.

Among the macrophytes *Chara* and *Nitella* sp. of algae *Marsilea quadrifolia* and *Azolla pinnata* of Pteridophytes, *Eichhornia carssipes, Ceratophyllum demersum, Hydrilla verticillata, Apanogeton monostachyon* and *Potamogeton crispus* of angiosperms are dominant.

Among zooplankton rotifers rank the first followed by copepods. Rotifers as *Keratella tropica, K. cochlearis, Brachionus calyciflora, B. caudata, B. diversicornis, Filinia longiseta, F. terminalis, Polyarthra* spp. and *Asplenskna* sp. are very common, copepoda like cyclops, *Mesocyclops leuckart, M. hyalinus* and Cladocera like *Bosmina langisostris, Bosminapsis* sp., *Moina branchiata* and *Alona* sp. are observed frequently in marginal water. Arthropods belonging to Decapoda like shrimps, prawns and crabs are encountered in this zone. Insects like nymphs of may fly, stone fly, dragon and damsel fly and aquatic beetles (Ranatra water scorpion), dinettes (Gyrinidae), whirligig beetle and dytiscus (Predatory diving beetle) are very common. Aquatic mollusks are transitory in nature preferring different zones of aquatic bodies in different seasons and different stages of their life cycle. In general all freshwater gastrophods live in marginal water zone for food, shelter and reproduction and only for a short period they come to margin of the river. It may be related to their ciliary mode of feeding habit. Freshwater gastropod like Pila viviparous, Lymnaea, *Melanoides* and *Terebia* are very common. Important fish nekton of marginal water of river are mullet fish like *Rinomugil corsula* and *Channa* spp. and *Barbus* spp.

Aquatic snakes such as *Enhydris enhydris* and *Netrix piscator* are found moving in and out of the marginal water. One species of Dolphin of river Ganga known as *Platynista gengetica* is now kept under conservation programme. This aquatic mammal also come to marginal water in search of fishes. Besides these consumers, the migratory birds come at the top of food-web. *Hilsa ilisha* and other clupeoid fishes are very common on the Ganges.

Open Water Zone

In this zone of river very few organisms are able to withstand the high velocity of currents. As such very few diatoms such as *Melosira ambiguans, Synedra ulna, Navicula* sp., *Chlorococcales* such as *Pediastrum simplex, Scenedesmus, Actinastrum* and desmids such as *Cosmarium* and *Closterium* encounter in the middle of the river. About 99 per cent of the total population consists of phytoplankton in this zone. *Hilsa ilisha* and other clupeoid fishes live in the open waters of the river Ganga.

Profundal Zone

The consumers or higher animals are the dominant inhabitants of profundal zone of river. But annelids like burrowing *Tubiflex* and bivalve mollusks like *Parreysis favidens* and *Corbicula striatella* are also found in this zone, of which *Novaculina gangetica* is restricted to Ganga river only. Other bivalve mollusks such as Lamellidens are observed except during winter season.

In addition to carp fishes all the Siluroid fishes such as *Wallago, Mystus* are the inhabitants of this zone. Aquatic reptiles *Kachuga kachuga, Kachuga tectum, Trionyx gangeticus* and Crocodiles are encountered in river system. The dolphin *Platynista gangetica* is the only aquatic mammal found in the river Ganga.

B. Lakes and Reservoirs

Lakes and reservoirs are large aquatic bodies of water enclosed by land and occupies a basin. These are grouped under lentic systems and are larger in depth and areawise then a pond. Lakes are geologically transitory. They are usually born due to catastrophes or geological activities occurring in nature and die quietly in course of time (Hutchinson, 1957). Thus, a lake lives through youthful stages to maturity, senescence and death when its concave basin is finally filled up.

In addition to the natural lakes, man is building up artificial lakes usually called as reservoir or impoundments all over the world including areas where there are no natural lakes. While not exactly born of catastrophe, man-made lakes are probably also transitory in the geological sense. Artificial lakes vary according to the region and to the nature of the drainage. Generally these are characterized by fluctuating water levels and high turbidity. Production of benthos is often less in impoundments than in natural lakes. The heat budget of impoundments may differ greatly from that of natural lakes depending on the design of the dam. If water is released from the bottom as with the case of dams designed for hydroelectric power generation, cold nutrient rich but oxygen-poor water is exported downstream while warm water is retained in the lake. The impoundment then becomes a heat trap and nutrient exporter in contrast to natural lakes, which discharge from the surface and, therefore, function as nutrient traps and heat exporters.

Structure of Lakes

Littoral Zone

This is the peripheral shallow part of lakes and ponds with light penetration upto the bottom and occupied by rooted and emergent plants. It is subjected to variations in temperature and erosion of shore materials. In certain lakes, sometimes wave action is so much that large aquatic angiosperm are absent and only algae, as benthic organisms are noticed.

Limnetic Zone

The open-water zone (beyond the marginal vegetation) to the depth of effective light penetration, called as the compensation level where photosynthesis just balances respiration. In general this zone will have a depth at which light intensity is about 1 per cent of full sun-light intensity. The community in this zone is composed of plankton, nekton and sometimes neuston. This zone is absent in small shallow ponds. The term euphotic zone refers to the total illuminated stratum including littoral and limentic where primary producers fix inorganic carbon to manufacture organic substances.

Profoundal Zone

The bottom and deep water zone which is below the depth of effective light penetration where respiration and decomposition predominate. It is also called tropholytic region of lake, where breaking down or dissolution of organic compounds or cells take place. The community in this zone is composed of mainly animal benthos. This zone is absent in most of the ponds.

Origin of Lakes

There are different reasons attributed for the origin of lakes.

Glacial Action

These are formed by the action of glaciers, through the deposition of morain and debris. For example (a) the Dal and Wular lakes of Kashmir are of this type and (b) by excavation of substratum namely the lakes of Pongong valley in Kashmir.

Tectonic movements

It is concerned to the earth crust instability due to warping (twisting), faulting, fracturing, folding, thrusting and quaking. There is deformation and adjustment of the earth's crust. The two great ancient lakes of Baikal in Siberia with a depth of more than 1.7 km and Tanganyika of East Africa are formed by tectonic phenomena. Their basins are formed lying between faults and adjacent highlands.

Land Slides

These are formed due to the impoundments on stream valleys by rock slides, mudflows, or other mass movements of soil and rock. The Kumaon lakes in Uttar Pradesh are formed due to land slides.

Volcanic Actions

Volcanoes have influence and shaped the landscapes in many areas of the world and continues to do so today. In Japan many lakes are formed by volcanic activities. Tropical countries like Java, Sumatra and Bali have many such lakes. The deepest lake in the United States is the Crater lake of Oregon. This is an extinct volcanic crater, formed when underlying magma (molten rock) has flown out.

Dissolution of Substratum

Lakes are produced by dissolution and removal of material which hold water to become lakes or ponds. Usually the carbonates of calcium and to a lesser extent magnesium are the solutes. Nainital Lake of Uttar Pradesh has been produced by the collapse of surface caused by the removal of limestone by the solvent action of underground water. Many small lakes in the limestone tracts of the Khasi and Jaintia hills of north eastern region are examples of the type.

Wind Action

In arid and semiarid zones, the erosive action of wind removes sand particles from some place and thus lake basins may be formed in the depressions. Further deposition of these removed particle from one place may block the existing stream of other place giving rise to a lake. The salt lakes of Rajasthan (like Sambhar lake) come under this group.

River Action

Rivers systems create standing water habitats of various types. Flood plain lakes are examples which are again categorized as Levee lakes and ox-bow lakes. Levee lakes are shallow often elongate bodies of water that lie parallel to the river bed and are separated from it by strips of higherland. The levee is a natural embankment built up by river sediments deposited at times of high water. The second type of flood plain lake is the ox-bow lake formed from isolated loops of meandering rivers. These crescent shaped basins are usually deeper than the levee lakes. In the lower reaches of the Ganga and Brahmaputra rivers many such levee and ox-bow lakes are formed.

Biotic Action

The North European lakes are the depressions in peaty deposits. Dams of sphagnum and other bog plants have been reported by Hutchinson (1957). Beaver-damming and man's similar activity are well known. In both instances streams are impounded by putting stream side vegetation or bamboos to trap fish. With the passage of time, reservoirs or man-made tanks are constructed for multiple objectives, namely for irrigation, hydroelectric power generation, flood control, public supplies, recreation, and development of fisheries. Reservoirs are created by constructing stone masonry on concrete bundhs across a stream. These have penstocks, irrigation release tunnels and spillway as safety measures against floods. Important reservoirs of India are Maithon (Bihar and West Bengal), Hirakud (Orissa), Gandhinagar (M.P.), Ukai (Gujarat), Nagarjunasagar (A.P.), Stanley (Tamil Nadu), Tunga-Bhadra and Kali river dams (Karnataka), Bhakra Nangal (Punjab) and Rihand (U.P.).

Lake Characteristics

In case of temperate area the lakes during summer become warmer on the top zones than the bottom waters and as a result only the warm top layer circulates. It does not mix with more viscous cold water, creating a zone with steep temperature gradient in between called the thermocline. The upper, warm circulating water is the epilimnion and the colder, non–circulating water is the hypolimnion or under lake (Fig 3.2). The magnitude and direction of the water movement depends upon the velocity and duration of the wind. Steady vigorous shore winds may set up return currents which extend to the opposite side of lake. Encountering the colder denser water of the thermocline region, this down flowing returning current is diverted in a horizontal

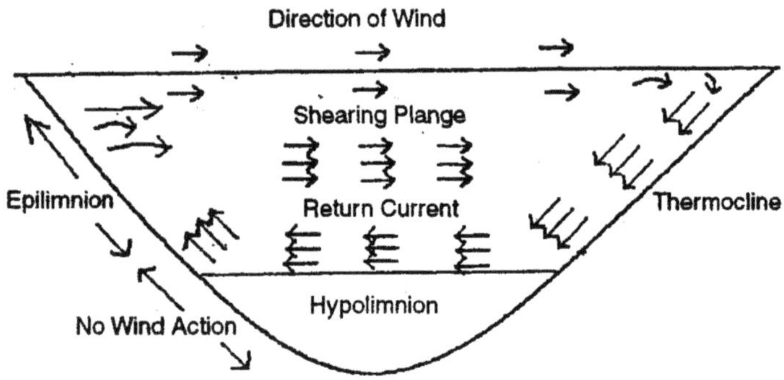

Fig. 3.2: Wind driven water currents in a lake in the epilimnion with a well defined thermocline.

direction and flows toward the opposite shore, maintaining a level above the thermocline. In such a circulation, the upper stratum of the epilimnion flows toward one shore, the lower-stream (the return current flows toward the opposite shore and the stratum which is known as the shearing plane is practically without motion. The name of the region in between epilimnion and hypolimnion was described as thermocline by Bronsted and Wesenberg-Lund (1911) and accepted by Hutchinson (1957). In the middle thermocline zone temperature drops at least 1°C with each 1 m increase in depth (Fig 3.3). Thus, with a typical direct stratification there are three compartments the upper epilimnion, a middle metalimnion or mesolimnion and a lower hypolimnion.

The thermocline, is not a broad region but an imaginary plane within the lake or in any lentic water body. It is located at a level intermediate between the two depths where the temperature decrease is greatest. In Fig. 3.3 the strong drop in temperature of the thermocline region in the hot months of summer has been shown on the right side. If the thermocline is below the range of effective light penetration as is often the case, the oxygen supply becomes depleted in the hypolimnion, because both the green plant and the surface source is curtailed.

During the cold weather conditions the temperature of epilimion drops until it is the same as that of the hypolimnion. The water of the entire lake begins circulating during the "full

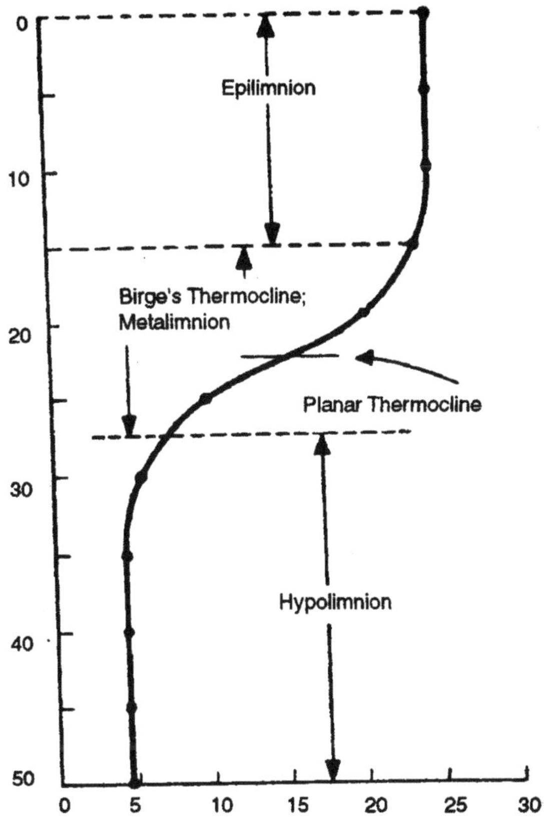

Fig. 3.3: Vertical Stratification of Temperature in a Lake (After Munshi and Munshi, 1995).

overturn". As the surface water cools below 4°C, it expands, becomes lighter, remains on the surface and freezes during winter stratification. In winter the oxygen is usually not greatly reduced because bacterial decomposition and respiration of organisms are not so much at low temperatures and water holds more oxygen. An exception to this generalization may occur when snow covers the ice and prevents photosynthesis, resulting in oxygen depletion and winter kill of fishes.

During the spring season, ice melts and water becomes warmer, it becomes heavier and sinks to the bottom. Thus, when the surface temperature rises to 4°C, the lake so to say takes another "deep breath" with the spring overturn. The classic pattern of two seasonal overturns is typical for many lakes in America and Russian countries. In general, the deeper the lake, the slower is the stratification and thicker the hypolimnion.

The rate of depletion of oxygen levels in the hypolimnion during summer stratification depends on the amount of decaying matter and on the depth of the thermocline. Productivity rich eutrophic lakes are subjected to greater oxygen depletion during the summer than poor oligotrophic lakes, because the 'rain' of organic matter from the limentic and littoral zones into the Profoundal zone is greater in the eutrophic lakes. It has been found that cultural eutrophication hastens oxygen depletion in the profoundal zone. Thus, fishes which are stenothermal and low temperature tolerant can survive only in less productive lakes in which cold bottom waters are not become depleted of oxygen. Lower organisms in contrast to fish of the profoundal zone are adapted to withstand low oxygen levels efficiently for long time.

Thermal Stratification in Tropical Lakes

Surface temperature of subtropical lakes never fall below 4°C. They show a distinct thermal gradient from top to bottom but experience only one general circulation period in winter. Tropical lakes as studied by Ruttner (1931) in the reservoirs of Java and Sumatra have shown high surface temperature exhibiting thermal stratification in the columns of water in summer. Though the temperature gradient from surface to bottom is little in comparison to temperate lakes but this is enough to make density difference of water resulting to thermal stratification.

Classification of Lakes

Different criteria are followed for classification of lakes. It is not easy to classify lakes on certain characteristics and there is a overlapping of characters between the lakes. Hutchinson (1957) classified the lakes on the basis of water circulation:

(1) **Dimictic (mictic=mixed) lakes:** Two seasonal periods of free circulation or overturns.

(2) **Cold monomictic lakes:** Water never above 4°C (polar regions), seasonal overturn in summer.

(3) **Warm monomictic lakes:** Water never below 4°C (Warm temperature or subtropical) one period of circulation in winter.

(4) **Polymictic lakes:** More or less continually circulating with only short, if any stagnation periods (high altitude, equatorial).

(5) **Oligomictic lakes:** Rarely (or very slowly) mixed (thermally stable) as in many tropical lakes.

(6) **Meromictic lakes:** Chemically stratified lakes. Meromictic denotes partly mixed. In contrast to most lakes in which bottom and surface waters mix-periodically (i.e. holomictic or wholly mixed lakes) some lakes become permanently stratified by the intrusion of saline water or salts liberated form sediments, creating a permanent density difference between surface and bottom waters. In this the boundary between circulating and non-circulating layers is a "chemocline", instead of a thermocline. Free oxygen and aerobic organisms will be absent in bottom waters of such lakes. Big soda lake in Nevada comes under this group.

Oligotrophic-eutrophic Series of Lakes Based on Productivity

Thienemann (1925) observed that conditions in deep lakes differed from those of the shallow lakes, even though they are located in the same climatic and edaphic region. The summer oxygen supply in the hypolimnion of a deep basin of oligotrophic lakes in contrast to eutrophic ones remain always abundant to support bottom fauna of diverse nature. By contrast the dissolved oxygen becomes critical in the shallower eutrophic lakes and only a few species are found in the profundal benthos. Features contrasting oligotrophic and eutrophic lakes are given in Table 3.2.

Rodhe (1969) reported different rates of inorganic carbon fixation by photosynthesis in these lakes. According to him 7-25 g $C/m^2/yr$ ot lake surface indicate oligotrophy : 75 to 250g $C/m^2/$ year indicate natural eutrophic lakes and 350 to 700g $C/m^2/$year in polluted lakes. Mesotrophy lies somewhere between annual primary production of 25 to 75g carbon fixation. But extremely low rates of carbon fixation are also known in certain oligotrophic lakes. Goldman *et al.* (1967) reported about 0.014g of carbon per square metre for a average summer day in an Antarctic lake.

Table 3.2: Differences in the productivity of oligotrophic and eutrophic lakes.

Oligotrophic Lakes	*Eutrophic Lakes*
1. Deep and steep-banked.	1. Shallow, broad littoral zone.
2. Epilimnion volume relatively small as compared to hypolimnion.	2. Epilimnion/hypolimnion ratio greater.
3. Blue or green water, marked transparency.	3. Green to yellow or brownish green, limited transparency.
4. Water poor in nutrients.	4. Nutrients and Ca abundant.
5. Sediments low in organic matter	5. Sediments high with organic matter.
6. Oxygen abundant at all levels at all times.	6. Oxygen depleted in summer hypolimnion.
7. Littoral plants limited, rosette type.	7. Littoral plants abundant.
8. Phytoplankton quanti-tatively poor.	8. Abundant phytoplankton.
9. Water blooms of blue-greens lacking.	9. Water blooms are common.
10. Profundal bottom fauna diverse.	10. Profundal benthos species poor.
11. The ratio E/H (Epilimnion / Hypolimnion) small.	11. The ratio E/H (Epilimnion / Hypolimnion) large.

Special Lake Types

There are seven special types of lakes classified based on their origin.

Dystrophic Lakes

These have high levels of humic acid in water and these are bog lakes, dystrophic in nature with low pH slowly develop into peat bogs and marshy wetlands.

Deep Ancient Lakes with Endemic Fauna

The Lake Baikal in Russia is the most famous of ancient lakes. It is the deepest lake in the world and was formed by earth movement during mesozoic era. Ninety-eight per cent of 384 species of arthropods are endemic (found nowhere else), including 291 species of amphipods, 36 species of fish are endemic. This lake is often called the "Australia of freshwater", because of its endemic fauna (Brooks, 1950).

Desert Salt Lakes

These are present in arid climates where evaporation exceeds precipitation resulting in salt concentration. The Salt Lake, Utah harbour community composed of a few species which are able to tolerate high salinity. Brine shrimps (Artemia) are characteristic animals of such lakes.

Desert Alkali Lakes

They occur in igneous drainages in arid climates with high pH and concentration of carbonates. *Examples:* Pyramid Lake, Nevada.

Volcanic Lakes

They are acidic or alkaline lakes associated with regions of active volcanic activity. They have restricted biota. *Examples:* some Japanese and Philippine lakes.

Meromictic

Chemically stratified lakes, not very common.

Polar lakes

Surface temperatures remain below 4°C, or rise above it for only a brief period during the summer when circulation of water can take place. Plankton population grow rapidly during this period.

Vertical Distribution of Dissolved Oxygen

The main sources of oxygen in the lakes are photosynthetic aeration of water by phytoplankton and the other is through surface air diffusion into water at the orifice. The deep layers are away from the both of these sources and light does not penetrate into it. As a result the respiration and decomposition processes remove oxygen from the water. In dimictic lakes during the summer and autumn overturns of water makes oxygen distribution more or less uniformly to all the surface and bottom strata. If a curve is drawn based on oxygen levels in relation to depth of the water body the line will be nearly straight and is referred as orthograde curve.

When thermal stratification occurs during summer when the lake is no longer homogenous throughout, the tropholyte zone (Hypolimnion) becomes isolated from the upper water. Now oxygen begins to be consumed there. In lakes with large hypolimnion volumes and relatively little production of organic matter in the eplilimnion above, the demands on the oxygen in the tropholytic zone are so slight that it shows an appreciable decline. This is characteristic of oligotrophy. The biomass, the ratio of epilimnion volume to hypolimnion volume, and the hypolimnion temperature interact to produce vertical oxygen curves. Under favourable environmental factors the production of a large epilimnion biomass of flora and fauna, the situation will be quite different during summer stratification. Great quantities of dead and decaying organic matter consume large amount of oxygen in hypolimnion water.

Important Reservoirs of India

The most important reservoirs of India are constructed mainly by impounding river systems. A reservoirs of the Ganga river system are Rihand, Matatila, and Sardar sagar in Uttar Pradesh and Govind Sagar in Madhya Pradesh.

Reservoirs of the Indus river system in Punjab and Himachal Pradesh are Beas dam and Govindsagar respectively. Reservoir of Mahanadi river system in Orissa (Sambalpur) is one of the longest dam of the world called Hirakund. Reservoirs of Krishna river system are Nagarjunsagar dam and Nizamsagar dams in Andhra Pradesh and Tungabhadra Kali, Krishanraj sagar and Bhadra reservoir project in Karnataka state. Bhavanisagar and Mettur dam (Stanley reservoir) in Tamil Nadu are important reservoirs of Cauvery river system.

Case Studies on Reservoirs

Idukki Reservoir

There are approximately three million hectares of river impoundments in India, spread over different types of soils exposed to different climatic and anthropogenic environments. Among them small number of aquatic systems are investigated and most of them are fragmentary. Though a considerable amount literature is spread in various journals, their scientific reviews are very meager. This lacunae made non–availability of sufficient books on the limnology of Indian water systems.

Idukki reservoir is one among the biggest fresh water bodies of our country. It is formed by three dams viz., Idukki double arch dam, Cheruthoni dam and Kulamavu dam. The reservoir is situated at an altitude of 693 m above MSL. It is roughly V shaped having serrate margins. The length of Ayyappancovil arm is 17 km and Kulamavu arm is 15 km. Total area of the reservoir is 59.8 Sq.km. The average depth normally varied from 50 to 80 m, while maximum depth near Idukki double arch dam is around 125 m.

This reservoir has a catchment area of 640.35 km, and a gross storage capacity of 19996 Mm^3. The Periyar river empties its water into the reservoir at the Ayappancovil. Besides it is fed by Erattayar tunnel and a number of streams in the catchment area. The bulk of water is used for power generation drawn through a siphon located near Kulamavu dam.

Physico-chemical Features

A study conducted by Dr. T.C. Khatri of Zoological Society of India revealed that an interesting variation in the diurnal characteristics of water quality are recorded. The air temperature varied between 23.5 to 30.6. Maximum temperature was recorded at 1500 hours on the first day and 0600 hours of the next day of 15th and 16th February, 1983, and the water temperature varied between 25 to 26.6°C at 5m depth, 23.4 to 24.1°C at 10 m depth, 23.1 to 23.5°C at 30 m depth and around 23.0°C at the depths below 50 m respectively. The pH varied between 6.7 to 7.4 at 5 m depth and 7.0 at further lower strata. The carbon dioxide showed increasing trend at the depth up to 30 m and remained more or less static thereafter. The dissolved oxygen showed a declining trend towards the bottom up to the depth of 30 m and increased thereafter.

The air and water temperature followed a definite trend. The water was isothermal throughout the observation period. The upper strata up to 30 m deep showed wide variations in temperature than the lower strata which indicate the solar influence up to 30 m deep. Water was alkaline at the upper 5 m and almost acidic in rest of the water mass and the pH fails to show any definite trend with the diurnal variations.

Free carbon dioxide was recorded through out the dial cycle and showed an increasing trend with the depth of 30 m deep. This might be due to the presence of plankton up to that depth. From 50 m downwards there was a erratic variations in carbon dioxide levels. Presence of free carbon dioxide at the bottom layers indicate benthic decomposition activity by the decomposers. The total alkalinity showed increasing trend during night hours which may be attributed to respiratory release of carbon dioxide by the planktonic organisms. An increase dissolved oxygen at 50 m depth onwards may be due to mass sinking or eddy like physical gaseous diffusion processes (Tables 3.3 and 3.4). He has also listed the productivity of various lakes and reservoirs reported elsewhere (Table 3.5). The member of phytoplankton recorded in the reservoir water are showed in Table 3.6. Besides algae there are several other organism to be researched and reported.

Case Study No. 2

Lakhotia Lake, Rajasthan

It is a manmade, rainfed perennial lake, in the northern side of Pali district (25.8° latitude 73.3° E longitude). The lake is roughly triangular, having a maximum length of 1825 m east-west and a breadth of 950 m north-south with a maximum depth of 45 m. It is surrounded by Umaid textiles in the north, Bathing ghats, Dharmashala and temples on the east, a spillway in the west. Its littoral zone is dominated by the growth of a natural scavenger plant often employed as biological filter in the biological waste treatment methods (*Example : Scirpus tuberosus*). The water colour is yellowish green due to the muddy basin and plankton blooms. The water quality of the lake are shown in Table 3.7.

Biological Factors

The phytoplankton population showed three peaks one each in summer (June), Monsoon (October-November) and winter (January). The summer and winter peaks were mainly contributed by Chlorophyceae and monsoon peak by Cyanophyceae. The phytoplankton population mainly comprised of 25 genera, out of which 15 belong to Chlorophyceae, 6 to Cyanophyceae and four

Table 3.3: Stratification of physico-chemical parameters in the Idukki Reservoir during February 1983.

S = Surface, B = Bottom.

Parameter	*Depth*	*Time (hours)*								
	(m)	*0900*	*1200*	*1500*	*1800*	*2100*	*2400*	*0300*	*0600*	*0900*
Air temperature (C)		28.5	28.8	30.6	26.5	26.0	25.1	25.0	23.5	27.3
Water temperature (C)	S	27.0	28.6	28.5	28.0	27.6	27.5	27.0	25.5	26.1
	5	26.6	26.5	26.5	26.5	26.2	26.1	26.0	25.0	25.2
	10	24.1	24.1	23.9	23.8	23.6	23.5	23.5	23.4	24.0
	30	23.5	23.5	23.4	23.3	23.3	23.3	23.2	23.1	23.3
	50	23.2	23.2	23.1	23.1	23.1	23.1	23.0	23.0	23.1
	B	23.1	23.1	23.1	23.1	23.1	23.1	23.0	23.0	23.1
Transparency (cm)		250	260	270	210				205	235
	S	7.5	7.2	7.1	7.4	6.7	7.2	7.2	7.2	7.2
	5	7.4	7.3	7.0	7.4	6.7	6.9	6.9	6.9	6.9
pH	10	6.9	6.8	6.7	6.9	6.5	6.9	6.9	6.8	6.6
	30	6.8	6.6	6.6	6.6	6.5	6.5	6.6	6.3	6.5
	50	6.8	6.9	6.9	6.7	6.6	6.6	6.6	6.2	6.3
	B	6.7	7.1	6.8	6.9	6.6	6.7	6.3	6.7	6.4
Free carbondioxide	S	2.0	1.9	2.2	2.6	2.8	2.9	2.9	2.8	2.3
(ppm)	5	2.2	1.6	1.7	2.5	2.8	4.5	2.7	2.2	2.0
	10	4.7	5.2	6.1	5.6	4.5	4.5	3.7	3.6	5.7
	30	5.8	7.4	8.1	6.8	5.9	5.4	4.5	6.8	5.9
	50	4.7	6.5	7.5	7.0	4.7	6.5	5.6	7.0	6.7
	B	5.5	6.3	7.8	5.2	5.5	6.3	7.5	5.2	6.0
Bicarbonate alkalinity	S	23	17	22	22	18	24	20	21	22
(ppm)	5	23	20	26	20	19	14	15	16	17
	10	20	20	22	22	18	18	18	16	17
	30	18	16	23	29	19	18	18	19	20
	50	30	28	22	27	20	19	20	21	22
	B	17	18	20	30	27	28	29	30	32
Dissolved oxygen	S	7.66	7.13	7.53	7.53	7.40	7.66	7.40	7.40	7.63
(ml/litre)	5	7.06	7.00	7.13	6.80	6.80	6.86	6.86	6.86	7.03
	10	4.33	3.73	4.00	4.00	3.86	3.63	3.93	3.93	4.13
	30	4.13	2.23	2.26	3.26	2.40	2.26	3.33	3.33	3.66
	50	3.43	3.26	2.26	3.33	3.26	3.33	3.43	3.43	3.43
	B	3.93	3.93	4.06	4.46	3.60	3.60	3.73	3.73	3.73

Source : Khatri T.C. 1985.

genera are Bacillariophyceae. Only one genus *Ceratium* of Dinophyceae was recorded. The Zooplankton fauna of the lake include Cladocera (*Moina, Daphnia, Ceriodaphnea*), Ostracoda (*Cypris*), Copepoda (Cyclops, *Diaptomus* and Nauplii larvae and Rotifera (*Brachionus*).

Case Study No 3

Dr. N.R. Birasal, Faculty member of Zoology Department, Gudleppa Hallkeri Science College, Haveri, has studied the limnological aspects of Supa reservoir of Karnataka.The supa reservoir is built on Kali river which rises at an elevation of 945 m in the Western Ghats near Dandeli. This reservoir with a surface area of 129 km^2 has flooded 10,781 hectares of land cleared from the tropical forests. The ecological succession of organisms and fish productivity during first filling stage and later were studied by the author.

Table 3.4: Variations in the physicochemical and biological parameters in the Idukki Reservoir, Kerala.

Parameter	*Range*
Water level	15-100 m
Air temperature	21.2°-29.8°c
Water temperature	21.8°-26.3°C
Transparency	55-540 cm
pH	6-9
Free carbon dioxide	nil-70 ppm
Bicarbonates	7-42 ppm
Dissolved oxygen	1.232-9.856 ml 1^1
Silica	95-199 / µg at 1^1
Phosphate	nil
Nitrate	1.7-18.7 µg at 1^1
Phytoplankton	nil-175 cell x 10^2 1^1
Chlorophyll-a	nil
Primary production	nil
Zooplankton	2-122 an 1^{-1}

Source : Khatri T.C. 1985.

Table 3.5: Annual gross production in different lakes reported by various workers

Name	Climate	Annual Gross Production (gCm^{-2}year^{-1})	Reference
Padamsagar	Tropical	4138.91	Bohra (1976)
Ranisagar	- do -	3451.63	Bohra (1976)
Ooty lake	- do -	3144.84	Sreenivasan (1963)
Amaravathi reservoir	- do -	2422.00	Sreenivasan (1965)
Othakadai pond	- do -	2000.20	Sumitra (1971)
Lakhotia lake	- do -	1956.03	Khatri (1984)
Yercaud lake	- do -	1136.24	Sreenivasan (1963)
Yanmali pond	- do -	857.60	Sumitra (1971)
Teppakulam pond	- do -	835.90	Sumitra (1971)
Aliyar reservoir	- do -	730.00	Sreenivasan (1970)
Bhavanisagar reservoir	- do -	919.80	Sreenivasan (1974)
Kodaikanal lake	- do -	273.75	Sreenivasan (1963)
Indiana lake	Temperate	600.00	Wetzel (1966)
Esromso and Furoso	- do -	200.00	Jonasson & Methiesen (1959)
Clear lake	- do -	160.60	Goldman and Wetzel (1963)
Lake leake	- do -	9.00	Croome and Tyler (1975)
Tome lake	- do -	9.00	Croome and Tyler (1975)

Source : Khatri, 1984.

Table 3.6: Abundance of phytoplankton in Idukki Reservoir

1. less than 5 per cent 2. less than 20 per cent
3. 20-70 per cent 4. more than 70 per cent

Phytoplankton	*Abundance*	*Maximum population Cells X (10^2 per litre)*	*Per cent*
		Chlorophyceae	
Athordesmus sp	(2)	2	a
Ankistrodesmus sp.	(4)	122	69.71
Coelastrum sp.	(1)	1	7.14
Cosmarium sp.	(2)	14	13.59

Contd......

Phytoplankton	*Abundance*	*Maximum population Cells X (10^3 per litre)*	*Per cent*
Chalmydomonas sp.	(1)	8	47.05
Microspora sp.	(3)	84	70.58
Mougeotia sp.	(1)	36	85.71
Scenedesmus sp.	(1)	5	31.25
Staurastrum sp.	(3)	12	11.88
Ulothri sp.	(1)	3	10.00
Volvax sp.	(1)	5	13.51
Chlorella sp.	(1)	1	3.33
		Bacilliairophyceae	
Navicula sp.	(2)	7	6.79
Syndra sp.	(1)	a	
Pinnularia sp.	(1)	a	
		Cyanophyceae	
Aphanozapsa sp.	(1)	5	11.90
Myxasarcina sp.	(1)	3	30.00
Oscillatoria sp.	(1)	a	
Spirulina sp.	(1)	a	
Algal group	**Premonsoon**	**Monsoon**	**Post monsoon**
Chlorophyceae	75.6	96.6	96.6
Bacillariophyceae	18.3	3.4	3.0
Cyanophyceae	6.1	nil	1.4

Source : Khatri, 1987.

Table 3.7: Physico-chemical and biological features of Lakhohita lake in Pali

Parameter	*Range*	*Season (month) of Minimum*	*Season (month) of Maximum*
Abiotic			
Physical			
Water level	2.25-4.5 m	Winter (Jan)	Monsoon (Aug)
Air temperature	22.6°-37.6°C	- do -	Summer (Jun)
Water temperature	17.0°-33.0°C	- do -	- do -
Transparency	51-105 cm	Summer (May)	Monsoon (Sep)

Contd.....

Parameter	Range	Season (month) of Minimum	Maximum
Chemical			
pH	6.3-9.0	Monsoon (Oct)	Winter (Feb)
Free carbon dioxide	0-15.0 ppm	*	Monsoon (Oct)
Carbonates	0-30 ppm	**	Winter (Feb)
Bicarbonates	4-158 ppm	Monsoon (Jul)	Summer (Jun)
Dissolved oxygen	1.575-8.368 ml/l	Monsoon (Oct)	Summer (Jun)
Silica	0.185-5.251 mg Si/l	Summer (Jun)	Monsoon (Aug)
Phosphate	0.15-4.5 μg/l	Monsoon (Sep)	Monsoon (Jul)
Nitrate	1.5-36.0 μg/l	Winter (Nov)	Summer (Jun)
Biotic			
Phytoplankton	106-379 units x 10^3/l	Summer (Apr)	Summer (Jun)
Bacillariophyceae	8-154 units x 10^3/l	Monsoon (Sep)	Summer (Jun)
Chlorophyll-a	0.532-6.959 mg/m^3	Winter (Feb)	Summer (Jun)
Gross primary production	0-6.301 gc/m^3/day	Monsoon (Oct)	Summer (Jun)
Zooplankton	4-12 nos./l	Monsoon (Oct)	Monsoon (Sep)

* Absent in most of the observations.
** In presence of free carbon dioxide.

Source : Khatri, 1984.

Case Study 4

Shanti Sagar lake, Shimoga

Shanti Sagar lake is one of the largest freshwater lake systems of Karnataka state. This case study was taken from the doctoral work of Mrs. Hina Kauser (1988). The lake embodies the study on the water quality of the lake for a period of two years in terms of physico-chemical factors and aquatic algae. However, there are no efforts made in her thesis on drift animals and sedimentary ecology and other biota. Hence the work is again found to be fragmentary. Further study on the limnology of the lake system is called for.

According to her investigation the lake water is used for agriculture and inland fisheries by the Inland Fisheries Department. The water spread area of the lake is 64 sq.km. Total catchment area is 896.9 sq.km. Depth is 8.25 m. and the irrigated area is 2976 hectares. The lake water is free from pollution and is not eutrophic in nature. The detailed phycology revealed that the lake possess 7 genera 31 species of desmids, 17 genera, 9 species of diatoms, 3 genera 39 species of euglenoids, and 11 genera 23 species of blue greens respectively. The authors advocates that the lake system being free from pollution, it may used economically as one of the important tourist/sport be site/nature club to invite tourists.

C. Ponds, Swamps and Marshes

For the definition of ponds as compared to lakes and reservoirs the depth of water is of prime importance than the actual size of the water surface. The depth of a pond is usually around 2 m or rarely more. The biological processes in a pond are more intensive and vigorous than in a deeper lake. The littoral zone of ponds is often larger compared to that of lakes. In ponds limentic and profundal zones are small or absent and hence stratification has little importance. In contrast to the limentic zone of a lake the littoral zone is the main producer region in ponds. Ponds are present in all parts of the world where there is adequate rainfall and they are mostly rainfed. In arid and semiarid parts of tropical India these are the main sources of drinking water.

Classification of Ponds

Natural Ponds

Natural ponds are perennial or seasonal shallow water bodies formed in various ways. When a stream shifts in position it leaves behind an isolated body of lentic water which forms the "Ox-Bow" Pond. In the places where depressions are formed due to the solution of the underlying strata, the water gets accumulated either by flood water or rainfall and thus the natural ponds are formed. Sometimes the last remnant of a lake whose basin has become filled progressively by sedimentation in course of time is transformed into a natural pond.

Fish-Ponds

Most of the fish-ponds are artificial or semi-artificial in built. Many times these are constructed by cement masonry with a well designed plan. Some are created by erecting dams across a stream or basin. With few exceptions their water level can be regulated by inflow and drainage where pisciculture is practised. Fish-pond is a shallow body of water that can be drained completely. It is often supplied by running water but also by spring ground or rain water. Naturally the change from wet to dry conditions as well as various cultivation processes including fertilization with manures and fertilizers will affect the flora and fauna of a fish pond. Generally little water is allowed to flow through them.

Beaver Ponds

Such ponds are created by making holes in bank of streams passing through forests by beavers. Beaver ponds were common in North America in primeval days. A beaver pond is abandoned when the supply of plants in the vicinity become reduced. Beaver ponds act as water reservoirs and such ponds are absent in India.

Apart from the above, there are duck breeding ponds, and the various types of constructions by the Pollution Control Board authorities as sewage oxidation ponds, stabilization ponds, maturation and settling ponds. Some ponds are kept for ornamental purposes. About the sewage oxidation ponds is explained in detail in Chapter 10.

Pools and Puddles or Temporary Ponds

These are the depressions in the ground either at the margin of glaciers where they fill with meltwater or in the vicinity of river bed after the floods have receded. The water thus collected usually is very shallow and measures a depth maximum of a few feet only. Also prolonged rainfall may form temporary small pools and puddles. All these pools dry up in some part of the year, and as such organisms in these habitats must be able to survive in a dormant stage during dry periods and able to move in and out of the pools like amphibious and adult aquatic insects.

Case Study of a Fish-Pond

A typical fish-pond study carried out by Kiran *et. al.*, (2000) is incorporated into this book as a case study. The fish-ponds include hatchery and nursery ponds, spawning ponds, and rearing ponds.

The Study Area

The Fish Seed Farm located in the outskirts of city of Shimoga (latitude 13°42′ North and longitude 75°38′ East) in a placed called

B.R. Project, Karnataka State, India. The Bhadra Fish Seed Farm was established in 1965-66 by acquiring a land of 46.8 hectares near the left bank channel of Bhadra reservoir. Since then construction of ponds and other office buildings was undertaken. These ponds were used for seed production. The whole area is divided into five blocks and each block is divided into two units namely Bundh breeding and National fish farming units. The total number of ponds are 278. Out of these four cement ponds have been studied as an example for fish-ponds and are shown in Table 3.8.

Table 3.8: Details of the fish–ponds

Species of fish reared	*Period of study*	*Pond type and size*	*Stocking rate*
Fingerlings of common carp	18/10/96 to 12/10/97	cement ponds 4 m^2	1 lakh/ha

Physico-Chemical Properties of Water

The mean water temperature in the surface layer varied from 19.5° to 35.0° during the period of the study. The transparency was maximum to the extent of 92.0 cm showing highest light penetration in the aquatic body. pH of the pond was always in alkaline side. High dissolved oxygen values were observed around 10.0 ppm.

Algal Blooms

Blooms of algae occurred in the summer season which continued up to rainy season. The phytoplankton members consist of Chlorophyceae, Bacillariophyceae, Cyanophyceae and Euglenophyceae members. The list of algae commonly recorded in the fish pond are listed in the Table 3.9.

Zoo-Biota

The zooplankton exhibited seasonal periodicity in their abundance similar to that of phytoplankton. In ponds with thick deposits of organic matter, there was an abundance of cladocerans, but fewer copepods and rotifers. The zooplankton organisms are of general importance in the productivity of fish-ponds, as they

Table 3.9: Phytoplankton composition of fish-ponds at B.R. Project

Chlorophyceae	*Cyanophyceae*	*Bacillariophyceae*	*Myxophyceae*	*Euglenophyceae*
Ankistrodesmus falcatus	*Oscillatoria sps.*	*Navicula rhomboides*	*Aphanocapsa*	*Euglena polymorpha*
Coelastrum microporum	*Nostoc sps.*	*Navicula hustedtii*	*Microcystis aerugenosa*	*Euglena limophylla*
Micractinium pussilum		*Navicula acicularis*		
Selenastrum gracile		*Navicula palea*		
Scenedesmus quadricauda		*Navicula spicula*		
Tetraedon caudatum		*Nitzschia palea*		
Pediastrum simplex		*Pinnularia gibba*		
Pediastrum duplex		*Anomoenieis brachysira*		
Ulothrix sps.		*Amphora cofformis*		
Zygnema sps.		*Fragillaria construens*		
		Gyrosigma kutzingii		
		Melosira granulata		
		Cymbella tumida		
		Gomphonema gracile		

from the main food of young fish. Oliogochaetes, such as *Nais, Chaetogaster, Dera, Autophorus* were found among vegetation in the bottom mud. Crustaceans were represented by prawns, shrimps, and crabs. Insects both adults and larval forms of Mayflies (Ephemeroptera), Caddisflies (Trichoptera), Midges (Diptera), Mosquito larvae, Chironomids, water bugs like Notonecta, Nepa, Ranatra and Cybister-Beetles were abundant. Other macrofauna were represented by fishes, frogs, snakes and birds. Birds specially common crow, egrets, little cormorants and snake birds occurred as predators on fish larvae. Many workers have studied the ecology of fish ponds and the literature available is mostly on the pond ecology. But still there is no much information available on pond sediments.

Case Study 2

Padmavathi and Durgaprasad (1997) have studied the ecology of two fish-ponds in environment of Kolleru lake, Krishna district, Andhra Pradesh. A comparative study of two ponds A and B is made and average of their physico-chemical factors, plankton, culture period, stocking density were recorded and the yield of carps and unwanted fish species is given in Table 3.10.

Plankton

In pond A total of 46 genera of plankton were identified and their seasonal abundance was studied. Phytoplankton was numericaly abundant than zooplankton. Algae were represented by 25 genera belonging four classes viz., Cyanophyceae, Chlorophyceae, Bacillariophyceae and Euglenophyceae. The genera recorded in these groups are given in Table 3.11. Zooplankton were represented by rotifers, copepods and cladocerans in order of dominance whereas the other zooplanktonic forms were almost negligible. About 30 species belonging to 21 genera of animals were recorded. In the fish ponds in addition to the cultured carps, 21 genera of fishes in order of dominance, *Puntius sophore, P. chola, Salmastoma bacaila, Amblyharyngodon mola, Oreochromis mossambica, Rasbora daniconis, Barilius bola, Chela labuca, Chanda nama, Esomus danricus, Puntius ticto, Etroplus suratensis, mystus vittatus, M. guilo, Channa punctatus, Channa marulius, Heteropneustes fossilis, Clarius batrachus, Wallago attu, Mastacembalus armatus* and *Macrognethus aculatus* are recorded.

Table 3.10: Comparative tabulation of physico-chemical factors, plankton and fish yield in the fish ponds near Kolleru lake

		Pond A	*Pond B*
A.	Area (ha)	6.4	1.6
B.	Duration of culture period (Days)	238	320
C.	Physico-chemical parameters		
	a. Air temperature (°C)	30.30	30.50
	b. Water temperature (°C)	30.28	30.00
	c. Transparency (cm)	33.50	20.30
	d. Rainfall (mm)	164.00	156.74
	e. Dissolved oxygen (mg/l)	4.96	8.77
	f. pH	8.06	8.34
	g. Total alkalinity (mg/l as $CaCo_3$)	149.50	338.20
	h. Total hardness (mg/l as $CaCo_3$)	161.88	311.80
	i. Conductivity (umhos/cm)	695.60	2080.00
	j. Total solids (mg/l)	461.50	1598.00
	k. Chlorides (mg/l)	90.72	521.60
	l. Nitrate-N (mg/l)	0.88	1.42
	m. Nitrite-N (mg/l)	0.13	0.09
	n. Ammonia-N (mg/l)	0.80	0.45
	o. Orthophosphate (mg/l)	2.36	2.40
	p. Iron (mg/l)	0.37	0.54
D.	Plankton (Individuals/l)		
	a. Phytoplankton	5.338	44.924
	b. Zooplankton	1.224	4.714
	c. Total	6.563	49.638
E.	No. of fish stocked (No./ha)	2617	5238
F.	No. of fish harvested (No./ha)	2570	5156
G.	Gross yield (carp)	3.75 t/ha/238 days or 5.75 t/ha/year	12.83 t/ha/320 days or 14.63 t/ha/year
H.	Net yield (carp)	2.92 t/ha/238 days or 4.48 t/ha/year	11.99 t/ha/320 days or 13.68 t/ha/year
I.	Unwanted fish yield	1.00 t/ha/238 days or 1.53 t/ha/year	—

Table 3.11: List of various algae and zooplankters recorded from fish-ponds near Kolleru lake.

I. Phytoplankton	
A. Cyanophyceae	*Anacystis, Agmenelum, Gomphosphaeria, Anabaena, Raphidiopsis, Oscillatoria* and *Lyngbya.*
B. Chlorophyceae	*Eudorina, Pandorina, Actinastrum, Pediastrum, Tetraedon, Ankistrodesmus, Scenedesmus, Crucigenia* and *Coelastrum.*
C. Bacillariophyceae	*Synedra, Nitzschia, Navicula, Amphora, Calonies* and *Cyclotella.*
D. Euglenineae	*Euglena, Phacus* and *Trachelomonas.*
II. Zooplankton	
A. *Rotifera*	*Branchionus calyciflorus, B. angularis, B. forficula, B. falcatus, B. caudatus, B. rubens, B. budapestinensis, B. patulus, Keratella tropica, K. cochelaris, Asplanchna sp., Polyarthra sp., Hexarthra sp., Lecane (lecane) papuna, Monostyla bulla, Euchlanis diletata* and *Filinia longiseta.*
B. Cladocera	*Diaphanosoma excisum, Ceriodaphnia cornuta, Moina micrura, Bosminopsis dietersi* and *Alona pulchella.*
C. Copepoda	*Mesocyclops thermocyclopoides, M. aspericornis, Thermocyclops crassus, Microcyclops varicans, Ectocyclops phaleratus, Heliodiaptomus cinctus, Allodiaptomus raoi* and *Pseudodiaptomus binghami.*

In a similar situation in the Salem district of Tamil Nadu Marimuttu and Krishnamurthy (1985) have studied phytoplankton ecology of two tropical ponds and reported a detailed list of algae (Table 3.12).

Swamps and Marshes

Swamp and Marshes are wetlands having standing waters which form links between terrestrial and aquatic ecosystems. These have been defined by Reid (1961) as shallow water areas with luxuriant growth of macrophytic vegetation. Beadle and Lind (1960) defined the swamp as most shallow slow moving water which permit the growth of particular vegetation. Hora and Pillay (1962) and Welch (1960) pointed out that ecologically, the swamps through their different phases of dereliction, are considered to be

Table 3.12: List of phytoplankters recorded in two ponds from Salem district, Tamil Nadu

Algae	*Pool A*	*Pool B*
Chlorophyceae		
Chlamydomonas kleinii	+	+
Chlamydomonas sp.	+	+
Chlosterium arcuatum	+	+
Cladophora sp.	+	-
Chlorella vulgaris	-	+
Docidium truncatulum	+	-
Golenkinia adiata	+	+
Oedogonium sp.	+	-
Pandorina morum	+	-
Pediastrum simplex	+	-
Scenedesmus quadricauda	+	+
Spirogyra maxima	+	+
Staurastrum cristatum	-	+
Cyanophyceae		
Aphanocapsa pulchra	-	+
Aphanothece saxicola	+	+
Chroococcus coharens	+	+
Cylindrospermum indicum	+	+
Lyngbya contorta	+	-
Merismopedia punctata	+	+
Nostoc sphaericum	+	-
Oscillatoria tenuis	+	+
Raphidiopsis indica	+	+
Bacillariophyceae		
Cyclotella meneghiniana	+	-
Gomphonema lanceolatum	+	+
Melosira granulato	+	-
Navicula laterostrata	+	-
Pinnularia sp.	+	+
Synedra ulna	+	+
Terpsinoe musica	+	-
Euglenophyceae		
Euglena acus	+	+
Phacus acuminatus	+	+
Dinophyceae		
Gymnodinium sp.	+	+

the final stage of the evolution of land from aquatic bodies. Many workers have characterised these water bodies as water logged, shallow water areas with loose peaty bottom, rich in decaying organic matter, retaining water either periodically or shrinking or drying up completely during summer months.

The shallowness of water and growth of vegetation are undoubtedly the characteristic features of existence. In temperate region, swamps and marshes are often shallow wet depressions, where macro-vegetation is largely of emergent and floating type spreading over the entire water surface. Technically a swamp contains persistent standing water among the vegetation, whereas marshes contain water saturated sediments with no or little standing water among the vegetation. On the basis of extent, magnitude of dereliction and relationship with water sources, these are categorised into two main types.

Permanent Swamps

These are found where shallow moving or stagnant water remains perennial and protected from wave action and wide changes in seasonal levels. Such swamps are most common all along the west coast of Karnataka and Kerala states starting from Mangalore to Kanyakumari. These not studied from limnological point of view. Similarly in the regions of North-Bihar Districts (Katihar, Purnea, Samastipur and Begusarai) many such swamps exist. These are larger and smaller tributaries of complex river systems of Kosi, Bagmati, Gandak etc. in North-Bihar with shallow profiles having many slow moving run off channels infested by water hyacinths, Pistia, Lemna sp. and other floating or submerged vegetation. The bottom silt is very rich with organic matter and other nutrients. Thick coverage of water hyacinth and permanent algal bloom make these water bodies unsuitable for fishes.

Seasonal Swamp

These are periodically inundated or is subjected to water logging. The seasonal swamps are of two types: (a) Sensescent swamps, which are wetted grasslands commonly known as "Chaurs" with a few metre deep water attaining acute stage of dereliction. These are most common in Darbhanga, where a few square kilometre area is inundated with water during rainy season either by spilling over of neighboring rivers or water received from its catchment areas. Air breathing teleosts migrate to breed in these shallow waters during the rainy seasons. (b) Semi-Sensescent swamps, are those with shallow water bodies or other low lying areas, manipulated for commercial use by harvesting aquatic cash crops such as "Makhana" (Euryale ferox) and "Singhara (*Trapa bispinosa*). These activities prevent the water areas from becoming swampy and thus remain in semi-sensescent condition.

Physico-Chemical Conditions

The ambient temperature and rain fall are the limiting factors to control the water level of swamps. The thick aquatic vegetation of swamps cause heavy water loss through evapo-transpiration. The temperature, it is known that smaller is the body of water, the more quickly it reacts to changes in the air temperature. The presence of macrophytes influence greatly the water temperature. The diel and seasonal temperature variations in swamps are influenced by the day length, air temperature, and wind velocity and wave action of water.

The swamps are very poor in dissolved oxygen levels and specially the lower layers are often completely deoxygenated. The presence of macrovegetation in water usually induces a hypoxic and hypercarbic aquatic conditions, except during period of peak photosynthesis. There exists difference in the levels of dissolved respiratory gases in waters infested with free floating, rooted floating or emergent types with combination of submerged vegetation. Generally dissolved oxygen was lower, dissolved carbondioxide higher, pH lower and temperature fluctuations lower under the water hyacinth and *Pistia* and *Lemna* covered areas than in the open water. These factors vary according to changes in diurnal and seasonal conditions of the area. Due to extreme hypercarbic and hypoxic conditions mostly air-breathers and supplemental air-breathers are found living under the coverage of water hyacinth.

Low pH (6.25 to 7.00) of water together with high atmospheric and water temperature, heavy accumulation of organic matter and hypoxic conditions of soils and water were responsible for accumulation and occurrence of major obnoxious gases viz., hydrogen sulphide, methane and free ammonia.

The free CO_2 levels in swamps control the pH and bicarbonate carbonate alkalinities. Because free CO_2 forms carbonic acid (H_2CO_3) with water which dissociates into H+ and HCO_3 ions. This brings a change in the pH of water, as hydrogen ions are set free. HCO_3 reacts with calcium or naturally occurring minerals to form bicarbonates which are soluble. If at this stage free CO_2 is not available calcium bicarbonate gets converted into $CaCO_3$. However, if CO_2 is in surplus, calcium can be retained in the form of calcium bicarbonate. That amount of CO_2 which inhibits the conversion of bicarbonates into carbonates, is referred as "equilibrium carbondioxide" Ruttner (1953).

Nutrient Flow

The species composition of swamps is mainly influenced by nutrient flux. Nitrogen occurs in natural waters in elemental, inorganic and organic forms. The elemental nitrogen in water is derived mostly from the atmospheric and bacterial denitrification of nitrates, nitrites and ammonia. Phosphate and nitrogenous compounds are usually released into water either by the decomposition of organic matters in water and sediments or by rain, soil and sewage water run-off. These nutrients are absorbed and stored by plants present in the swamps.

Biotic Factors

The biotic component of swamps are of varied type, which include, macrophytes phytoplankton, zooplankton, periphyton and macroinvertebrate and air-breathing fishes. Four characteristic vegetational communities were recognized in swamps. Those are free floating type (Plate 3.1) e.g., *Eichornia crassipes, Trapa bispinosa, Lemna minor, Lemna major, Salvinia natans* and *Salvinia molesta,* bottom rooted floating type e.g., *Euryale ferox,* submerged type (Plate 3.2) e.g., *Hydrilla verticillata, Potamageton crispus, Naias graminea, Ceratophyllum demersum* etc., and Emergent type (Plate 3.3) e.g., *Cyperus sp., cattail, Sweet flag, Spike rash* etc.

The freshwater swamps have always higher macrophytic productivities than any other natural community (Thompson, 1975). Productivity of submerged macrophytes, though considerably high, are always at stressed condition in such habitat, where large floating and emergent aquatic weeds grow. The submerged vegetation was most sensitive to physico-chemical changes. The problems due to infestation of aquatic weeds are great as they affect the general aquatic weeds are becoming

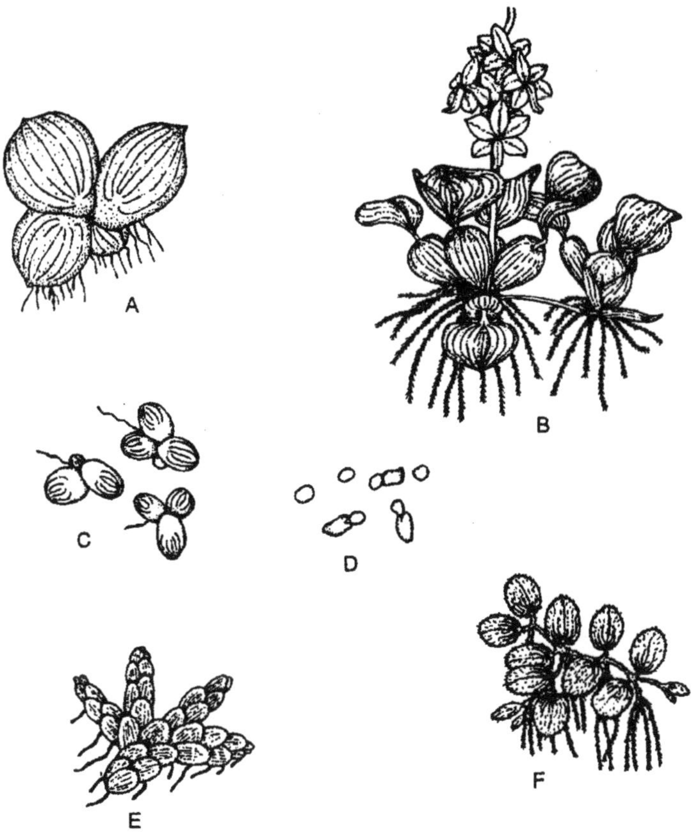

Plate 3.1: Floating hydrophytes

Phylum: Spermatophyta

A. Great duckweed, *Spirodela*

B. Water hyacinth, *Eichornia*

C. Lesser duckweed, *Lemna*

D. Watermeal, *Wolfia*

Phylum: Pterydophyta

E. Water velvet, *Azolla*

F. Water fern, *Salvinia*

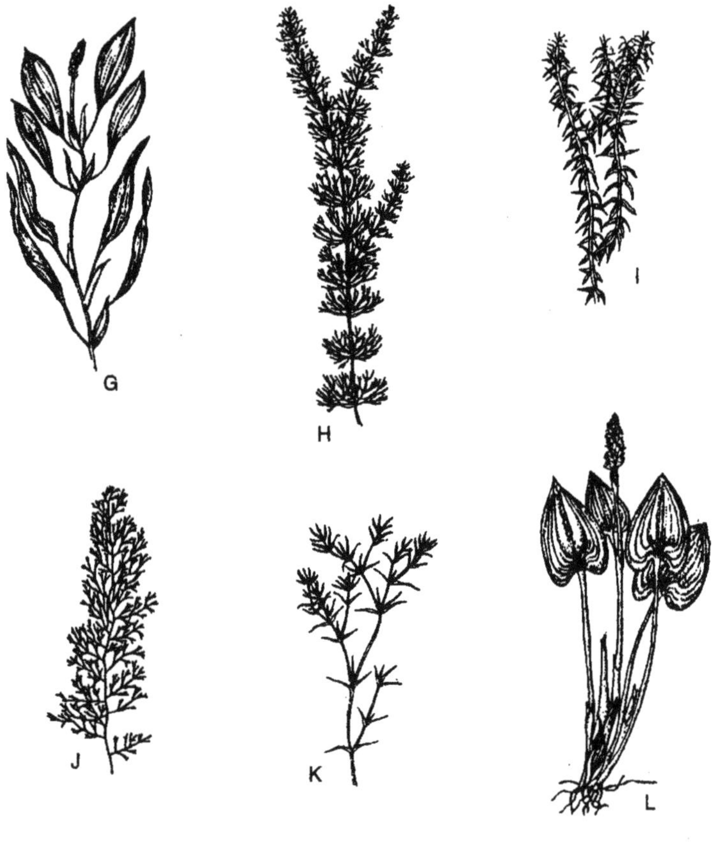

Plate 3.2: Submerged and emergent fresh water hydrophytes.

G. Pond weed, *Potamogeton*

H. Coontail, *Ceratophyllum*

I. Waterweed, *Elodea*

J. Water millfoil

K. Naid, *Najas*

L. Pickerelweed, *Pontederia*

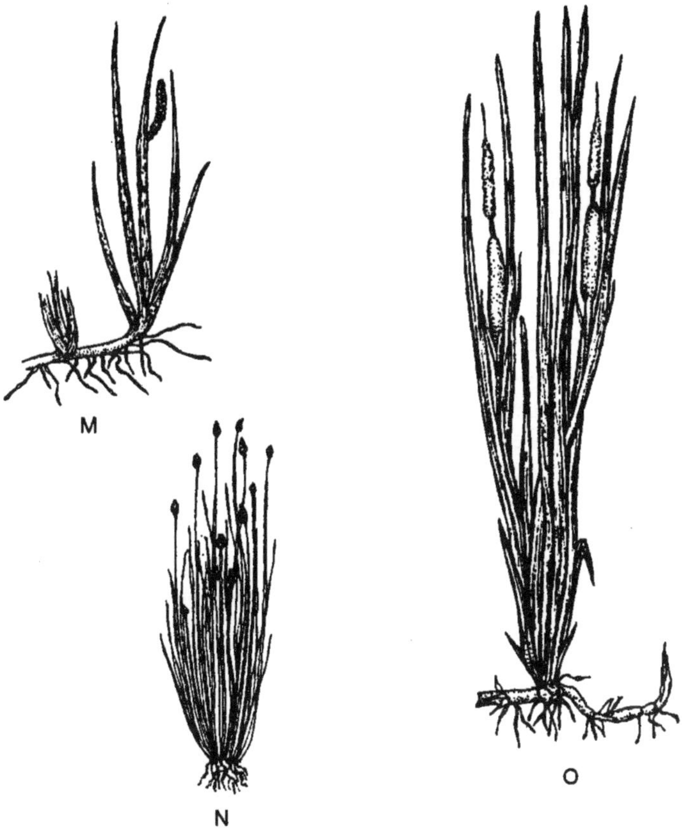

Plate 3.3: Emergent type hydrophytes

M. Sweet flag, *Acorus*

N. Spik rush, *Eleocharis*

O. Cattail, *Typha*

important and are used as filtering beds in the treatment of polluted waters. They are said to absorb potential harmful heavy metals like cadmium, nickel, and mercury and phenolic substances. Dr. Trivedy from the Envrionmental foundation, Karad, Maharasthra, recommends the use of hydrophytes to remove nutrients from wastewater prior to its discharge into the river.

Phytoplankton

The phytoplankton communities of swamps are mainly represented by three groups of algae: Cyanophyceae (Blue green algae), Chlorophyceae (Green algae) and Bacillariophyceae (diatoms). Usually the phytoplankton were highest during winter and early summer, while lowest during monsoon seasons. Bacillariophyceae dominate among the phytoplankton communities throughout the year. In India both in north and south, swamps usually show two-three seasonal peaks of algae.

Zooplankton

General zooplankton communities of swamps belonged to Rotifera, Cladocera and Copepoda. The rotiferan population always dominate over other zooplankton communities in swamps. The swamps are characterised by rotifer genera like *Brachionus, Filinia, Keratella* and *Polyarthra* in India.

Periphyton

The periphyton communities or attached forms of the swamps mainly include algae and testaceous rhizopods (e.g. *Euglypha, Diffugia, Centropyxis, Arcella* etc.) Usually the periphytic species prefer rich dissolved oxygen, high pH and alkaline conditions. Hard testaceous shells of rhizopods are mainly composed of fine sand particles of silica and decomposed diatom cells. The abundance of Bacillariophyceae and silicate contents in swamps is commonly observed.

Macroinvertebrates

Generally the macroinvertebrates of swamps belonged to Annelida, Gastropoda, Odonata, Ephemeroptera, Diptera, Hemiptera, Coleoptera and other invertebrates. The aquatic insects and gastropods dominate the weed infested swamps. The shallow standing water with macrophytes provide a variety of niches for several insect larvae adapted for boring into stems and leaves of plants and substrate for benthos. Floating and submerged macrophytes support large invertebrate fauna than emergent vegetation. Depending upon the degree of association of macroinvertebrate with aquatic macrophytes, they can easily be differentiated into two major groups. The fauna closely associated with submerged macrophytic parts of vegetation (annelids, chironomids, odonata and ephemeroptera) and others comparatively less associated or generally free moving types (Gastropoda, Hemiptera and Coleoptera).

Air Breathing Fishes

Considerable quantities of fish food material including plankton, perphyton, bottom fauna, insects and decaying organic matter are available in different combinations in swamps, but their fish fauna is limited. This is mainly due to the adverse physico-chemical conditions of these water systems, where fast growing, non-predaceous and herbivorous fishes are unable to survive. However, certain slow growing hardy fishes with carnivorous habit and low sensitivity to harmful obnoxious gases and hypo and hypercarbic conditions survive. The air breathing teleosts with their predaceous food habits, accessory respiratory organs for aerial respiration and other biological and physicological adaptations can live well in these swamps. Most common air-breathing teleosts inhabiting these habitats include four species of *Channa, Anabas testudinesus, Heteropneustes fossilis, Clarias batrachus* and *Monopterus cuchia*.

The availability of these should enable interesting ecological conclusions to be drawn about animals inhabiting swamps infested with macrovegetation. With the high temperature, when the oxygen under macrovegetation depletes rapidly to zero especially in the summer, the oxygen demand of most aquatic organisms is increases. As such most of the vertebrates found in the macrovegetation areas are either entirely dependent upon aerial respiration (snakes, tortoises), or supplemental air breathers (frogs, air breathing fishes like the murrels, climbing perches, mud-eel).

Case Study on Savalanga Pond, Shimoga

The water quality of Savalanga pond was studied by Shivappa *et al.* (2000). The water quality of savalanga Tank was studied for pH, DO, BOD, CO_2, carbonates, chlorides, calcium, magnesium and total hardness. All the parameters analysed for the water quality from three sampling stations revealed that the parameters comply with drinking water quality standards. Eleven species of algae belonging to 2 blue greens, 2 greens and 7 diatoms were recorded. The dominance of diatoms indicated the purity of water. Savalanga tank is located in a village called Savalanga about 24 km away from shimoga city on the Shimoga Honnavar state highway in the Karnataka State. The total area of the tank is 80.7 hectares. It lies 13°27′ to 14°39′ North latitude and 74°38′ to 76°11′ E longitude. The water from this tank is used for drinking purpose for the village. The inland fisheries Department is practising carp culture in the pond. Additionally the Forest department established a plant nursery adjacent to the pond.

The water quality is characterised by dissolved oxygen average values 4.0, 6.0 and 7.2 mg/l at the reaches S_1, S_2 and S_3 respectively. The rise in aerobicity from station S_1 to S_3 indicates increase in aerobicity. Comparatively high amounts of DO at the S_3 is attributed to enrichment of nutrients by slight human activity and thereby increase algal production and hence increase in oxygen levels (Ganapathi, 1974; Adoni, 1985). The carbon dioxide was recorded average values 0.1 mg/l (S_1), 1.0 mg/l (S_2) and 1.1 mg/l (S_3) respectively (Table 3.13). Due to continuous release of carbon dioxide through respiratory activities and decomposition process at the bottom layers and also atmospheric diffusion of the CO_2 formed combines with calcium and magnesium is responsible for the hike in alkalinity of the water. Hence all the water samples during the study showed alkaline pH values (Table 3.14).

In the pond systems most of the CO_2 produced is utilised in the process of photosynthesis by phytoplankton (Saha and Pandit, 1986). An inverse relationship between free carbon dioxide and carbonate alkalinity was reported in freshwater ponds by Kant and Raina (1990).

The five day biochemical oxygen demand indicates the amount of organic load in the water samples (Hosetti *et al.*, 1985). The BOD levels were minimum in all the samples. As per the drinking water quality standards BOD should be less than 6.0 mg/l in the raw freshwater and the data hitherto obtained also comply with above statement. The water quality of Savalanga pond is most suitable for drinking purposes. It is interesting to note that why the chloride levels is high in all the three samples is not clear. Maximum chlorides were recorded at S_3 and this require further investigation. The pond water is also represented by high levels of calcium and magnesium. These are the two elements keep

Table 3.13: Physico-chemical characteristics of Savalanga water Tank (Average values ± SD)

Parameters	*Stations*		
	S_1	S_2	S_3
Dissolved Oxygen	4.0 ± 0.07	6.0 ± 0.07	7.2 ± 0.16
Biochemical Oxygen demand	2.0 ± 0.07	2.8 ± 0.07	3.2 ± 0.07
Carbon Dioxide	0.1 ± 0.07	1.0 ± 0.15	1.1 ± 0.11
Bicarbonates	84.0 ± 1.11	120.0 ± 1.22	138.0 ± 1.87
Chlorides	52.2 ± 0.15	52.2 ± 0.12	82.8 ± 0.07
Calcium	40.08 ± 0.01	40.08 ± 0.03	40.08 ± 0.04
Magnesium	51.92 ± 0.01	53.92 ± 0.02	39.93 ± 0.02
Total Hardness	92.0 ± 0.12	94.0 ± 0.14	80.0 ± 0.16

All the parameters are expressed as mg/l

Table 3.14: Physical parameters of the Savalanga tank water (Average values)

Sampling stations	*pH*	*Air T°C*	*Water T°C*
S_1	7.2	25.0	25.0
S_2	7.2	26.0	26.0
S_3	7.6	27.0	27.0

the pond water always towards alkaline range by precipitating into carbonates and bicarbonates. The potable waters are classified into 3-4 categories based on the total hardness as soft, medium, hard and very hard and the Savalanga pond water is medium in hardness.

Among the eleven species algae recorded 7 were diatoms, 2 green and 2 were bluegreen respectively (Table 3.15). Dagaonkar and Saxena (1992) have reported 25 species of algae and 26 species of drift animals from a temple pond at Gwalior. The temple pond was dominated by green and bluegreen algae indicating pollution and eutrophication of the water body. In the present study on Savalanga pond, the diatoms were dominant organisms which indicates that the water was free from organic pollution and eutrophication. Hence the pond is most suitable for drinking water supplies.

Table 3.15: List of algae recorded in the pond water

	Diatoms	Green algae	Blue greens
1.	*Pinnularia major*	1. *Spirogyra sp.*	1. *Ocillaotoria sp.*
2.	*Stauroneis phoenicentron*	2. *Chlorella sp.*	2. *Microcystis sp.*
3.	*Anomenieis sphaerophora*		
4.	*Anomenieis brachysira var. thermalis*		
5.	*Navicula husted*		
6.	*Gomphonema tenellum*		
7.	*Gomphanema subpicatum*		

D. River Ecosystem

A river with its tributaries forms an unique aquatic ecosystem which cover different types of climatic zones, landscapes and biogeographic regions. It is the natural drainage system of the earth which moves continuously. The river water flows in a directed by the topography of the area and the underlying rock layers. The water current cuts the water course wider and deeper while moving downwards increasing its drainage area. Near the origin the river is small, straight and swift while in the downstream the velocity of water decreases and enters into the plains. As the velocity of water current gets reduced the load it carries is deposited as silt, sand and mud. When a river flows into a lake or sea the velocity of the water current is suddenly checked and the load of sediment is deposited in fan-shaped area at the inlet point to form a delta. Here the course of water is broken into a number of channels which are blocked or opened with subsequent deposits. In course of time the sediments build up new land surface. The Gangetic plains of India has been built up like this.

The continuous swirling and churning of stream water results in a greater contact with the atmosphere and leading to increased oxygen levels in the water is high at all levels and often is near saturation point for the prevailing temperatures. Only in deep holes or in polluted areas dissolved oxygen show significant decline. Most of the carbon dioxide in flowing water occurs as carbonate and bicarbonate salts.

Functional Aspects

The river is a heterotrophic and an open-system, as it depends for energy from outside in contrast to standing water and other terrestrial systems. The energy is supplied mainly by allochthonous sources in the form of leaf litter, twigs and branches carried by air and flood. Other inputs come by a geological pathway through sub-surface seepage which brings nutrients leached from the adjoining forest grasslands and agricultural or residential areas and from upstream flow carrying both dissolved nutrients and particulate matter. Many river inputs come from mechanical pathways through the dumping of urban and industrial effluents into the flowing water.

A minor biological autochthonous source of nutrients is primary production mainly by diatoms and algae growing on rocks and by rooted aquatic macrophytes. Even in slow streams where primary production may be substantiated by rooted hydrophytes provide immediate source of energy. The energy source for the huge population of consumers (Fishes) inhabiting the river mainly depend on allochthonous organic matter coming from various other sources. Energy is lost along two pathways in rivers, geological pathway through streamflow to down stream systems and biological pathway from heat loss through respiration.

Important River Systems of India

The five major river systems of India are the Ganga, Brahmaputra and the Indus river systems in the north and the Peninsular East Cost river system comprising the Mahanadi, Godavari, Krishna, the Cauvery and the West Coast river system in the South comprising the Narmada and Tapti rivers.

The Ganga river system is regarded as one of the largest river systems of the world. It drains the southern slopes of the Central Himalayas and cover the States of Haryana, Uttar Pradesh, Bihar, West Bengal and parts of Rajasthan and Madhya Pradesh. In the plains the water get warmed up and becomes biologically more productive. The Ganga river system harbors the richest fresh water fish fauna of India ranging from mahseers and the torrestial fishes of the hills to the culturable carps, and migratory Hilsa and a wide array of other species of fishes, some are of great commercial value. The Brahmaputra river system drains the northern slopes of the Central and Eastern Himalayas, Assam, Bhutan, Sikkim and parts of North-West Bengal. It has a rich fish fauna of torrential streams and several species of carps, catfishes and air-breathing fishes. The anadromous Hilsa is of considerable commercial importance.

The Indus river is a massive system of five rivers of Punjab and Pakistan. Only a small segment is left in the present-day India where the important rivers are the Beas and the Sutlej. These rivers harbors the exotic rainbow and brown trouts in their upper reaches and variety of indigenous carps and catfishes in their lower section. The East coast system of rivers of Peninsular India comprises the Mahanadi, Godavari, Krishna and Cauvery having a combined length of about 6,437 km. This system drains the entire peninsular India in including east of Western Nagpur Hill ranges. The Mahanadi has all the Indian major carps common with the Ganga system. The other rivers of the East coast system besides their own indigenous fish fauna of several carp species have catfishes, murrels, perches, prawns etc., which might have been migrated or intentionally introduced from the north. These species have established themselves and contribute significantly to the fisheries of these rivers. Tributaries of the Cauvery from the Nilgiris have cold water fish like trout and masheer.

The West coast river system in the south drains the narrow belt of peninsular India originated from the Western Ghats. The Narmada and Tapti are the longest rivers of the system and are faunistically quite rich. All the other rivers of this region arising from Western Ghats are short and though many are perennial most of them are like torrential streams only. The length of the river of this region is about 3,380 km and their over-all fish fauna consists of carps, catfishes, mahseeers, murrels, perches, prawns etc. A case study carried out by Datta Munshi (1990) is chosen for the study.

Case Study 1

Kosi River System

The Kosi river system comprises seven rivers called Saptakosi, draining the southern slopes of the Himalayas between 25°20′ to 29°0′ N lat. and 85°20′ to 89°0′ longitude covering a large catchment area of 70,400km^2. The three main streams Sun Kosi, Arun Kosi, and Tamur Kosi originate from the glaciers of

Mt. Gosaithan (8,013m), Mt. Everest (8,848m) and Kanchanjungha (8,579m). The other four streams Indravati, Bhote, Thamha and Dudh join the Sun Kosi. All these rivers join together to form the mighty Himalayan Saptakosi of Nepal. The Saptakosi assumes its proper name Kosi while passing through Chatra in Nepal.

The Kosi drainage system in the plains covers an area of about 58,570 km^2. In comparison to the river Ganges (2575 km) it has got a much shorter length of 965 km. It arises from a height of 5490 meter in the Tibetan plateau and has a different water quality than the river Ganges and has turbulent water current.

Drainage of Kosi in North Bihar

Flowing down the high mountains and low Siwalik hills of the Himalayas the Kosi enters into the plain of Nepal terai at Chatra. It runs through its two main branches for some distance before uniting again near Hanumannagar (Bihar-Nepal border). Now the Bhimanagar barrage controls the flow of Kosi in the plains of Purnea, Saharsa and Darbhanga. It then meets the "Ghugri Kosi" near Mansi, North-East of Munger. Soon after it enters into the plain it becomes about a mile broad and flows through many network of channels of braided pattern. They bifurcate through interlacing the Bhimnagar barrage to as far South of Koparia (Saharsa district) the Kosi is well known for its profuse branching into many channels called Dhars. Due to its ever shifting nature the river system is of braided nature. During its course the bed is nearly 4-5 miles wide with diarah land formation known as Kosidiarah with embankments and depressions alternating each other at a number of places through out its course.

Further South below Koparia and Dhamaraghat the shifting tendency gets reduced and it runs almost parallel to the Ganga river in a well defined channel till its confluence with the latter where the depth of the river bed increases considerably. Quite a large number of rivers meet the Kosi river system at one point or other. They are Bagmati, Burhi Gandak, Kareh, Jibach, Kamala, Balan, Tiljuga, Dhemura and others.

Glaciers, Rainfall and Discharge Characteristics

Many of the flooding and other related problems of Kosi are chiefly due to the effects of glaciers and heavy rainfall in the Himalayan terai regions and high siltation rate. Kosi discharge is second highest among the Indian rivers. Its average discharge in normal years is about 1,75,00 cusecs. But very high discharge have been recorded 7,05,00 in 1927; 4,78,442 in 1948; 8,55,000 in 1954 and 9,25,000 cusecs in 1968 respectively.

Siltation

Geologically speaking the upper catchment basin is composed of rocks like granites, crystalline limestones, fossiliferous limesstones, boulders, sandstones, limestones, less compact pebbly grit and silt (coarse, medium and fine), and loam with calcareous nodules. Soil of Kosi river basin is alluvium non-calcareous and non-saline. The slope of the river not only induces rapid erosion on hill slopes but also allows the river to carry sand and silt down the hill to the low land in the plain. The various streams, tributaries and other (small water courses active during rain season) of the Kosi carry enormous amount of sand causing siltation got the highest capacity of siltation amongst the rivers of world including the Yellow river of China (Mookerjee, 1971). From the analysis of the data it appears that about 95 per cent of the silt load comes down the river during the monsoon (June-October) and only 6 per cent during the rest of the year. Everywhere the process of silting and scouring the river bed continues. The river deposits the silt somewhere and washes it out from other places. This characteristic of silting of Kosi is the main cause for its shifting nature and formation of sand bars in the river course. The silt deposition is rendering barren vast areas of land estimates at 5,20,000 to 8,00,000 hectares in Bihar and 80,000 to 1,30,000 hectares in Nepal.

Flood

The flood of very serious concern have become almost annual feature of the Kosi region. The devastating flood water carry enormous amount of sand and deposits them on the fertile lowlands making them unproductive for longer periods. Besides its catastrophic flood affects the lives of about two million people of North Bihar every year. Conditions that obtained are more or less the same as those in the Hwang-Ho river basin of North China plain. On account of its heavy sand and silt discharge flood of damaging proportions and calamities brought in their wake and above all its serious impact on the land people and economy of the region it has earned the name as the Sorrow of Bihar.

In order to mitigate the miseries of the people the authorities were prompted to harness the "Sorrow of Bihar" by constructing parallel flood embankments (western eastern) tied up to a masonry barrage (Bhimnagar barrage) at Hanumannagar and by diverting large part of the discharge through sluices as well as numerous irrigation canals for flood protection checking westerly transition, irrigation and generation of hydal power.

Shifting Courses of Kosi

Due to the physiographical nature the river Kosi is well known for its profuse branching into many dhars and for the unstable nature of its course. It may be mentioned that in early part of the 18th century, the Kosi flowed below the town of Purnea but it has gradually moved westward across 120 km of the countryside as is evidenced by the deserted channels. The left out channel at Purnea is known as KaliKo. These dhars (streams) have lost their connection with the river Kosi hence are less powerful today and normally go dry except during the rains. Besides these a number of sandy wastelands, marshy chaurs, water reeds, river bluffs separated here and there by creeks and minor trickles spread over whole of the Kosi belt are still noticeable.

Impact of Embankments

Bundhs, barrage and canals were constructed during 1954 to control flood and river shifting. This checked the process of formation of new waterbodies. Due to these barriers the existing dhars, swamps, and chaurs have shrinked up or dried up depleting the freshwater resources of Kosi belt.

Physico-chemical Factors

The physico-chemical parameters of Kosi river water passing through different zones along with ambient temperatures are given in Table 3.16. The pH of the river water was mostly alkaline. The fluctuation in the oxygen content of water along the longitudinal zone is very little. Carbondioxide is present in the form of free state as well as bicarbonate state, carbonate being nil. High content of silicate in the water is remarkable in all the sampling stations.

Biotic Factors

About 27 species of diatoms were reported from the river system. The remarkably high content of silica probably makes the

Table 3.16: Phyisco-chemical parameters of Kosi River water along its course at different places.

Parameters	*Units*	*Bharda (Nepal)*		*Birpur (India)*		*Supaul*	*Dumari-ghat*	*Kur-sella*
		Up.St.	*Dn.St.*	*Up.St.*	*Dn.St.*			
At.temp.	(°C)	26.33	29.66	24.66	25.50	24.83	26.50	26.30
Wt.temp.	(°C)	22.33	23.50	22.30	22.16	23.00	24.00	23.83
pH		7.23	6.97	7.1	7.2	7.13	7.2	7.22
Turbidity	(NTU)	165.0	113.33	103.30	96.67	76.67	221.67	136.67
Conductivity	(μMNos)	143.33	180.0	150.0	163.33	170.0	150.0	213.33
DO_2	(ppm)	7.03	6.72	6.53	7.76	7.6	7.07	7.2
FCO_2	(ppm)	2.33	2.80	1.90	1.87	2.2	2.2	1.6
Co_3 alk.	(ppm)	N	I	L	—	N	I	L
HCo_3 alk	(ppm)	53.67	50.33	46.87	47.47	54.67	58.00	68.00
Chloride	(ppm)	27.52	29.37	33.08	27.52	32.81	22.56	22.49
Total Hardness	(ppm)	46.00	53.67	46.33	43.33	52.33	52.67	44.00
Ca-Hardness	(ppm)	38.67	41.33	40.00	33.00	42.67	38.33	37.67
Ca^{++}	(ppm)	15.47	16.53	16.00	13.20	17.07	15.30	15.07
Mg^{++}	(ppm)	1.789	3.009	1.545	2.520	2.358	3.497	1.538
Na^{++}	(ppm)	2.0	13.0	5.533	4.033	3.366	3.497	1.538
K^{+}	(ppm)	1.4	9.0	1.8	1.8	9.0	1.4	1.4
Silicate	(ppm)	19.09	16.33	17.68	14.84	17.76	18.10	21.13
No_3N	(ppm)	1.74	6.52	4.82	9.62	1.65	2.16	1.95
Po_4P	(ppm)	0.205	0.18	0.276	0.30	0.325	0.230	0.155

dominance of this group in the river. There is important and dominant biota of Kosi river are as follows.

(a) **Chlorophyceae:** *Spirogyra polymorpha, Hydrodictyon reticulatum, Vaucheria sp., Pediastrum sp., Hormidium sp., Cosmarium reniforme, Oedogonium sp., Monostroma quaternariwn, Mesotacnium sp., Chactophora sp., Cladophora insiguis, Draparnaldia sp., Eudorina sp., Crucigenia terrapedia, Botrydium granulatum, Clolsterium aerosum, Cylindrocystis sp., Sirogonium sp.*

(b) **Myxophyceae:** *Spirulina major, Spirulina sp., Phormidium ambiguum, Phormidium sp., Anabaena orientalis, Anabaena circularis, Ankistrodesmus sp., Nostoc sp., Oscillatoria limosa, Microcystis sp.*

(c) **Bacillariophyceae:** *Cymbella cistula, C. turgida, Cymbella sp., Fragilaria intermedia, Fragilaria sp., Gomphonema sp., Melosira ambiguans, Navicula viridula, Navicula viridis, Nitzschia sp., Syndra ulna, Synedra acus, Pinnularia nobilis, P. gibba, Pinnularia sp., Mastogloia danseii, Denticula sp., Rhopalodia sp., Neidium affinis, Frustulia rhomboides, Calones amphisbaena, Anumoeoneis sp., Amphora avalis, Diatomella halfouriana, Gyrosigma kwzingii, Rhzosolenia sp., Surirella sp.*

(a) **Protozoa:** *Euglena sp., Difflugia sp.,*

(b) **Rotifers:** *Filinia longiseta, Braclionus caudatum, B. falcatus, B. forficula, Keratella tropica, Keratella cochlearis, Polyarthra sp., Hexarthra sp., Horaella brehmi, Lecane luna, Cephalodello auriculata, Trichocerca similis* and *Synchaeta sp.*

(c) **Copepods:** *Diaptomus sp., Cyclops sp., Cladocera, Moina sp., Diaphanosoma sp., Bosmina sp., Bosminopsis sp.* and *Nauplius.*

(d) **Nematoda:** *Prismatolamus sp.* and *Aphelenchoides sp.,*

(e) **Mollusca**

(I) **Gastropoda:** *Bellamya bengalensis (Lamarck), B. dissimilis (Muller), B. bengalensis, B. bengalensis, Pila globosa (Swainson), Thiara scabra (Muller), T. (Tarebia) lineata (Gray), Brotia costula (Refinesque), L. acuminata, refescens (Gray), Lymnaea accuminata, L. acuminata patula (Lamark), L. luteola, Indoplanorbis exustus (Deshayes), Achatina fulica fulics (Bodich)* and *Cryptozona ligulata (Ferussac).*

(II) **Bivalvia:** *Lamellidens corrianus, L. marginalis, Parreysia (Radiatula) caerulea, P. favidens (Benson), P. favidens var. pinax (Benson), P. favidens var. maruns (Benson), P. favidens var. deltae (Benson), P. olivaria (Lea), P. pachysoma* and *Corbicula striatella.*

Case Study 2

River Cooum Madras (Chennai)

The water quality of almost all rivers in India are influenced by manmade activities. Specially the reaches near by the urban centres are mostly degraded and most of the rivers have become sewers. No river is left free from industrial effluents and sewage pollution. However, during monsoon and flood conditions, the rivers are comparatively free from pollution due to dilution, at least in the upstream reaches from urban centres. The case studies chosen for this topic are also polluted from sewage.

Another problem commonly met with any limnologist interested in river ecology is that it is very difficult to find an aquatic system (freshwater) free from pollution and eutrophication and thus the biota is different from the desired ones. Many time a condition is reached that the biota of polluted reaches its confused with normal biota.

The literature survey on the case studies of freshwater systems of India revealed that majority of the reseraches/papers are dealt partially, that all the reports are mostly fragmentary. Many time holistic approach is lacking. Hence it becomes very difficult for a freshwater biologist to review the subject and select suitable case studies in Indian rivers.

Salvanayagam *et al.*, (1988) have studied the water quality of Cooum river near Madras city in relation to aquatic hemipterans. This case study deals with changes in physico-chemical parameters along the river course and their influence on the occurrence of hemipteran insects. The sampling stations are printed in Table 3.17. And the river course is showed in Fig 3.4. The chemical factors showed wide variation from upstream to towards the downstream

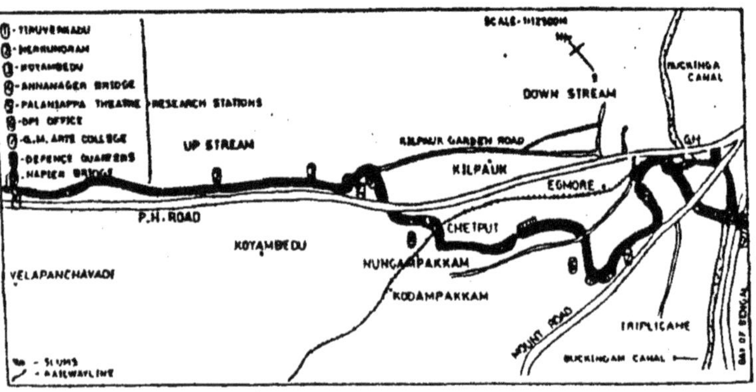

Fig.3.4: The Cooum River Course

Table 3.17: Names of sampling stations, distance and source of pollution

Sampling stations	Names	Distance from sea (Km)	Source of pollution
1	Tiruverkadu bridge	17 Km	Water clear, useful for domestic purpose.
2	Nerkundram bridge	14 Km	Water clear, bathing and washing taking place and no appreciable pollution.
3	Kovembedu bridge	12 Km	Human waste, domestic sewage and waste from small factories.
4	Annanagar bridge	10 Km	Treated sewage (5 million gallons/day) from Madras metropolitan water supply and sewerage board along with domestic and human waste.
5	Nungampakkam bridge	8 Km	Small sewage outfalls waste from Dhobi ghats, cattle washing and the wastes from small factories situated adjoining the Nelson Manikam Road.
6	Anderson bridge	5.5 Km	Buffalo waste and washing, four main sewerage outfalls along with waste from other establishments.
7	Harris bridge	3.5 Km	Treated sewage from Greams road sewage pumping station, two major sewer outfalls, waste from slums and other establishments in the heart of city.
8.	Quaid. E. Millat bridge	1 Km	Waste from north Buckingham canal, Southern Railways, General Hospital, Central Jail, slums and other domestic and human waste with five major outfalls.
9	Napier bridge	sea-mouth	Waste from transport corporation workshops, Trade fair and other Government establishments and University of Madras.

Clear water : Zone 1 to 3, Polluted : Zones 4 to 9

Table 3.18: Physico-chemical characteristics of water in the nine sampling stations of the river Cooum

Physico-chemical character	*1*	*2*	*3*	*4*	*5*	*6*	*7*	*8*	*9*
I. Temperature (0°C)	30.1	30.8	31.7	30.3	30.2	29.8	29.8	29.7	29.4
II pH	6.5	6.5	7.0	7.5	7.0	7.5	8.0	7.5	8.0
III Salinity (ppt)	1.91	2.86	1.91	2.19	2.32	6.73	15.36	26.41	27.75
IV Dissolved Oxygen(mgl^{-1})	10.09	9.06	7.78	0.0	0.96	0.0	0.0	0.0	0.0
V Free CO_2 (mgl^{-1})	4.0	3.0	11.0	21.0	20.0	21.0	18.0	30.0	22.0
VI BOD (mgl^{-1}) 5 day, 20°C	10.0	10.0	18.0	20.0	25.0	38.0	45.0	78.0	73.0
VII COD (mgl^{-1})	36.0	42.0	123.0	147.0	158.0	279.0	330.0	517.0	429.0
VIII TOC (mgl^{-1})	8.0	6.0	16.0	14.0	24.0	36.0	45.0	58.0	61.0

Clear water : Zone 1 to 3, Polluted zones 4 to 9

Table 3.18. The first three stations are considered as unpolluted zones and latter one is a polluted zones. Hence only clear zones considered in this Chapter.

Clear Water Zone

The first three stations Thiruverkadu, Nerkundrum and Koyembedu bridge stations are included in the clear zone. In the first two stations the water is completely free from pollution and from station three onwards it began to pollute.

There was slight variations in the chemical parameters except salinity levels. The DO was in the range 7.78 mg/l–10.1 mg/l and it was recorded comparatively low levels in the third station. The CO_2, BOD, COD and TOC were also little high in the third station. These parameters were recorded in the range 4 to 11 mg/l, 10 to 17 mg/l, 36 to 128 mg/l and 8 to 16 mg/l respectively. The domestic sewage accruing from the city is the main culprit for pollution in station three. Among the three clear water stations the third station Koyambedu had abundant hemipterans (Table 3.19). The slight pollution in this has not affected the insect population instead, it enhanced the population of insects in the area. The aquatic vegetation namely *Vallysneria spiralis,*

Table 3.19: Distribution of aquatic Hemipterans in the nine selected stations of River Cooum

	Aquatic bugs	*Zones*								
		1	*2*	*3*	*4*	*5*	*6*	*7*	*8*	*9*
I	Limnogonus fossarum fossarum	++	++	+++	+	+	+	+	–	–
II	Gerris (A) spinolae	+	+	++	+	+	–	–	–	–
III	Hydrometra sp.	++	+	++	+	+	+	–	–	–
IV	Corixide sp.	++	+	++	–	–	–	–	–	–
V	Anisops bredeni	++	+	++	–	–	–	–	–	–
VI	Anisops bouvieri	+	+	++	–	–	–	–	–	–
VII	Ranatra filiformis	+	+	++	+	–	–	–	–	–
VIII	Ranatra elongata	–	–	+	–	–	–	–	–	–
IX	Laccotrephes griseus	+	+	++	+	–	–	–	–	–
X	Laccotrephes robustus	–	–	+	–	–	–	–	–	–
XI	Diplonichus indicus	++	+	+++	+	–	–	–	–	–

+++ Abundant
++ More
+ Less
– Not found

I Zones 1 to 3 : clear water
II zones 4 to 9 : polluted water

Monochorion sp., and *Marsilia sp.* served as food and habitat for these insects. Nirmala kumari and Balakrishnan (1984) have observed that Dragon fly, damsel fly and hemipterans were present in large number in the aquatic weeds. The aquatic, hydrophytes provide suitable habitats for these insects. Similarly more number of aquatic insects like *Diplonichus indicus,* water striders, *Limnogonus fossarum,* hemipterans, *Anisops sp.*, and *Laccotrepjes sp.* were in the polluted zone of the river.

Case Study 3

River Tunga

A stretch of 20 km near Shimoga city the river water quality was studied by Dr. S. R. Patil for his doctoral work during the period 1995-1997. Basically he is a botanist and he could concentrate his studies on the routine analysis of water quality parameters and a detailed aspect of phycology was undertaken.

The river is polluted severely near the city limits for a reach of 2 km. The upstream and down stream of the city limit of the river is free from domestic pollution. Every year the pollution will be washed away during monsoon floods and polluted conditions develop in the summer as the river flow receeds and the basin opens. During summer months the river is characterised by unhygeinic scene, growth of floating hydrophytes like *Pistia, Lemna, Eichornia* and other weeds. Additionally the nuisance of flies and midges is common problem. Open defecation by local slume dwellers, Lorry cleaning, cattle wading, and dumping of garbage is also observed. However, the author has emphasized his investigation on the unpolluted part of the river. The study reveals that it is rich in zoo and phytoplankton, fishes and invertebrates.

The phytoplankton include 18 genera 11 species of desmids, 23 genera and 19 species of diatoms, 16 genera and 20 species of blue greens, 13 genera of euglenae and a single species of xanthophyceae were recorded. Relative abundance of various algae and the pollution index was studied based on Patrik's formula and the values were in the range of 1.33 to 4.0. The data on pollution index revealed that the lower the index, higher is the organic pollution. It was 1.32 to 2.0 in the stretch near the city limits and around 11.0 in the stretches away from pollution sources at the outskirts of the city. He also reported protozoa (12 genera), rotifera (12 genera), crsutacea (11 genera), aquatic insects (6 genera) and fishes (21 genera and 25 species) respectively.

E. Thermal Springs

Thermal springs are those aquatic bodies where mean annual temperature is higher than the atmosphere (Gilber, 1875). The thermal springs generally occur in the following geologically characteristic areas.

(1) Along faults, fissures and shear zones.

(2) Along the nose of folds.

(3) Along the contact plane of two adjacent rock type.

There are three distinct causes of thermal environment, which include, solar heating, combustion process, geothermal activity and radioactive decay. Solar heating can lead to soil temperature as high as 60°C as is generally seen on the surface of a desert soils at mid-day. In aquatic environments solar heating leading upto 40°C-60°C has been recorded. In a stratified pond near to gulf of Eilath (Israel) temperature as high as 50°C has been reported as a result of solar heating. Combustion processes may be of biological or non-biological origin. Thermophilic micro-organism, like fungi and bacteria grow on dumped up organic waste material, decompose and generate heat. Geothermal activity is probably responsible for the creation of high temperature environment. The temperature in active volcanoes is too high for living organisms. Molten lava can have a temperature more than 1000°C.

Geothermal Water Source

The atmospheric water from the surface descends into the interior of earth through some extended fissures or along the surface of an impermeable layer. It goes below the ground water level and is made to circulate there under the combined action of gravity and capillary forces. The temperature of earth crust increases with depth roughly at the rate of 1°C per 31.7 m. Suess (1902) hypothesized that inside the lithosphere, water is formed by O_2 and H_2 at high temperature and under great pressure. The atmospheric O_2 and H_2 which have penetrated long ago into the crust of earth remain in contact with the volcanic magma and the synthesis results. The constant temperature of the thermal spring is maintained by the mixing of the two types or water, magmatic (Juvenile) and meteoric (rain) water at some depth below the ground water level, where the quantity of the surface water is fairly constant being dependent on the pore space, cracks and crevices of the rock bed and the mean annual rainfall.

Based on the extensive work carried out on the thermal spring areas of Yellowstone National Park, U.S.A., Allen and Day (1935) have concluded that thermal springs are essential circulating ground water of surface origin, which got heated and augmented by steam originally present in a super heated state in contact with the magma. As the rain water descends down in the interior of the earth, the temperature increases with the depth regularly. This water is driven upwards by thermal expansion pressure exerted by the containing gases and by density between descending cold and ascending hot water and is finally discharged to the surface as a thermal spring. The water of the hottest springs of Bihar is possibly coming out from depths over 6-8 kms.

Sources of Thermal Energy

Heat from the underlying magma is transferred to the spring water either through rock conduction or by direct contact. High content of CO_2, H_2S, Li, B, F, in such waters is indicative of the fact that considerable amount of heat is contributed by the active participation of the magmatic fluid. Another source of heat seems to be the exothermic reaction of certain chemical and mineral transformations. It is found in the disintegrating radioactive elements (Bhimbundh thermal spring) and the decomposition of kyrite in the generation of H_2S which characterize many such springs as Surajkund and Rajagir. Heat also evolves in the zones where faulting of rocks occur. Many thermal springs are seen to emerge out along active faults and fissures.

Most of the thermal springs are associated with volcanic rocks and many scientists think that the source of heat in a thermal spring is due to volcanic activity. The steam produced by magma is the main source of heat in thermal springs. Probably geothermal energy is responsible for the creation of high temperature environments. The largest concentration of hot springs are found in Yellowstone National Park (U.S.A.) Iceland, New Zealand, Japan and the U.S.S.R. The geothermal heating is a common phenomenon throughout the earths surface and in most areas the heat production is so small that it remains unrecognized.

Mineral in the Thermal Springs

Chemical studies on mineral springs were initiated by Robert Boyle in the 19th century. The chemical composition of hot spring varies widely. The pH of hot springs varies and values ranged between 1.0 to 11.0. Many hot springs have significant amounts of hydrogen sulfide. Many acidic springs are rich in ferrous ions. The concentration of such interesting elements as fluoride, arsenic, rare earths and gold varies very much. Many springs are highly radioactive, whereas other have no more radioactivity than normal ground water levels. The early geochemical work is reviewed by Allen (1934).

Thermal Springs of India

There are more than 200 thermal springs scattered throughout the country and occur either as solitary or in groups at different altitudes ranging from sea level to nearly three thousands metre above the sea level. Generally these occur only in those area which are volancially active. The radioactivity and therapeutic value of several springs have been evaluated by different workers in early part of the present century. Some of the springs of Maharashtra are permanently radioactive. However, most of the Indian thermal springs are radio-emanative and not permanently radioactive. Nag (1931) has studied the radioactive properties of Rajagir thermal springs. Chatterjee and Chatterjee (1958) have found strong radioactivity in the thermal springs of Bakreswar, Swetaganga and Agnikund in West Bengal.

In India thermal springs are found in four main broad belts, those are south Bihar belt, west coast of India, Sind-Baluchistan area and the Himalayan belt. Thermal springs are abundantly located in the states of Bihar, West Bengal, Andhra Pradesh, Maharashtra, Gujarat, Tamil Nadu, Karnataka and Orissa. In Sind and Baluchistan, Rajasthan, Punjab and Haryana the thermal springs are more in numbers. In the Himalayan region the thermal springs are present in Jammu and Kashmir, Himachal Pradesh, Assam, West Bengal, Uttar Pradesh and Bhutan.

Surajkund

It is the hottest thermal spring of Bihar having a temperature of 85°C. It is situated about 23 km west from Bagodar Police Station, Hazaribagh district. Water is accumulated in a kund. There are also three small kunds near the Surajkund named as Bharatkund, Ramkund and Sitakund, but they do not have water of high temperature.

Rajagir

It is the most popular hot spring spot of Nalanda district, 266 km south-west of Bhagalpur and 100 km from Patna. There are six different sources and all of them have been covered in the form of water jets, temperature ranging between 40-42°C. The thermal springs of Bhimbandh (Munger) have been studied extensively (Saha 1980; Saha and Datta Munshi, 1980 and 1982).

Physico-Chemical Factors

Generally the water temperature varied between 48.5 and 52.5°C and lower temperatures are associated with the winter months and higher temperatures during the summer months. The water was acidic and the pH varied between 1.4 and 5.8 ppm.

The gases come out along with water at different sources. The gas samples were collected from the thermal springs of Bhimbandh (63°C) and were analysed (Datta Munshi, *et al.*, 1984) the presence of rare gases in the thermal springs is quite remarkable. The biological characteristics of thermal water depend upon several factors of which temperature is most potent one. Life exists mostly in the form of flora at higher temperature gradient. Animal life appears only at lower range of temperature. The diversity of the species increases as the temperature decrease along the thermal gradient of the hot spring.

The flora of hot spring along the thermal gradient of Bhimbandh mainly comprises algae, occurring both as phytoplankton and periphyton. They play key role in the productivity of the thermal stream. The algal flora is mainly represented by myxophyceae, bacillariphyceae and chlorophyceae. Observation revealed that myxophyceae constituted 100 per cent population at higher temperature range of 51-63°C. Only one species *Synechococous liveidus* occur at 63°C. At lower temperature of 59°C a few more members appear viz., *Synechococcus elongatus*, Phormidium africanum, *Oscillatoria amphibia*. The growth of these algae was most luxuriant between 55-50°C, where thick mat was formed exhibiting varied pattern of greenery along the stream. Rich accumulation of algal mat at this temperature was due to total absence of herbivores. Additionally large number of bacteria were also recorded.

The list of flora recorded at different gradients include the algae belonging to different classes.

Cyanophyceae: *Synechococcus lividus, S. elongatus, Phormidium aftricanum, Oscillatoria amphibia, O. jasovensis, Lyngbya spirulinoides, Scynechocystis sp., Nostoc commune, Mastigocladus laminosus, Anabaneba sp.*

Chlorophyceae: *Oedogonium, Ulothrix moniliformis, Closterium, Cosmarium trichyderum, Cylindrocapsa sp. and Mougeotia sp.*

Fungi: *Syncephalastrum racenosum, Aspergillus flavus, A. fumigatus, Curvularia lunata, C. affinis, Fusarium oxysporum, Trichoderma sp. and Chaeotonium sp.*

Chapter 4
Mangrove Swamps

Introduction

Some plants grow and complete their life cycle in habitats with a high salinity are known as salt plants or halophytes. Such plants are commonly found near sea shores where mesophytes and fresh water hydrophytes can not grow well. Although these plants grow in the areas which are well saturated with water yet they cannot avail of the water because of high levels of salts in the soils. Thus, the halophytes, are also known as mangroves which are plants of physically wet but physiologically dry habitats. According to Bhosale (1978) the critical level of salinity for plant is 0.5 per cent of the dry weight. Mangroves are partially submerged, salt-tolerant aquatic plants subject to tidal influences found mainly in tropical and subtropical intertidal regions.

These may be small group of trees, shrubs, and climbers. Their habitat is shallow coastal waters, estuaries and lagoons. The mangrove swamps are specialized tropical marine ecosystem. The biological productivity in these systems is exceptionally high. The food web is detritus based. The mangrove vegetation is represented by various families of flowering plants that have adapted themselves to the intertidal environment by modifying their morphological, physiological, anatomical and other characters.

These plants cope with the problems of salinity in various ways, some of them avoid salinity. Salinity avoidance is usually accomplished by limiting germination, growth and reproduction to specific seasons of the year as well by secretion of excess salts.

In the salt tolerants, the protoplasm functions normally and endures a high salt concentration without apparent damage.

Classification of Halophytes

According to Champan (1976) halophytes have been classified into the following categories:

(i) Miohalopytes – Plants growing in the habitats of low salinity (salinity below 0.5 per cent NaCl)

(ii) Euhalophytes – Plants of high Saline habitats.

Coastal angiosperm halophytes have been divided into the following groups.

(i) Submerged marine halophytes – hydrohalophytes

(ii) Low coast plants – hydrohalophytes

(a) Swamp halophytes, the mangrove

(b) Marsh Halophytes

(iii) High coast plants – aerohalophytes

The total area of mangroves in the country is about 6740 sq. km. which is nearly 7 per cent of the world's mangrove. The mangrove areas of Sunderban (West Bengal) and Andaman and Nicobar Islands constitute about 80 per cent of the Indian mangroves. The rest are scattered in the states of Andhra Pradesh, Tamil Nadu, Orissa, Maharashtra, Gujarat, Goa and Karnataka. Areawise distribution of mangroves of India are shown in Table 4.1.

In India, we find mangrove forests on the East and West coasts and the tidal swamps of the deltas of the large rivers. For example, the Gangetic delta and deltas of Mahanadi, Godavari, Krishna and Cauvery, and on the coast of Bombay and its suburbs, including the Elephanta Islands, Uttain Islands, Mora Islands, Madh Island, Ghor Bunder and many other Islands.

Untawala *et al* (1982) classified the saline regimes of Goa as follows with their characteristic mangrove vegetation.

1. **Euhaline zone** : 20-40 per cent Patchy growth of mangroves.
2. **Polyhaline** : Salinity 18-30 per cent majority of vegetation are mangroves belonging to *Rhizophora*

mucronata, Sonneretia alba, Avicennia marina, Bruguiera pariflora, B. cyanorhiza, Ceriops tagal, Dumnitzera racemosa and *Acanthus ilicifolius.*

3. **Mesohaline:** Salinity 5-18 per cent. The flora include, *R. apiqulata, Avicennia offiucnalis, Kandelia candel, Sonneretia alba* and *Aequiceros corniculatum.*

4. **Oligohaline :** Salinity 0.5-5 per cent. The mangrove plant species include, *Sonneretia cassolaris, Acrostichum aureum, Cyperus sp.* etc.

Plants occurring in mangrove vegetation belong to several families (Table 4.2). Common mangrove plants are *Rhizophora murconata, Kandelia, Rheedi, Ceriops roxburghiana, Bruguiera gymnorhiza, Carallia lucida* and many other plants belonging to the family Rhizophoraceae, *Avicennia officinalis, Avicennia alba* of the family Verbinaceae, *Acanthus ilicifolius, A. volubiles* (Acanthaceae)

Table 4.1: Areawise Distribution of Mangrove Vegetation in India

	States/Union Territory	*Areas in sq. km. (Values rounded to the nearest 10 sq. km.)*
1.	Andaman and Nicobar groups (including Nicorbar Islands)	1,190
2.	West Bengal (the Sunderbans)	4,200
3.	Orissa (All deltaic and coastal mangroves, including the Mahanadi area)	150
4.	Andhra Pradesh (the Godavari and the Krishna)	200
5.	Tamil Nadu (the Cauvery and adjacent coastal stretch)	150
6.	Karnataka (Coondapur and Malpe, Karwar .... Patches)	60
7.	Goa	200
8.	Maharashtra (inclusive of Ratnagiri, Vijayadurg, Malvan, Devgad, etc. inclusive 85 sq. km.)	330
9.	Gujarat (including the Narmada – the Tapti – Gulf of Khambhayat mangroves of 105 sq. km.)	260
	Total	**6,740**

Aegiceros majus, Ardesia humiles of family Myricinaceae. *Cassula species* (a very common plant of rice field in northern India belonging to the family Compositae), *Cometes surathensis* (Caryophyllaceae) which is commonly found in Western Asia, *Avicennia, Sechium edule* (Aroidea) – an Indomalayan plant, *Excoecaria agallocha* (Euphorbiaceae), *Phoenix paludosa* (Palmae) and a number of other plants belonging to the families Palmae, Ficoideae, Chenopodiaceae, Plumbaginaceae.

Mangrove vegetation of Gangetic estuary, particularly of Sunderban region, according to Prain (1903) are localized in the following three geographical zones:

(i) Southern Coastal strip and South-Western Part
(ii) Central Zone
(iii) North-Eastern Zone

Table 4.2: The principal mangrove species of India

Family	*Species*
Sterculiaceae	*Heritiera littoralis*
Tiliaceae	*Brownlowia Lanceolata*
Meliaceae	*Xylocarpus granatum syn.* *Carappa granatum syn.* *C. obovata*
Leguminosae	*Cynometra ramiflora*
Rhizophoraceae	*R. mucronata* *R. stylosa, Ceriops tagal* *(syn. C. mandollaeana,* *(Bruguiera conjugata (Linn)* *B. parviflora.*
Combretaceae	*Lumnitzera racemosa* *(Syn. L coccinena)*
Lythraeceae	*Sonneratia acida, S. alba*
Rubiaceae	*Scyphiphora hydrophyllacea*
Apocyaneae	*Cerbera manghas*
Myrsinceae	*Aegiceros corniculatum (syn.)* *A. majus*
Avicennlaceae	*Avicennia officinalls*
Euphorbiaceae	*Exeoecaria agallocha*
Palmae	*Nypa fruticans Phoenix paludosa*

The Central Zone is characterised by climax forest of Heritiera. Ceriops is one of the dominant species.

Mangrove vegetation the Western sea coast of India is represented by the following plant communities.

(i) *Avicennia alba* and *Avicennia officinalis*
(ii) *Acanthus ilicifolius* and *Suaeda fructicosa*
(iii) *Salvadora* and *Sesuvium portulacastrum*
(iv) *Aeluropus repens*

Godavari delta in Andhra (South India) shows characteristic mangrove vegetation of *Avicennia alba, Avicennia marina, Rhizophora, Ceriops, Bruguiera, Acanthus ilicifolius, Myriostachya weightiana*, etc. Some important plant and animal species associated

Table 4.3: Approximate estimates of mangrove associates in India

Taxonomic Group	*No.*
Bacteria	20
Fungi	40
Algae	65
Bryophytes/Ferns	35
Lichens	2
Monocotyledons	73
Dicotyledons	110
Protozoa	30
Copepoda	50
Sponges	5
Ctenophara and Coelenterata	10
Bryozoa	20
Non-Polychaete worms	13
Nematodes	50
Polychaetes	11
Crustaceans	229
Insects and Arachnids	500
Molluscs	212
Echinoderm	1
Fish	300
Reptiles	22
Amphibians	2
Birds	177
Mammals	36

with mangrove forests in Asia with special reference to India are shown in the Table 4.3. The taxonomic distribution of principal vegetation is shown in Table 4.4.

The ecological conditions which are essential for the development of mangrove vegetation are

(a) Shallow water with thick mud.
(b) Water logged saline soil or sandy or loose soil or heavy clays containing large amount of organic matter.
(c) High rainfall, and
(d) High humidity in the atmosphere and cloudy weather.

Table 4.4: Taxonomic distribution of mangrove species

Family	*Genera*	*Total No. of species known*	*Total No. of species known from Indian Ocean Pacific region & (Indo-W.Pacific)*	*No. of species in India*	*Remarks*
1	2	3	4	5	6
1. Rhizophoraceae	1. Rhizophora	7	5	4	*R. stylosa* (A&N Islands and Pichavaram) West Coast
	2. Brugiera	6	6	4	*R. lamarckii* known from Pichavaram
	3. Ceriops	2	2	2	
	4. Kandalia		1	1	*K. candel. K. rheedi* occur mainly in Goa-Ratnagiri
2. Avicennianceae	5. Avicennia	11	6	3	(Described as *B. coccinea* in Nicobar)
3. Combretaceae	6. Lumnitzera	2	2	1	
4. Sonneratiaceae	7. Cymnometra	–	1	1	
	8. Sonneratia				
5. Rubiaceae	9. *Schyphiphora*	5	5	1	*S. caseolaris* known only from Goa

Contd......

1	2	3	4	5	6
6. Plumbagianceae	10. Aequitilis	2	2	1	
7. Meliaceae	11. *Xylocarpus*	10	8	4	Southern Limit: Pichavaram (Tami INadu), West Coast
8. Myrsinaceae	12. *Aegiceras*	2	2	1	
9. Acanthaceae	13. *Acanthus*	3	3	2	
10. Euphorbiaceae	14. *Excoecaria*	1	1	1	
11. Papilionaceae	15. *Dalbergia*	2	2	2	
	16. *Derris*	2	2	2	
12. Palmae	17. Nypa	1	1	1	
	18. *Phoenix*	2	2	2	
13. Sterculicceae	19. *Heritiera*	2	2	2	
14. Bombacaceae	20. *Camptostemon*	2	2	2	Recorded less from East Coast also (Orissa) now confined to the West Coast.
15. Salvaqoracede	21. *Salvadora*			1	
16. Apocyanaceae	22. *Cerbera*	2			
17. Aizoaceae	23. *Sesuvium*			1	
18. Chenopodiaceae	24. *Arthrocnemum*			1	
19. Polypodiaceae	25. *Salicornid*			1	
	26. *Suaeda*			3	
20. Filicices (Pteridophyta)	27. *Acrostichum*			1	
21. Poceae	28. *Porterasia*			1	
	29. *Aelruropus*			1	
	30. *Urochondra*			1	

Important Characters of Halophytes

As the water in the habitat is not such that can easily be absorbed by the plants, the halophytes develop almost all important xerophytic devices in them for H_2O economy.

1. Habit

A great majority of halophytes in the tropical and sub-tropical regions are shrubs, but a few of them are herbaceous. For example, *Acanthus ilicifolius*. In temperate zones, halophytic vegetation is purely herbaceous.

2. **External Morphology**

(a) **Roots:** Halophytes develop many shallow normal roots. In halophytes, in addition to normal roots, many stilt or prop roots develop from the aerial branches of stem for efficient anchorage in muddy or loose sandy soil. These roots grow downwardly and enter the deep and tough strata of the soil. In some plants for example, *Rhizophora mucronata*, the stilt roots may be strong and extensively developed, but in others they may be poorly developed (*Rhizophora conjugata*). In some plants, the stilt roots may not at all develop.

Sometimes, a large number of adventitious root buttresses develop from the basal parts of tree trunks. These root buttresses provide sufficient support to the plants.

The soil in coastal region is poorly aerated and it contains very little percentage of oxygen because of water logging. Under such conditions, the roots of halophytes do not get sufficient aeration. In order to compensate this lack of soil aeration, the hydrohalophytes develop special type of negatively geotropic roots, called 'pneumatophores' or breathing roots. The pneumatophores usually develop from the underground roots and project in the air well above the surface of mud and water. They appear as peg like structures. The tips of these respiratory roots may be pointed. They possess numerous lenticells or pneumatophores on their surface and prominent aerenchyma enclosing large air cavities internally. The gaseous exchange takes place in these roots through the lenticells. The aerenchyma helps in the conduction of air down to the submerged roots.

In some plants like *Bruguiera* the horizontal roots grow above the surface of mud and then again bend downwardly and enter deep in the mud. In this way, they form knee like structures. The aerial surface bears a number of pores which facilitate the exchanges of gases. Pneumatophores do not develop in some species of *Rhizophora*. In those cases, the upper aerial parts of descending stilt roots probably take up the respiratory activity.

(b) **Stem:** Stems in several halophytes develop succulence. *Salicornia herbacia, Suaeda maritima* may be quoted as examples for it. According to Arnold the Succulence depends on the ratio of absorbed free ions in the plant cells rather than absolute amounts of sodium chloride or sulfate present. Succulence is induced only after the accumulation of free ions in an organ which increases above a critical level.

(c) **Leaves:** The leaves in most of the halophytes are thick, entire, succulent, generally small sized, and are often galley in appearance. Some species are aphyllous. Stems and leaves of coastal aerohalophytes show additional mode of adaptation to their habitats. Their surfaces are densely covered with trichomes. Leaves of submerged marine halophytes are thin and have very poorly developed vascular system and frequently green epidermis. They are adapted to absorb H_2O and nutrients from the medium directly.

(d) **Fruits, Seeds and their dispersal:** The fruits and seeds are generally light in weight. Fruit walls have a number of air chambers and the fruits, seeds and seedlings which can float on the water surface for pretty long time are dispersed to distant places by water current. A littoral species of *Spinifex quarrosus* a member of Graminae commonly found growing in the sandy saline sea shores in Andhra, shows peculiar type of fruit dispersal. In this plant, female inflorescence is spherical in shape and consists of many spikelets. A number of stiff bractiolar bristles of the inflorescence help in the dancing and some resulting of the inflorescence when the seeds mature the globular and hairy inflorescence becomes bodily detached from the creeping plant and trails on the sandy substratum dropping its seeds at places. Finally, the inflorescence with rest of the fruits becomes buried in the mud.

(e) **Viviparous mode of seed germination:** Mangrove plants growing in the tidal marshes are met with the phenomenon of 'Vivipary' which is defined as the germination of seeds while the fruits are still attached to mother plants. In Rhizophoraceae plants, when the embryo reaches to advanced stage of development, the massive club-shaped hypocotyl and terminal radical pointing downwardly emerge out of the fruit. When the hypocotyl attains a length of several centimetres (about 50-80 cms) the seedlings falls, vertically down. Thus, the radical and a part of hypocotyl become fixed in the mud and the remaining upper part of hypocotyl along with other embryonal parts, such as plumule and cotyledons remains above the surface of mud or water, within a few hours the radical develops a tuft of roots and plumule also starts growing rapidly. Sometimes seedlings fall in deep water and they float on the surface of water vertically with hypocotyl pointing downwards. On reaching to shallow areas, the radical becomes fixed in the soft mud and the plants starts growing very rapidly. It is noted that high degree of salinity in the soil or water checks the germination of seeds. So, the viviparous germination is a very significant adaptation in these plants to avoid the retarding effects of salinity on seed germination. Species of *Rhizophora, Aegiceros, Avicennia, Cassula, Ranansatia vivipara* are some of the common examples of vivipary.

3. Anatomical Features

Halophytes possess specific and typical structural characteristics which make them distinguishable from other groups of plants. There are as follows:

(1) Large cells and small intercellular spaces.
(2) High elasticity of the cell walls.
(3) Extensive development of water storing tissues.
(4) Smaller relative surface area.
(5) Small and fewer stomata.
(6) Low chlorophyll content.

Anatomy of halophytes reveals a number of xerophytic features in them. They are as follows:

(1) Presence of thick cuticle on the aerial parts of the plant body. The epidermis of succulents and coastal halophyes is characterised by a cover of waxy layers in addition to thick cuticle.

(2) Leaves may be dorsoventral or isobilateral. They develop protected stomata which are not deeply sunken. Epidermal cells are thin-walled. The palisade consists several layers of narrow cells with intercalated tannin and oil cells.

(3) Stems in the succulent plant possess thin-walled water storing parenchyma cells in them. Mucilage cells may be found in abundance. Epidermal cells of various mangrove species contain large quantities of tannins and oil droplets. Cortex is fleshy, several cells thick and in old stems it may become lacunar. Salinity causes extensive lignification of stele.

(4) The leaves and stems of coastal halophytes are abundantly covered with various types simple and branched trichomes, giving the plants a grayish appearance.

The trichomes may exert a protective function in plants by:

(a) Affecting water economy;
(b) Affecting the temperature of their leaves, and
(c) Preventing sea water droplets from reaching the live tissues of leaves.

(5) Leaves of many species of mangrove are dotted with local cork formation "Cork warts". Leaves of *Sonneretia, Aegiceros, Nitrasia* (A desert shrub), and *Suaeda monica* contain well developed aqueous tissue. Salt secreting glands may be found in some halophytes.

Pneumatophores develop a number of lenticells on their surface. The cortex is spongy and consists of extensively developed aerenchyma enclosing large air chambers. Highly developed air chambers are continuous with the stomata of leaves and with the cortex and primary phloem of the stems.

4. Physiological Adaptation in Halophytes

To overcome the physiological drought condition:

(i) they show high rates of transpiration;
(ii) they show exrudation of sap that contains dissolved salts, and

(iii) they develop many shallow absorbing roots.

The importance of mangrove ecosystem is summarized as follows:

(i) The mangrove ecosystems intercept and filter surface water run-off, thereby preventing large scale nutrient losses to the open area. A bi-product of this filtering action is the maintenance of water quality.

(ii) The complex root strength helps to accumulate organic and inorganic materials. The systems also act as mechanical energy buffer to strong winds and storm tides, and thus prevent periodic devastations of the coast line.

(iii) Mangroves have a high net productivity, which is largely exported into estuaries and forms the basis of detritus food chain. This net production can also be evaluated in terms of the fisheries it supports.

(iv) The shallow waters around mangroves provide breeding and feeding place for many estuary dependent fishes.

(v) Mangroves support large bird population by providing rookeries adjacent to productive estuarine feeding areas.

Succession of Mangrove Vegetation in Sea Coast

The distribution of halophytic community appears to be limited by salinity and depth of water as well as by competitive ability of the members of the next community in the halosere. The aggressiveness of plant communities in saline habitat is due to the changes in the salinity level. In coastal region, nature of vegetation is greatly affected by the gradation of sea coast. Succession of mangrove species in coastal regions may take place very slowly in the following sequence:

(i) In the deep water generally true mangroves, like the species of *Avicennia* grow.

(ii) When the bottom of the sea is slightly raised up, *Avicennia, Rhizophora, Ceriops, Bruguiera* etc. form mixed mangroves vegetation in the shallow water.

(iii) As the ground is exposed, true mangroves disappear and other halophytes for example species of *Aegicera,*

Excoecaria etc. gradually invade the land within short period. These halophytic communities are interspersed by salt tolerating succulents and the grasses make the soil fit for the cultivation of some crop plants. Several varieties of paddy, such as *Oryza Sativa* var, achra, wild species of *Oryza coarctata* grown very well in these area.

Succession of angiospermic halophytes varies also in each habitats in accordance with other ecological conditions besides salinity. Thus, in reality, there is no general trend for development of various halophytic plant communities around the world and local variations are encountered in each specific site.

Navalkar (1951) found that in mangrove vegetation, succession is governed by disintegration of rocks, soil salinity, soil humidity and biotic factors. He recognized seven stages of plants in the succession.

Ecological Role of Mangroves

These plants perform stabilisation of shore lines and prevention of shore erosion.

The dense network of prop-roots, pneumatophores and stilt root not only give mechanical support to the plant, but also trap the sediment. The rate of sedimentation is much higher. Its detritus helps in feeding and provides breeding and nursery grounds for the juveniles of many commercially important shrimps and fishes.

The major primary production in mangrove ecosystem is from trees. However, only a fraction of this production is consumed by herbivores. The remainder enters the mangrove water as litter form. The decomposition of the litter formed detritus which in turn is colonized by heterotrophic organisms thus, enhancing its nutritive value. This detritus besides forming a food source for suspension and deposit feeders, is also consumed by the juveniles of a variety of bivalve, shrimps, and fishes, which migrate into the mangrove environment in their life cycle for better feeding and protection. In fact there is a direct correlation between the extent of mangrove forests along coast line and fishes as well as shrimps catchers from the coastal waters. The high rate of production and easy accessibility of the mangroves render them as ideal sites for aquaculture of shrimps.

Mangroves in Karnataka are of fringing type found in the inter-tidal region along estuaries, back water islands and other protected areas. In India, mangroves have been subjected to reckless exploitation by anthropogenic activities and also by other biotic interference. Around 50 years ago, North of Bombay 2,500 Sq. Km. of mangrove once existed. A figure that is down to between 350-500 Sq. Km. of undistributed mangrove, the thickest being around the Popli, Dattiwara and Vaitarana area.

The finest mangroves of the west coasts of India can still be found in and around Piram Island and Aliya between 70 Km. off Maygon and Bhadhup, the last estuarine villages which form part of the estuarine zone of the Narmada. These mangroves have been processing sediment and building up the shores of the Gulf of Khambaat undistributed for the last 60 million years. When they vanish, a thriving fishing industry plus the cotton, gram and peanut crop of the estuary will become marginal.

In and around Bombay, the mis-conception of '**Khar Zameen**' as useless waste land has caused ignorant and zealous land-filling for buildings. If this continues it will destroy all over what was once a natural fish nursery, a coast-extending natural mechanism beyond the brainscope of most builders and a filtration system for industrial waters. Because mangroves can partially process all kinds of heavy metals, oxides and salts, the destruction of creeks and bounded rice paddies, has already started to play havoc with our fish, fruit and vegetable source, besides allowing the discharge of factory discharges to go into the sea virtually untreated. So as we systematically fill in these swamps, we are destroying the most effective natural system of protecting and actually extending our coastlines.

Economic Aspects

Mangroves forests form a unique system comprising of land, water, vegetation, animals and also shelter to human beings. The natural combination results in the multiplicity of goods and services. The following points are suggested as management packages for the mangrove area:

1. Any tourism or economic development should be ecologically sustainable for the mangrove ecosystem.

2. Any landscape change or conversion of precious resource base to alternative uses involving laying of roads for tourist traffic or causing disturbance in the channels or estuaries to the wetland mangrove vegetation through boats should be properly planned.
3. The goods and services produced by this ecosystem may be divided into three categories. The goods and services produced within the ecosystems, for example in the Pichavaram mangroves have 200 species of fish. 30 species of mollusks and crabs. The largest honey is collected from Sunderban forests.
4. Potential medicinal values of the plants from the mangrove ecosystems.
5. Mangrove serving as buffer protection against cyclone threat or as a sanctuary to rare and endangered flora and fauna.

Management Aspects

The forest resources of India, apart from Eastern and Western Ghats also include Mangrove forests on the East and West coasts. These mangroves stabilise our shoreline and augment our continental shelf fisheries. The mangrove are considered as secondary wood sources used in the manufacture of carpenters tools, handles, handicrafts, newsprints, paper pulp and cardboards etc. The management of mangrove forests should aim primarily to go for judicious management and it should become a perennial beneficial source for the nation.

The aspects which need immediate management plan of action are given below:

1. Protection when endangered and preservation and conservation of resources.
2. Improving the quality by planned regeneration of plants.
3. Continuous recruitment of fishery resources without resorting to large scale landscape alternation.
4. Optimum utilization for economic development, without degrading or depleting and endangering the resources and the whole ecosystem complex.

The mangrove may be maintained as scenic spots for recreation and care should be taken that they will not be disturbed by the tourists. The vanishing wild life of mangrove like Bengal tiger, barking dear, cheethal, crocodile, birds, the monitor lizard and phython etc. should be carefully protected from further depletion. It also includes wild ass of run of Kuchh and wild fauna of Sunderban. Every effort should be taken to protect the existing wild life of the area in their, natural environment. The natural regeneration occurs in the area where conditions are favourable and under social forestry scheme the degraded mangroves should be developed.

To the extant possible establishment of heavy industries in the sensitive coastal and adjacent mangrove areas should be avoided. The research on various aspects of mangrove with particular reference to their conservation and management is called for. There are about fourteen major centres in India, engaged in the mangrove research. The important fields of study includes survey, distribution, remote sensing, ecobiology, fisheries, physiology, soil, pollution, productivity, afforestation, wildlife etc.

A National Mangrove Management Committee (NMMC) has been constituted by the Ministry of Environment and Forestry in 1979. According the committee the following areas R and D activities are recommended.

1. Nationwide mapping of the mangrove areas.
2. Quantitative survey of the total mangrove area, climatic regime, rate of growth and seasonal variation in the environmental factors.
3. Location of suitable areas to announce for reserve forests under conservation plan.
4. Afforestation of degraded forests and research activity on the mangrove species.
5. Floral and faunal study, microbial role, detritus production and the biochemistry of organic sediments.

Conclusion

Mangrove swamps are the very interesting and distinctive shallow water marine communities of tropical and sub-tropical waters. They are important land-builders which helps to form islands and extend shores. Mangrove are among the few emergent land plants that tolerate the salinity of the open sea. Different species form zones in the intertidal region or even beyond. The mangrove has extensive profits which reduce tidal currents, causes extensive deposition of mud and slit and provide surfaces for attachment of marine organisms. Its seeds sprout while still on the tree. The seedlings drop off and float on the water until they lodge in shallow water where well developed roots may take hold, perhaps to start a new island. Mangroves are not only important in extending coasts and building islands but also in protecting the coasts from excessive erosion which might otherwise be produced by fierce tropical storms. Leaf detritus from mangroves has been shown to constitute a major energy input into fisheries and detritus food chain.

Chapter 5
Intertidal Organisms

Intertidal zone is the smallest area of all on the worlds oceans, existing as an extremely narrow fringe, a few metres in between high and low tides of water. The intertidal zone is perhaps the best known and studied of all areas of sea because it is the most accessible to humans. Despite its extremely restricted area, the intertidal has within it the greatest variations in environmental factors of any marine area, and these can occur within centimetres of each other. Coupled with this is a tremendous diversity of organisms, which is greater than that found in the more extensive subtidal areas.

Environmental Conditions

The tremendous array of environmental factors found in the intertidal zone is due to the fact that this zone is exposed to sunlight for a certain amount of time during every day of the year and most physical factors show greater range in air than in water.

Tides

The periodic, predictable rise and fall of the level of the sea over a given time interval is called a tide. Certainly, this is the most important environmental factor influencing life in the intertidal zone. Without the presence of the tide or some other means of inducing a periodic rise and fall in the water level this zone would not exist as it is, and many of the other factors would cease to influence. With very few exceptions, most shore areas of the world experience tides. The only major seas that virtually lack tidal action are the Mediterranean and the Baltic. In these areas, the water level on the shore line fluctuates primarily due to the effects of wind action pushing water.

Tides occur due to the interaction between the gravitational attractions of the sun and the moon on the earth and centrifugal force generated by the rotating earth and moon system. As a result of these forces, the water in the ocean basins is pulled into bulges. Gravitational attraction of one body for another is a function of the mass of each body and their distance apart. The gravitation generating force of the moon on the earth is about twice that of sun, even though the sun is many times more massive than the moon. This is due to the great distance of the sun from the earth as compared to the moon.

The earth and moon form an orbiting system that revolves around them common centre of mass. Because of the large size of the earth relative to the moon, this point is located inside the earth. The revolution of the earth and moon creates a centrifugal force (acting outwards) which is balanced by overall gravitational force acting between the two bodies. However, the gravitational force is much stronger that centrifugal force on the side of the earth than on the side opposite. As a result the side facing the moon has the water pulled into a bulge (high tide). On the opposite side of the earth, the gravitational force of the moon will be the least and the stronger centrifugal force will pull water into a bulge away from earth (another high tide). Thus, we have the two high tides. These then circle the earth following the position of the moon as the earth revolves on its axis once every 24 hours. Low tides are about halfway between the high tides.

The system as described above would give two high tides and two low tides of equal magnitude each day. As we know, however, tides are not equal, nor are these always four tides per day. How do these discrepancies arise? In part, this is due to the fact that the earth is vertical with respect to its plane of orbit around the sun. It actually is inclined 23 and half degree from the vertical. As a result during the earth's rotation on its axis, a given point on the earth's surface is subjected to different tidal heights. Tides are also influenced by the rotation of the moon as it is elliptical but not circular.

The effect of the sun is seen in spring and neap tides. Spring tides are tides that show the greatest range (both high and low) and result when the moon and sun are directly aligned so these two forces are combined. Neap tides, on the other hand, are tides showing minimum range and result when sun and moon are at right angles to each other and thus counteract each other.

Differences both in number of high and low tides per day and in height in various parts of the world are due to peculiarities of the various ocean basins in which the tides occur. The well-known extremely high tides of the Bay of Fundy or Cook Inlet in Alaska are the result of the basic tide forces acting in the geometry of the basin.

Tides that expressed as single low and high per day are called diurnal tides. Tides that have two high and low each per lunar day are termed as semidiurnal tides. Where there is a mixture of diurnal and semidiurnal, we have what are called as mixed tides. The heights of the high and low tides vary from day-to-day as the positions of the sun and moon relative to each other change, each being alternatively fully aligned and 90° out of alignment every 14 days.

Perhaps the most significant effect of the tide with respect to intertidal organisms and communities is that is subjected them to periodic exposure to the air, with its considerably wider range of physical parameters. It is this exposure to aerial conditions that necessitates most of the adaptations required by intertidal organisms in order to survive in this zone.

Tides in combination with time can, however, have two significant direct effects on the present and organization of intertidal communities. The first effect results from the time that a given area of the intertidal is exposed to the air relative to the time that it is submerged in water. It is the duration of exposure to air that it is most important, because that is when the marine organisms will be subjected to the greatest temperature ranges and the possibility of desiccation (water loss). The longer the duration of exposure, the greater the chance for encountering a lethal temperature or desiccating beyond tolerable limits.

Temperature

Because of its inherent physical characteristics water, particularly large bodies such as the oceans, show a minimum range of change in temperature. Furthermore, this range rarely exceeds the lethal limits for the organisms. Intertidal areas are, however, regularly subjected to aerial temperature for varying periods and these temperatures have a much wider range, on both daily and seasonal basis. The range may often exceed the lethal limits of the marine organisms. The low tide occur when the air temperature is at a minimum (cold temperate, polar) or a maximum (tropics), the lethal limits may be exceeded and the organisms will die. Even if death does not occur immediately, the organisms may be so weakened by the extreme temperatures that they cannot resume normal activities and will die due to secondary causes. Temperature also has an indirect effect on the organisms. Marine organisms die by desiccation. Desiccation may be hastened by increased temperatures.

Wave Action

It is in the intertidal zone that wave action exerts the most influence on organisms and communities of any area in the sea. This influence is manifested directly and indirectly. Wave action affects shore life directly in two major ways. In the first place, it has a mechanical effect, which acts to smash and tear away objects with which it makes contact. One can appreciate this action by noting the many pictures and stories of destruction caused to man made structures due to storm waves of varying types. The action is similar in the intertidal zone, and hence, any creatures that inhabit this zone must be adapted to this persistent force in some way. On shores that are made up of loose sand or gravel, the wave action is even more profound in that it moves the entire substrate around influencing the actual shape of the zone. Wave exposure may thus prove to be limiting for certain organisms that are unable to sustain the force and necessary to others that cannot exist other than in heavy wave areas.

Secondly wave action to extend the limits of the intertidal zone. It does this by throwing water higher on the shore than would normally occur as a result of the tide alone. This continual splashing allows the marine organisms to live higher in wave-swept areas than in calm areas within the same tidal range.

Wave action has some other, perhaps minor, effects also. It acts to mix atmospheric gases into the water, thus increasing the oxygen content, so wave-washed areas never lack for oxygen.

Because of the constant interaction with the atmosphere and the formation of bubbles plus the stirring up of the substrate, wave-swept areas may have decreased penetration of light.

Salinity

Salinity changes may occur in the intertidal in two situations that may affect the organisms. In the first place, the intertidal may be exposed at low tide and subsequently flooded by heavy rains or run-off from land surface. This means they will be subjected to great dilution in salinities. In certain cases this reduction in salinity may exceed the limits and the organisms may die, as most intertidal organisms show a limited tolerance to decreased salinity. The second situation has to do with tidepools, those areas that retain sea water at low tide. These may either be flooded with freshwater run-off from heavy rains, thus reducing salinity, or they may show increased salinity when exposed to heavy evaporation during the day.

Adaptations

Since the intertidal organisms are primarily marine in origin, the adaptations that we observe have to do primarily with avoiding or minimizing the stress imposed by the daily exposure to the terrestrial environmental factors. The major stress from the marine environment is wave action.

Resistance to Waterloss

Soon after the removal from the water the marine organisms began to lose water content of their body in the air. If they are to survive in the intertidal, this loss must somehow be minimized and or the organisms must have body systems that will tolerate considerable water loss during the hours of exposure to the air.

The simplest mechanism for avoiding water loss is obvious in motile animals such as crabs. These animals simply move from the exposed surface areas of the intertidal into very moist cracks, craves, or burrows, where water loss is minimum or absent. Alternatively, they may seek refuge under the moist covering of algae. In other words, these animals avoid the adverse environmental conditions of the shore by actively moving to the suitable microhabitats. Similar situation occurs in some species of anemones such as *Anthopleura xanthogrammica* of the Pacific coast of North America. It is soft bodied, with no way of preventing water loss. However, it is normally found only among barnacles or in crevices where water loss is minimum. Physiological adaptation is not needed in this case.

Another simple mechanism is seen in several of the high intertidal algae such as *Porphyra, Fucus* and *Enteromorpha*. These plants cannot move and have no mechanism to avoid water loss. They simply are adapted to withstand a severe loss of water from their tissues. Specimens of these genera can be found that are dry and brittle after a long exposure at low tide yet will quickly absorb water and resume normal body processes when the tide returns. Kanwisher (1957) found that they can tolerate as much as a 60-90 per cent loss of water and still recover. A similar tolerance to great water loss is seen in some intertidal sedentary animals. Boyle (1969) for example, reports Chitons can tolerate a 75 per cent water loss and limpets, according to Davis (1969) from 30-70 per cent, depending on the species (Fig. 5.1).

Many species of intertidal animals, in contrast to the above, have mechanisms for prevention of water loss. These mechanisms may be structural, behavioural or both. Barnacles of many species are predominant members of the intertidal zone all over the world. These animals are sessile and avoid water loss by closing up within their shells at low tide. Here, the presence of an impermeable shell checks the evaporative water loss. Limpets belonging to the genera *Patella, Acmaea,* and *Coillisella* are also dominant animals of the rocky intertidal zone. Certain species of these limpets have a home scar into which their shells exactly fit. At low tide, they return to these home and by fitting into these grooves, greatly economize the water loss. Other limpets that have no scar simply clamp down tightly against the rock so no soft tissue is exposed. Other gastropods such as the *Littorina* have operacula which completely seal off the aperture to their shells. At low tide, they may pull into the shell, closing the aperture with the operculum and thus reduce water loss. Certain bivalves, such as *Mytilus edulis,* are able to survive into intertidal because of their ability to close their valves tightly to prevent water loss. Still other organisms, such as the Anemone, *Actinia* and the hydroid *Clava squamata,* produce mucus that prevent water loss. Many inhabitants of sand and mud simply burrow into the substrate to prevent desiccation.

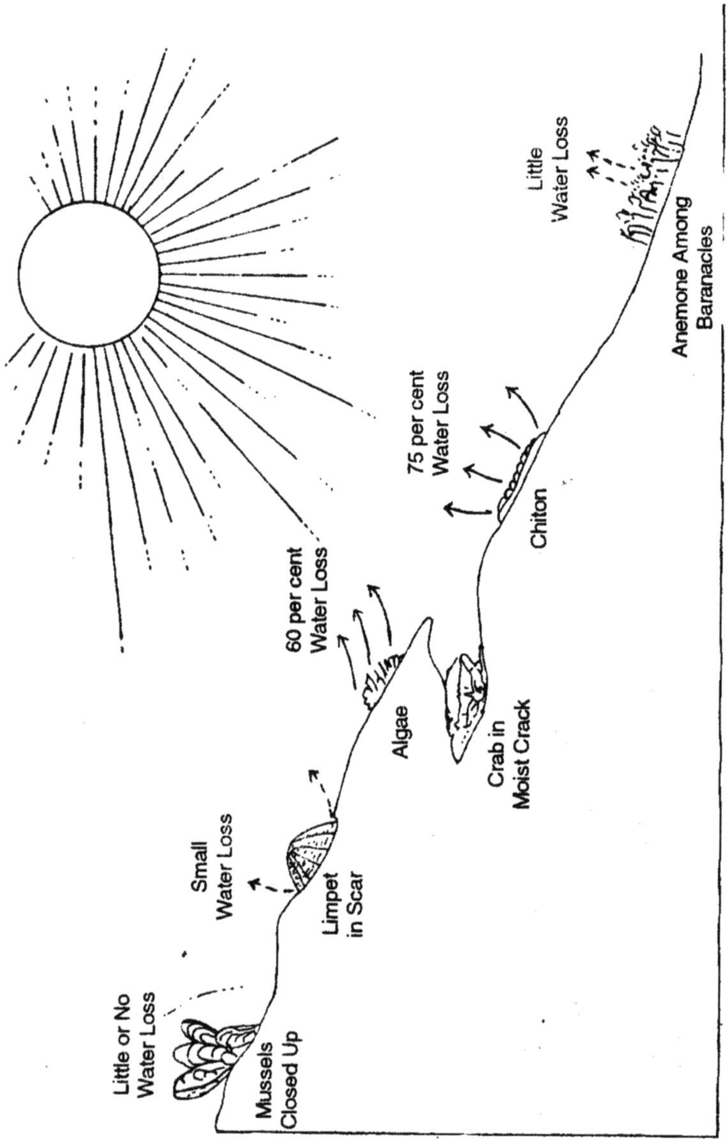

Fig. 5.1: Representation of water loss by intertidal organisms.

Maintenance of Heat Loss

Intertidal organisms are also exposed to extremes of heat and cold and show behavioural and structural adaptations to maintain their internal heat balance. Though deaths by freezing have been recorded for intertidal organisms, extreme low temperatures seem to be less of a problem to shore organisms than high temperatures. This may be due to the fact that organisms are often living much closer to their upper lethal temperature than to their lower lethal one. Hence, most mechanisms for heat balance concern the avoidance of too high temperatures. This may be accomplished by reducing heat gain from the environment and increasing the heat loss from the body of the animal. Heat gain from the environment is reduced in several ways. One way is to have a relatively large body size when compared to similar species, either lower in the intertidal or in the subtidal. Large body size means less surface area relative to volume and hence, less area to gain heat. At the same time, a large body takes longer to heat up than a smaller one. It has been recorded for gastropod mollusks such as *Littorina littorea* and *Olivella biplicata* that the larger individuals occur higher in the tidal zone and the smaller ones lower. Yet another are some shore animals that do have such a small area of attachment or none at all. These occur where wave action is not severe and in those animals that lose attachment. The ability to re-attach as the tide comes in is well developed.

Heat loss can be attributed to a number of causes. One mechanism, found in hard-shelled organisms such as mollusks, is greater elaboration of ridges and other sculpturing on the shell. These act as radiator fins and facilitate heat loss. Examples of such sculptured mollusks are best found in the tropics and include *Tectarius muricata* and *Nodolittorina tuberculata*. Enhancement of heat loss occurs when the organism is light in colour (Dark organisms gain heat by absorption. Many tropical and subtropical snails of the high intertidal such as *Nerita peleronta* (Caribbean) and *Littorina unifaciata* (New Zealand) are much lighter in colour than their lower level relatives, presumably to facilitate the heat loss (Fig. 5.2).

Mechanical Stress

It is necessary for any organism that lives intertidally to adapt in order to resist the smashing and tearing effects of waves. Wave

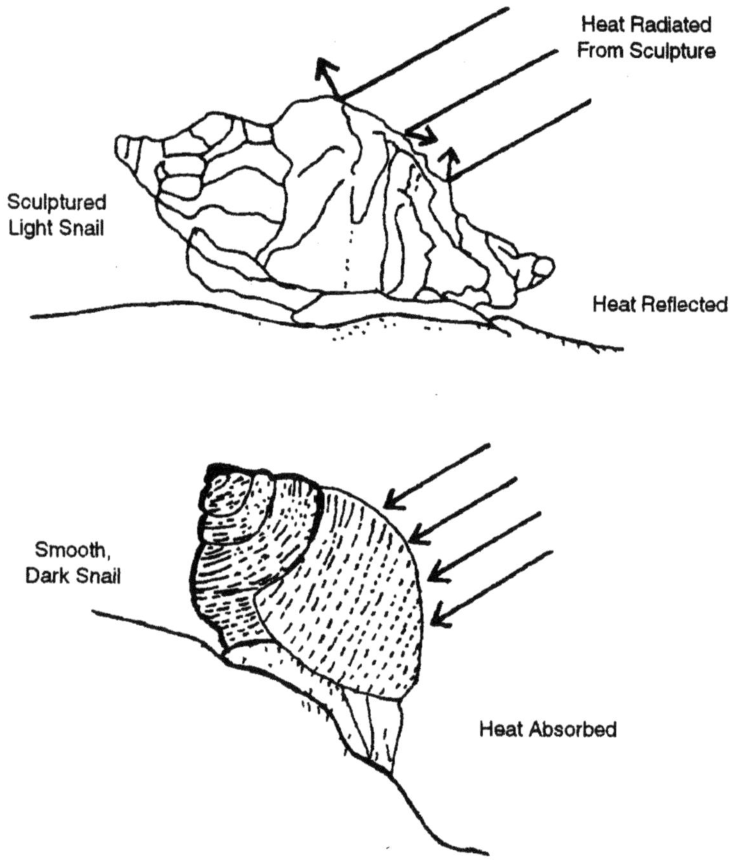

Fig. 5.2: Difference in the heat absorption by smooth dart shells and ridged weavy light coloured shells.

action has different effects on rocky shores as compared to sandy shores and requires different adaptations. In order to maintain position in the face of wave action, several adaptations have been developed by intertidal organisms. One such adaptation, observed in barnacles, oysters, and serpulid polychaete worms, is simply to become fixed to the substrate. Similarly, the majority of algae in the intertidal are fused to rocks via holdfast. Still other organisms develop attachments that are strong but not permanent thus giving limited movement. Such attachments are the abyssal threads of *Mytilus,* which anchor the animals to the rocky substrates but still can be broken and remade to allow movement on the nearby area. The dominant intertidal mollusks, such as various limpets and chitons resist wave action through a strong, enlarged foot which clamps them to the substrate.

Motile organisms such as crabs have no structural mechanism to resist being swept away and survive only because they seek shelter from the waves in crevices or under rocks. Most intertidal mollusks are adapted to wave shock by possessing thicker shells than their subtidal relatives and lack delicate sculpturing, which is easily broken in the waves.

Respiration

The animals inhabiting the intertidal have respiratory surfaces adapted to extract oxygen from water. Usually, these have extensions from the body surface. These respiratory organs are thus very susceptible to desiccation in air and do not function unless immersed in water. Such organs would be a distinct disadvantage in the intertidal. Among intertidal animals, there is a tendency to enclose the respiratory surfaces in a protective cavity to prevent them from drying. This is seen particularly in various mollusks where the gills are in a mantle cavity, which itself is protected by the shell. A similar situation prevails in barnacles where the mantle tissue acts as a respiratory organ. In order to conserve water loss during low tide most animals often close their operculum or clamp down (chitons and limpets) and others are quiescent.

All intertidal animals usually expose the fleshy parts of their bodies during feeding, which are susceptible to desiccation. As a result virtually all intertidal animals are active only during the time the tide is in and they are covered with water. This is true whether the animals are grazers, filter feeders, detritus feeders, or predators.

Salinity Stress

Some part of the year, the intertidal zone may be subjected to flooding by fresh water, which would create an osmotic stress on the intertidal organisms adapted to be submerged only in sea water. This is particularly true since most intertidal organisms do not show adaptations to tolerate salinity changes, certainly not like estuarine organisms do most have no mechanism to control the salt content of their body fluids and hence, are osmoconformers. The only adaptation observed are the same as those for preventing desiccation, such as closing up valves or shells in barnacles and mollusks. Perhaps this is why there are record of catastrophic mortality of intertidal organisms following heavy rain or run-off.

Reproduction.

Most of the intertidal organisms are sedentary or sessile, they must rely upon eggs and larvae that are free floating in the plankton for dispersal. Hence, we witness very high incidence of planktonic larvae among intertidal organisms. A second reproductive adaptation attributable to the intertidal position is that most organisms have a breeding cycle that is closely synchronized with the occurrence of certain tides, such as spring tides. An example is *Mytilus edulis*, in which gonads mature during periods of spring tides and spawning occurs on subsequent neap tides. In *Littorina neriotoides* the eggs are also spawned at time of spring tides.

Rocky Shores

Of all the intertidal shore, those composed of hard material, the rocky shores, are the most densely inhabited by macro-organisms and have the greatest diversity of animal and plant species. They stand in striking contrast to the almost barren appearance of sand and mud shores. It is these densely populated, topographically diverse, and species-rich rocky beaches that have fascinated marine biologists and ecologists for many years.

Zonation

One the most striking features of any rocky shore anywhere on the globe, when seen at low tide, is the prominent horizontal banding or zonation of the organisms. Each zone or band is set off from those adjacent by differences in colour, morphology of the major organisms, or some combination of colour and morphology. These horizontal bands or zones succeed each other vertically as one progresses up from the level of the lowest low tides to the true terrestrial environment. The zonation observed on intertidal rocky beaches is similar to the zonation pattern that one observes with increasing elevation on a mountain, where the different horizontal zones of trees and shrubs succeed each other vertically. Rocky intertidal zones very in vertical extent, depending upon the slope of the rocky surface, the tidal range and the exposure to wave action. Of course, the striking bands of organisms may be interrupted or altered in various place along a rocky shore wherever the rock substrate itself shows changes in slope, composition or irregularities that change its exposure or position is relative to the prevailing water movement.

It is a fact that the prominent zones can be observed on nearly all rocky shores throughout the world under many different tidal regimes. It is characterized by establishing the zones on the basis of the distributional limits of certain common groups of organisms and not on the basis of tides. It, thus, reflects the knowledge of any intertidal ecologists, that distribution patterns of the organisms and zones vary not only with tides, but also slopes and the exposure. Thus, under similar tidal conditions, there could be different band widths.

The Stephenson scheme has three main divisions of the intertidal area. The uppermost is termed the supralittoral fringe. Its lower limit is the upper limit of barnacles and it extends to the upper limit of snails of the genus littorina (periwinkles). The dominant organisms are the littorine snails and black encrusting lichens (Verrucaria type). Part of this zone is reached by extreme high water of spring tides, but its water should come from wave splash. Above this zone is the terrestrial supralittoral zone.

The middle part of the intertidal is termed as the midlittoral zone and is the broadest in extent. Its upper limit is coincident with the upper limit of barnacles, while its lower limit is that point where large kelps (*Laminaria,* etc.) reach their uppermost distribution. This zone is often subdivided and contains a host of different organisms. Perhaps the only universally present dominant group are the barnacles. The lowermost zone of intertidal is the infralittoral fringe, which extends from the lowest low tide up to the upper limit of the large kelps. This is an extremely rich zone composed of organisms that can tolerate only limited exposure to air. It is really an intertidal extension of the infralittoral zone or what we know as the sublittoral area.

Causes of zonation easy to recognize and measure the extent of the zones on a rocky shore, it is more difficult to come up with suitable explanations as to why organisms are distributed into these zones. The most obvious explanation for the occurrence of the zones is that they are as result of the tidal action on the shore and therefore reflect the different tolerances of the organisms to increasing exposure to the air and the resultant desiccation and temperature extremes. If one observes a whole series of such tidal curves such a series of breaks does become apparent. At this point the tide will cover the organism at least once every six hours. If, however, the organism moves up only a few inches the tide will not cover it for 12 hours. Thus, there is a great change in exposure time with a very short vertical movement. The effects of these topography and exposure differences suggest, however, that the real upper limits are set by certain physical factors namely desiccation and temperature.

Along with desiccation and often acting in concert with it temperature is the another major limiting factor. As noted previously ambient temperatures have ranges that may exceed the lethal limits and hence intertidal organisms may die from either freezing or "cooking" effect by upper limits of temperatures. The levels of high temperature promotes desiccation and the synergistic effect of these factors may be even more devastating than each acting alone.

In general the biological factors are complex often subtle and closely linked to other factors. The biological factors are competition, perdition and grazing (herbivores). Competition for a certain resource does not occur if the resource is available in plenty. In the rocky intertidal zone, only one resource is commonly limited is the space. This is perhaps the most restricted area in the marine environment in terms of physical dimension. As a result there is an intense competition for space. This has resulted in observed zonal patterns (Fig. 5.3).

A more complex example concerning competition is that among the mussel *Mytilus californianus* and several species of barnacles. In this case, studies by Dayton (1971) shown that *M. californianus* is the dominant space competitor on open coast shores. Given enough time and freedom from predators, the *M. californianus* eventually overgrow and outcompete all other macro-organisms and takeover the substrates of the mid-intertidal. *Mytilus californianus* takeover is however slow. Wherever open

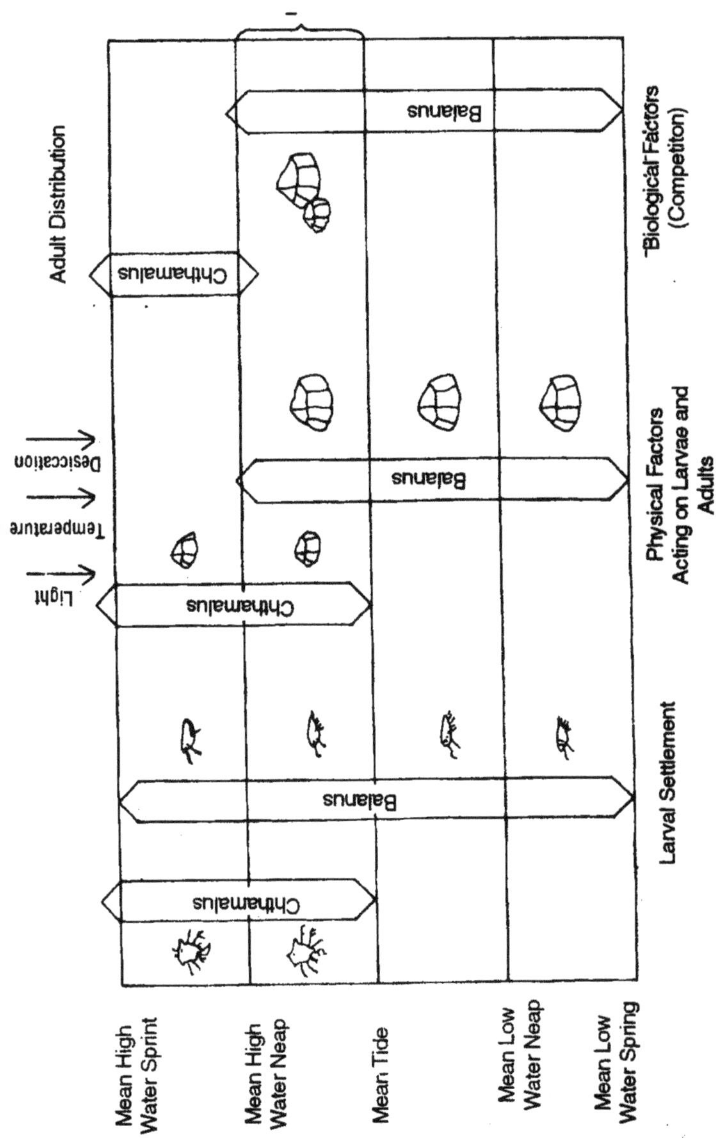

Fig. 5.3: Intertidal zonation due to interaction between Physico-biological factors for the larvae of barnacles (*Chthamalus stellaius and Balanus balanoides*)

space occurs, it may be rapidly colonized by other organisms, including three species of barnacles, *Balanus glandula, B. cariosus,* and *Pollicipes polymerus*. These further replace other rapidly growing algal species. The barnacles persist only until the mussels enter. The mussels outcompete and destroy the barnacles by setting on top of them and smothering them. Since nothing appears to be large enough to settle and smother the *M. californianus*, they remain in control of the intertidal space.

Lubchenka (1978) studies the tidepool algae of New England coast and indicated that *Enteromorpha interstinalis* was a dominant space competitor as compared to *Chondrus crispus*. In the absence of grazers, *E. interstinalis* would quickly outcompete *Chondrus* and takeover the space. The role of predators in determining the distribution of organisms in the intertidal zone is best documented for the Pacific coast of North America.

The dominant abundant intertidal animals on the coast other than the space dominating *Mytilus sp.* are the barnacles *balanus cariosus* and *Balanus glandula*. These latter two species occur abundantly in intertidal region, despite the fact that they are competitively inferior to *M. californianus*, because a predatory starfish, *Pisaster ochraceus*, preys upon it, thus preventing it from completely dominating the area. At the same time *Balunus glandula* is found primarily as a band of adults in the high intertidal while *Balanus carisus* occurs as scattered, large individuals or clumps in the mid-intertidal.

Physico-chemical Interactions

The high intertidal tends to have the distribution pattern of its characteristic species set by physical factors, whereas biological factors become more important in setting limits in the low intertidal. It is important to realize that these factors do not act in isolation and the final distribution pattern observed, either on a small or large scale, is often the result of the interaction among two or more factors. Persistence of *B. cariosus* in the mid-intertidal of the Pacific coast depends upon the predation of *Pisaster ochraceus*.

In the low intertidal of New England, the dominant space competitor is *Mytilus edulis* which can outcompete other space occupiers such as *Balanus balanoides* and the algae, *Chondrus crispus* (Fig. 5.4). On open exposed rocky headlands no predators exist

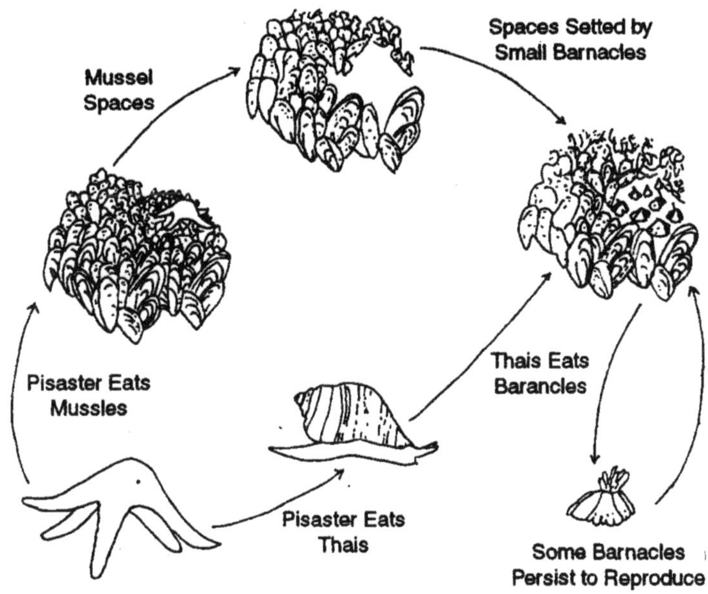

Fig. 5.4: Interactions between the mussel, *Perna viridis* and its predators on a rocky shore intertidal zone.

for *M. edulis* and hence it outcompetes the barnacles and algae to dominate the low intertidal. The only major herbivore in this system *Littorina littorea* prefers to graze on ephermeral algae that compete with *Chondrus*. Thus, the presence of the herbivore reduces competition by other algae on Chondrus allowing it to persist and dominate protected areas.

Distribution Patterns

The appearance of the various vertical bands or zones changes as one moves from areas of high wave exposure to protected and quiet water. This is true for all the other areas as well. Partly this is due to the action of water restricting certain species in or out of wave-swept areas. In other words some species of algae and invertebrates simply cannot tolerate the wave action and are absent from wave-swept areas.

Succession

Succession is defined as an ecological phenomena referring to the orderly progression of communities or series in which one community or association of species occupies a site and somehow modifies its conditions before the next one can takeover. This replacement of one association by another continues until reaching the last or climax community which persists indefinitely under a given climatic regime. Thus, in any one area when open space appears there is an orderly predictable sequence of communities that develops on that site which will be culminated by a known climax community. A fundamental requirement of this concept is that certain species must be present before others in order to modify the environment for subsequent ones. Does such a predictable sequence occur in the rocky intertidal is not clear.

Usually these intertidal species are those that we have termed opportunistic of fugitive and they are gradually replaced by the better competitors. In this sequence usually the first settlers on cleared areas are colonial diatoms or fast-growing small algal species. These are followed by barnacles and tube worms which are followed by the dominant large algae and mussels. The reason for the sequence is not however that the earlier species must be present or modify the substrate but simply that their larvae are more quickly produced and dispersed and hence arrive first. *Mytilus* for example is perfectly capable of settling on newly cleared areas.

Age Structure

In the marine environment especially the intertidal the long-lived species tend to be various invertebrate animals whereas the plants and various algal species tend to be short lived. We in turn tend to name zones in their intertidal on the basis of animals *Mytilus zone*, barnacles zone and so forth. As it is known that on the Pacific coast of North America the high intertidal has a band of *Balanus glandula* below that any of these species found are usually less than 2 years old. Because the harsh physical conditions this area the juvenile barnacles simply do not survive. As a consequence the barnacles establish from a single good year in which conditions allowed survival through the vulnerable juvenile state. *B. cariosus* populations in the mid-intertidal consist primarily of animals less than 2 years old. However, there are a fear that may be large and many years old. Again the majority are less than 2 years old because that is the time period during which *Thais* consume them.

Tidepools

Presence tidepools is a characteristic feature of a many rocky shores. These vary in size, depth and locations. Certain conditions affecting life of tidepools differ from surrounding intertidal zone. Tide pools can occur at any level in the intertidal, however, low level pools are slightly different from the surrounding sea and are essentially extensions of sublittoral.

Three major physical factors are subjected to variations in tidepools. The first one is the temperature. The ocean itself is a vast reservoir that heats up and get cooled very slowly and usually within very narrow limits, the same is not true in tidepools. These are relatively small bodies of water, undergo considerable more rapid changes in temperature. Shallow tidepools exposed to the sun on warm days quickly reach lethal or near lethal temperatures. Similarly tidepools in cool temperate or subpolar regions may have temperatures in the freezing range in winter. An additional problem in the tide pools is that the pool may either heat up or cool over a several-hour period while exposed to the air, but when the tide returns it will suddenly change the temperature of the whole pool. Thus, organism inhabiting such pools must be adapted to sudden fluctuations.

The second factor to vary in pools is salinity. During exposure to low tides tidepools may heat up. When that happens, evaporation occurs and the salinity increases. Under hot tropical conditions, the salinity increase can be enough in some instances to reach the point of precipitating out of salts. If opposite situation is the case when heavy rains occur at low tide and flood with fresh water, dramatically lowers the salinity. Again tidepool animals and plants may have to be adapted to wider ranges in salinity that typical marine intertidal organisms. As before, when the tide returns, the pool will be flood with sea water at some point and again, there will be a sudden abrupt return of normal conditions.

Since the substrate itself, at least the surface layers, is in constant motion and may be picked up and held for greater or lesser amounts of time in suspension, depending upon the strength of the waves; if follows that any change in wave intensity on a given beach will mean a change not only in grain size, but also in the profile or shape of the beach. This comes about when gentle wave action is replaced by heavier waves, which dig deeper into the beach, pick up more and heavier grains, and hold them in suspension longer, so they are deposited some distance away, usually offshore. This changed profile is a common seasonal occurrence on many temperate zone beaches, where a gentle slope of fine sand occurs during the summer months and is replaced by a steep, coarse beach during winter storms.

Because of this condition of ceaseless movement of the surface layers of the sediment, few large organisms have the capability of permanently occupying the surface of sand or gravel beaches. This is the reason for the barren appearance of such beaches. In contrast to rocky shores, sand beaches usually have a rather smooth, uniform profile and thus lack the great topographic diversity of rocky shores. There are, for example, no crevices, no slopes facing different directions and offering different moisture conditions, nor are there overhangs or tidepools. As a result, the environmental factors such as temperature, desiccation, wave action, and insulation act uniformly at each tidal level on the beach.

The final physical factor undergoing change in pools is oxygen concentration. Since the amount of oxygen that can be held in sea water is a function of temperature, it follows that those pools that heat up during exposure to the air will lose oxygen. Under normal conditions, this may not be serious enough to produce oxygen stress, but if the pool is crowded with organisms, it may produce a stress situation. For example, a pool filled with algae that was exposed with high respiration could reduce the oxygen level significantly.

Tidepools are thus areas of refuge from desiccation for intertidal organisms. But in turn these organisms suffer from rapid changes in temperature, salinity and occasionally oxygen, levels thus restricting the fauna and flora to those organisms that are able to tolerate such ranges.

Sandy Shores

Intertidal sandy beaches common throughout the world and well known than the rocky beaches to most human beings. It is because they are the substrate of choice for recreational activities. Usually the sand beaches lack the macrophyte vegetation. Invariably all organisms are buried in the soil.

Environmental Conditions

The most important factor governing the life of sand beaches is wave action and its effect on particle size. The beach profile and particle size varies in different beaches and the different seasons. The particle size of sand in a beach is function of wave action on that beach. Where wave action is light, the sand particle size are fine, but where the wave action heavy and strong, the particles are coarse, forming deposits called gravel. The importance of particle size to distribution and abundance of organisms rests with its effect on water retention and its suitability for borrowing. Fine sand through its capillary action tend to hold much water above the tide level in the interstices after the tide has retreated. Coarse sand and gravel, on the other hand, allow water to drain away quickly as the tide retreats. Since the organisms inhabiting the intertidal are aquatic, they desiccate in a coarse gravel beach. This makes the latter less hospitable. Fine sand is also more amenable to burrowing than coarse gravel.

A second important physical factor in sand beaches, which was not of concern in rocky shores, is also the product of wave action. That is wave-induced substrate movement. The particles or sand or gravel beaches are not large enough to be stable when waves strike the shore. As a result, with each passing wave the substrate particles are picked up churned in the water and redeposited. Particles are thus being continually moved and sorted. The reason that fine sand beaches occur only where wave action is light and coarse ones where it is heavy is that in heavy wave action the smaller particles remain in for any organism living in the sand simply because the sand is opaque to light and reflects or absorbs it in the surface layers. Desiccation is not a problem as long as the beach sand is fine enough to hold water by capillary action during periods of low tide. Thus, any organism burrowing into the sand will be constantly moist.

A final physical factor acting on beaches is oxygen content. Oxygen is never limiting in the water bathing the beach because the turbulence of wave action ensures constant saturation. Where it might become limiting is in the substrate itself. The water held in the beach and responsible for ameliorating the changes in temperature and salinity also contain oxygen which is available for use by the organisms. This supply, however, is used up by respiration of the organisms and must be replenished. This replenishment comes through interchange of the water with the sea above and is dependent on the fineness of the sediments. Fine sediments have a slow rate of exchange and coarse sediments a rapid exchange. Thus, in fine-grained beaches water interchange is slow and may result in reduced oxygen supplies.

Adaptations

The dominant environmental factor acting on sand beaches is wave action which creates the unstable constantly moving substrate. In order to inhabit this area organisms must first be adapted to tolerate these features. There are two routes that may be taken by organisms in adapting. The first is to burrow deeply enough into the substrate such that the organism is deeper than the depth of sediment affected by the passing wave. This strategy is employed by many large clams such as *Tivela stultorum* the *Pismo clam*. Such animals are usually also aided by developing a very heavy shell which helps in keeping the animal down in the substrate. The only difficulty with this type of adaptation is that a very severe storm may generate waves large enough to pull these animals out and throw them on the beach. This is the reason for the catastrophic destruction that occasionally occurs in these forms.

The second route of adaptation is the ability to burrow in very quickly as soon as the passing wave has removed the animal from the substrate. This appears to be the more common mechanism and is employed by many annelid worms small clams and crabs. A good example of this type of adaptation can be found in the various sand crabs of the family Hippidae found in many beaches around the world. Spider crab is a common burrower in the west coast of India. These animals have a short body with limbs highly modified passing wave they reburrow before the water motion can carry them off. Small clams of the genus *Donax* also do the same thing and razor clams of the genus *Siliqua* are extraordinarily fast at reburrowing. *Cerianthus Spp.*, of West coast of India are also good burrowers of the intertidal zone.

Other adaptations are correlated with the above. For example most burrowing mollusks tend to have very smooth shells that reduce the resistance to burrowing into the sand. Similarly shore echinoderms such as sand dwellers have much reduced spines to allow them to burrow into the sand.

Type of Organisms

The most noticeable group that is absent from sand beaches is the large plants. No macroscopic plants occur on open sand beaches presumably because there is no way for them to attach and maintain themselves in the wave action. Small benthic diatoms, however, may be occur on the sand grains. The second major group of organisms conspicuous by its absence is the sessile animals such as barnacles and mussels. The sand beach is dominated by representatives from three invertebrate classes. Polychaete worms, bivalve mollusks and crustaceans. Combinations of species from these three groups dominate sand beaches throughout the world.

Feeding

The absence of large plants means that there is very little primary productivity. Although diatoms are present is the sand ensures that even the diatom population is restricted to the surface layers. As a result there are no macroscopic herbivores on a sand beach. Since there is virtually no primary productivity the animals living on the beach must depend for food upon either the phytoplankton carried in sea water above the beach or the organic debris brought in by the waves or consuming the other beach animals. There are relatively few true carnivores among beach animals because to be such would require active movement across the substrate in search of prey. This is probably not feasible in the face of the heavy wave action. As a result most of the organisms in a beach are either suspension or detritus feeders.

Suspension feeding animals are those that filter particles out of the water column. These particles may be plankton or various organic particles resuspended from the bottom by the passing waves. The dominant group of suspension feeders on sand beaches are the bivalve mollusks.

Sand beaches tend to have lesser amounts of organic detritus than muddy shores but enough debris from various sources finds its way to the beach to be a reliable source of food for certain organisms. Since this detrital is often carried up and down the beach suspended in the wave wash rather than being deposited on the bottom the mechanisms employed by detritus feeders on the beach are often different from those employed by detritus feeders on muddy shores. The sand crabs of the family Hippidae such as *Emerita analoga* and *Blepharipoda occidentalis* of the pacific coast of America employ and a unique mechanism to trap debris in the wave wash. The crab, *Emerita* orient so that their heads point shoreward. They remain completely buried as the incoming wave passes over them shoreward. After passage as the water begins to run down the beach they stick out their very large second antennae. These are highly clothed in hairs forming a net that when spread into the backwash intercepts all particles within a certain size range. The particles are then wiped off by the mouth and ingested. As the tide rises and falls crabs take advantage of this to maximize their feeding time by being moved up and down the beach with the tide. In a similar manner as the tide recedes they emerge and roll down the beach before reburrowing and commencing feeding.

A similar method of feeding is employed by the gastropod mollusk *Olivella columellaris* on the coast of Central America. In this case a mucus net is produced by the animal and held across the wave backwash. This net acts like a filter and captures small particles. Since the animals feeds by consuming the entire net with particles it must produce a new net at frequent intervals. Most sand beaches also include one or more polychaetes which burrow through the sand ingesting the food and excreting out the sand particles. Much like earthworms on land, however, they are rarely abundant on open sand beaches because the amount of organic material is less than in mud shores.

Sand dwellars are the echinoderms common on sandy beaches are generally deposit or detritus feeding animals. These animals generally burrow through the sand. As they do so the finer organic particles but not the sand grains fall down between the short spines are trapped in mucus and carried to the mouth. Some species such as the Pacific coast *Dendraster excentricus* according to Timko (1976) have been modified this basic feeding process to remove suspended particles by projecting part of the body up into the water column.

Carnivores are rare but the few found usually are polychaete worms or gastropod mollusks. The polychaetes such as species of the genera *Nepthys* and *Glycera* actively burrow through the sand and seize small prey with the help of proboscis. The carnivores gastropods are represented most commonly by burrowing snails of the family Naticidae (moon snails). These animals burrow slowly just below the surface layers of the sand in search of bivalve mollusks which they consume by boring holes through the shell and eating out the contents. A number of transient predators also occur. At low tide various shorebirds actively feed on beach organisms and at high tide fishes invade the area.

Zonation

Since there are no organisms visible on the surface of the intertidal sand beaches these areas do not display the surface zonation patterns very obvious as observed on rocky shores. Furthermore since the various physical factors such as desiccation and temperature that act to limit organisms on rocky shores are of far less importance to organisms living in the sand, it might be suspected that zonation of shore organisms but it is neither clearly defined nor as well understood as that for rocky shores. The lack of clear zonation patterns in sandy beaches is due partly to the habit of some dominant organisms moving up and down the beach for feeding on abundant plankton make competition unlikely. Similarly there are relatively few indigenous invertebrate predators and are responsible for any major distributional pattern. The effect, however, of large vertebrate predators such as birds and fishes remains to be investigated. A generalised zonation for sandy beaches is shown in Fig 5.5 and that of zonation by dominant invertebrate is shown in Fig 5.6

Muddy Shores

Sharp boundaries exist between rocky shore and sandy shores such that they are easily defined and recognisable. A similar situation does not prevail when trying to differentiate between sandy and muddy shores. Indeed it is not possible to draw sharp boundaries between the two because as shores become more protected from wave action, they tend to become finer grained and accumulate more organic matter and, thus, become more muddy. Sand and mud shores are, therefore, the opposite ends of a continuum.

Since sharp boundaries do not exist and one grades into the other along a gradient of increasing protection from wave action. The fauna and flora also show a change from organisms typical of

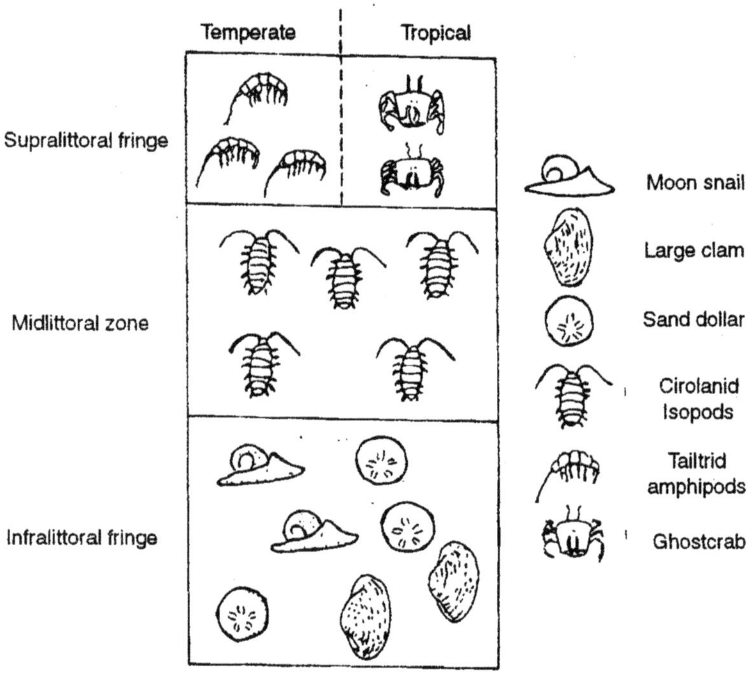

Fig. 5.5: A Generalised zonation pattern of intertidal organisms on sandy shores.

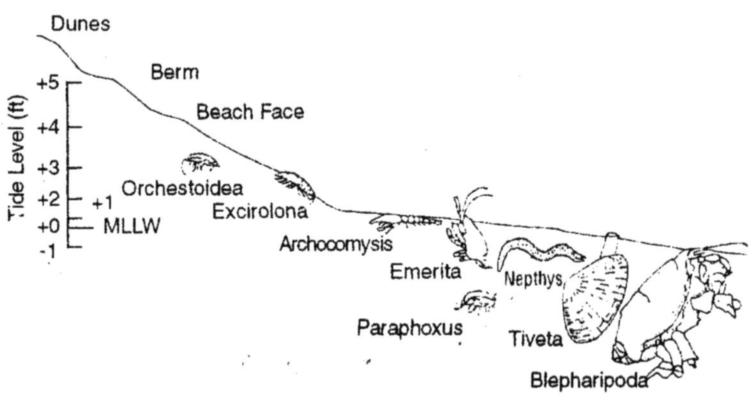

Fig. 5.6: Numerical dominance of invertebrates in sany beaches.

open sand beaches to those typical of muddy shores along the same gradient. It should be borne in mind that various transitional communities exist that will show mixtures of organisms from both extremes.

Muddy substrates are also characteristic of estuaries and salt marshes indeed most of the muddy shores of the world are associated with estuaries. Hence many of the types of organisms and adaptations are similar between the two areas.

Physical Factors

The major contrast with open sand beaches is that muddy shores cannot develop in the presence of significant wave action. Therefore, muddy shores are restricted to intertidal areas completely protected from open wave activity. Muddy shores are best developed where there is a source of fine-grained sediment particles. Muddy shores are located in various partially enclosed bays, lagoons, harbors and especially estuaries since these areas are developed where water movement is minimal and the slope of mud shores tends to be much flatter than that observed for sand beaches. Due to this reason these areas are often referred to as mud flats.

The very fine particle size coupled with the very flat angle of mudflats ensures that the sediments will not drain away and is held within the substrate. This long retention time for water coupled with a very poor interchange of interstitial water with the sea water above and a high internal bacterial population usually results in complete depletion of the oxygen in the sediments below the first few centimetres of the surface. Anaerobic conditions thus prevail within the sediment and this is one of the most important characteristics of the muddy shore. Muddy shores tend to accumulate organic material which means that there is an abundant potential food supply for the resident organisms, but the abundant small organic particles raining down on the mud flat also have the potential to clog respiratory surfaces.

Adaptations

Most organisms inhabiting these areas show an adaptation to burrowing into and through the soft substrate or else inhabit permanent tubes in the substrate. In contrast to sand beaches, however, the presence of the organisms in the mud flat is advertised on the surface by the presence of various holes of differing sizes and shapes. Thus, one of the primary adaptations of organisms in mud flats is the ability to burrow into the substrate or to form permanent tubes.

If organisms are to survive while burrowed into the substrate they must either be adapted to live under anaerobic conditions or else must have some way of bringing the overlaying surface water with its oxygen supply down to them. Most multicellular organisms cannot survive in the absence of oxygen. To obtain this oxygen rich surface water and food is carried to the various burrows, holes and tubes appear on the surface of the mud flat.

Most mud flat organisms are intolerant of completely anaerobic conditions. Many have adaptations that permit them to exist at lower oxygen tensions than similar forms that live on open sand beaches. The most common adaptation to low oxygen supply is the development of carriers (e.g. hemoglobin) that will continue to pick up oxygen at concentrations well below that of similar pigments in other organisms. Since wave action is essentially absent on these mud flats, there is no necessity for development of either rapid burrowing or heavy bodies to maintain positions as is the case for sandy shores.

Types of Organisms

In contrast to sand beaches muddy shores often support a substantial growth of various plants. On the bare mud flats the most abundant plants are diatoms which live in the surface layers of mud and often give a brownish colour to the surface of the mud at low tide. Other plants include large macroalgae such as species of *Gracilaria* (red algae), *Ulva Enteromorpha* (green algae). These large algae often undergo seasonal cycles of abundance becoming common in the warmer months and virtually disappearing in colder months. As a result of the occurrence of these various primary producers there is substantial primary productivity in the mud flats.

Mud flats contain large numbers of bacteria which feed on the abundant organic matter. Bacteria are the only abundant organisms found in the anaerobic layers of the mud shore and constitute a significant biomass. Among the dominant bacteria inhabiting this area most bacteria are capable of utilizing the potential energy of the various reduced chemical compounds abundant here. These chemosynthetic or sulfur bacteria obtain energy through the oxidation of a number of reduced sulfur compounds such as various sulfides (for example H_2S). These organisms are, thus, primary producers of organic matter analogous to green plants. They produce organic matter using energy obtained from the oxidation of the reduced sulfur compounds whereas plants produce organic matter using energy obtained from sunlight.

Since these autotrophic bacteria are located in the mud flats are unique among marine environments in that they have two separate layers in which primary productivity occurs on the surface where diatoms, other algae and marine grasses carry on photosynthesis and a deep layer where bacteria conduct chemosynthesis. The dominant macrofaunal groups on muddy shores are the same as those encountered on sand beaches namely various polychaete worms, bivalve mollusks and small and large crustaceans but of quite different.

Feeding Ecology

Because of the greater amount productivity due to both bacteria and plants there is vast food available on muddy shores than on sand beaches. This permits large organisms to live on mud shores and indeed mud flats may be very densely populated.

The dominant feeding types on mud flats are deposit feeders and suspension feeders a situation similar to that on sand beaches. Deposit feeders are particularly abundant due to the large amount of organic material and the large populations of bacteria in the sediments. Deposit feeding polychaetes are represented by genera such as *Arenicola* and *Capitella.* Both feed by burrowing through the substrate ingesting it and digesting out the organic matter (or bacteria) and passing out the undigested material through the anus. *Arenicola sp.*, however, form a U shaped burrow in which one arm of the U is a permanently open shaft to the surface while the other arm is filled with sediment that worm ingests. The worm consumes sediments from the shaft passes it through its gut and then moves up in the open burrow to defecate at the surface leaving characteristic mounds (Fig 5.7). Capitellids on the other hand form no permanent tubes but move like aquatic earthworms through the surface layers of the substrate ingesting it.

Deposit feeding bivalves are also common in these areas. Temperate zone mud flats often have large numbers of small tellinid clams of the genus *Macoma* or *Scrobicularia*. These bivalves have siphons that are separated. Each clam lies buried in the substrate with its siphons extended to the surface. There are intake siphon is moved over the surface like a vacuum cleaner ingesting organic particles and bringing them down to the clam for digestion.

Suspension feeders include various other species of clams some crustaceans and numbers of polychaetes. The fine sediment particles are easily stirred up on mud flats and remain resuspended in the water. Hence suspension feeders may possibly take in amounts of these particles in addition to various plankton organisms. Perhaps most of the suspension feeders on mud flats also partially take in resuspended sediments and hence are actually feeding on both deposited and suspended material.

In general deposit feeders are more common in fine-grained shores and suspension feeders become more abundant in coarser

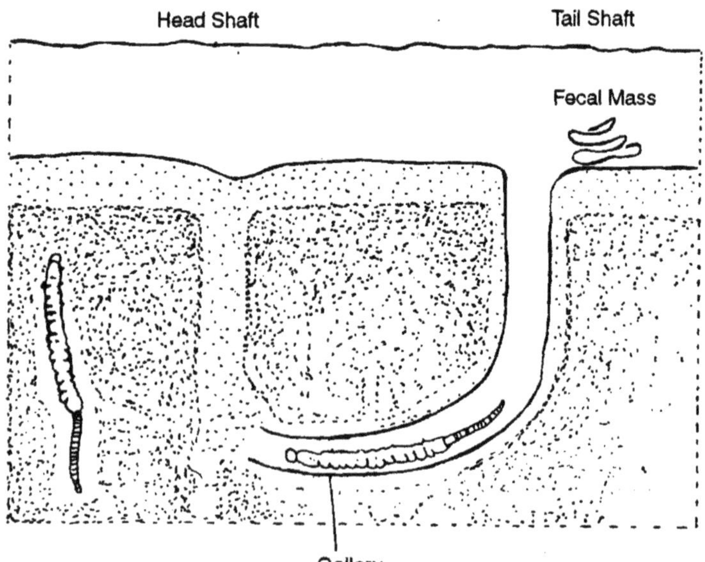

Fig. 5.7: Two polychaete worms in their U shaped burrow (*Arenicola*) and in the core substance sand (*Capitella*).

sediments where there is little organic matter. The major carnivores on mud shores are the fishes which feed when the tide is in and the birds feeding when the tide is out. Indigenous mud flat predators include a few polychaete worms such as *Glycera sp.*, moon snails (*Polinices* and *Natica*) and crabs.

Despite the relative abundance of plant material on muddy shores there are very few herbivores and most plant material finds its way into the food chain only after it has been broken down into small pieces to enter the deposit feeding food web. Thus, the trophic structure of a mud flat is often built up from two basis a detritus bacteria base and one based on plants. The detritus base web is derived from plants and other organic sources and includes the bacteria that live upon the detrital particles. Detrial particles can, thus, either be taken in directly by the large macrofaunal invertebrates or else be consumed by bacteria. An additional bacterial component at the base of this trophic pyramid is the sulfur bacteria.

Bacteria are consumed commonly by various nematode worms which, thus, occupy a position similar to that of herbivores in other areas. Nematodes and deposit feeders are in turn consumed by various carnivores including predatory invertebrates (moon snails *Glycera*) as well as birds and fishes. As second food web is based primarily on the microscopic diatoms as the autotrophic base. The diatoms are consumed by several different polychaetes, mollusks and crustaceans. These in turn are consumed by the large predatory birds and fishes.

Zonation

Little information exists regarding the zonation of mud flats. The very gentle slope of these areas means that the intertidal is often very extensive more so than rocky or sandy shoes. The upper areas the supralittoral fringe is inhabited by various species of crabs which burrow into the substrate. The extensive midlittoral area is the home of most of the common species of clams and polychaetes.

Intertidal Fishes

A large amount of research has been carried out on the ecology of invertebrates and plants of the intertidal zone on all three types of shores, there are very few accounts of the fishes of these areas or of the role they may play in community organization as grazers or predators.

Most intertidal fishes because of the turbulent environment are of small in size. The body shape is usually compressed and elongate (Blenniidae, Pholidae) or depressed (Cottidae, Gobiesocidae), which allows them to inhabit holes, tubes, crevices or depressions for protection against both desiccation and wave action. Most also lack swim bladders and are closely associated with the substrate. Many of these fishes are adapted to withstand greater ranges of salinity and temperature than their subtidal relatives. A few are even adapted to spend more time out of the water (*Periophthalmus*).

Most intertidal fishes in the temperate zone have been shown to be visual carnivores again suggesting a potentially significant role in intertidal community organization. The life history patterns of the few species that have been investigated are all largely similar. Eggs are demersal and laid on stones, rocks or submerged vegetation. Often the eggs are guarded by the male. The eggs hatch after a few weeks into planktonic larvae. The planktonic period varies depending on the species. It may be as long as two months. During this period the larvae gradually acquire the adult features and at the end become benthic. Life spans of the adults are generally short from two to ten years and sexual maturity occurs in the first or second year. Some intertidal fishes are migratory in habit move according to tidal, diurnal and seasonal variations.

Birds

At low tide a considerable variety of birds are often associated with the intertidal zone. Surprisingly the effect of these birds on the invertebrate communities is little known at the present time except for their role on estuarine mud flats. Most birds associated with the intertidal are carnivores or omnivores. Our knowledge of the diet of these birds indicates that they consume various intertidal invertebrates which in turn suggests they may have a potentially important role in the ecology of intertidal food webs.

Chapter 6
Interstitial Organisms

In addition to the large macrofaunal organisms that inhabit sandy shores, there is another world that exists here, comprising those organisms that occupy the microspaces between adjacent sand grains which are referred as the interstitial organisms. As interstitial fauna and flora assemblage has been found in intertidal and subtidal sandy substrates throughout the world, in both freshwater and saltwater. Major interstitial groups have been reported by Coull *et al.* (1977) down the depths as great as 5000 m. Studies of these associations are, however, unequally distributed. The marine interstitial fauns is better known than the fresh water, and the organisms and associations of the European seas are the best known of the various geographical areas. The pacific coasts of North America is practically unknown and unstudied with respect to interstitial fauna, whereas the Atlantic coasts of North America has been the site of numerous studies.

Definitions

The term interstitial is a general term that refers to the organisms that occupy the spaces between the sand grains. Psammon is a synonym for interstitial organisms, referring to all those that live between the sand grains. The term mesopsammon refers to all interstitial organisms associated with the fresh water or brackish water beaches while thalassopsammon refers to those associated with marine beaches and sand areas. Meiofauna is a term that is often used synonymously with interstitial or psammon but it has a more limited application because it refers to a size range whereas interstitial organisms include both the meiofauna and microfauna size groups.

Environmental Factors

The conditions that affect the interstitial fauna are somewhat different from those that affect the macrofauna in the same areas. Perhaps the most important factor influencing the presence, absence and types of interstitial organisms is grain size. Grain size is of paramount importance in determining the amount of interstitial space available for habitation. The coarser the grain size, the greater the volume of interstitial space and hence, the larger the interstitial organisms that can inhabit the area. Conversely, the finer the grain size, the lesser the space available and the smaller the organisms to inhabit the area. Grain size may thus act as a definite barrier to the dispersal of psammon organisms. Certain organisms will be confined to beaches of certain grain size or areas of a given beach simply because they are too larger to fit in between the grains of adjacent areas and, therefore, cannot disperse through the whole area. Psammon organisms show a definite zonation based on grain size.

Grain size is also of great importance because it controls the ability of beach to retain the circulate water. Interstitial organisms are aquatic and, therefore, require the presence of water in the interstices of the sand in order to survive. If the grain size becomes too coarse, when the tide ebbs the water will not be held in the sand by capillary action, but will drain away, leaving only a very thin layer coating the grains. On the other hand, fine-grained beaches are able to hold considerable water in the interstices through capillary action.

Circulation of water through the pore spaces in the sand is important because this water movement is responsible for renewing the oxygen supply. Circulation is best in coarse grained beaches and is reduced in fine-grained beaches. If the grain size becomes really fine, as in mud shores, the circulation practically ceases, and the result is the creation of anaerobic layers in the sediment. The maximum number of thalassopsammon have been found where sand grain diametres were between 0.175 and 0.275 mm. Fenchel (1978) reports that the interstitial fauna tends to disappear when the median grain size drops below 0.1 mm.

In addition to grain size, the mineral nature of the grains is also of importance in determining the composition of the psammon. Different association of organisms may be found in siliceous sand as opposed to carbonate sand. As we know the oxygen level is an important limiting factor in this environment. Practically all marine sediments have an oxidized layer at the surface, beneath which lies a completely anoxic layer. Thus, interstitial organisms living below a certain depth will encounter oxygen-free conditions. The thickness of the oxygenated layer depends on a number of factors such as grain size, amount of organic material, and water turbulence.

Temperature is another environmental factor of importance in determining presence or absence and distribution of interstitial organisms. Temperature ranges are most extreme in intertidal beaches and in the uppermost layers of the beach. Temperature changes are minimal in subtidal sands and in intertidal sands below 10-15 cm in depth. The surface layer of the sand on a beach acts as an insulator to the lower layers, however, the temperature may change markedly, depending on the air temperature, the effect of wind and rain, the amount of sunlight hitting the surface, and the temperature of the sea water. In cold temperate and polar regions, the temperature may be low enough to freeze the upper layers. Some interstitial organisms have apparently adapted to survive being frozen (Example – turbellarian–*Coronhelmis lutheri*, gastrotrich–*Turbanella hyalnia*, enchytraeid–*Marionina southerni*, and harpacticoid–copepod *Parastenocaris phyllura*). Because of the range of temperature that may occur in the upper layers, the fauna inhabiting these areas may be different from those of lower layers and includes those with a considerable tolerance for temperature change (eurythermal).

Another factor acting on interstitial organisms, particularly of intertidal beaches, is salinity. Reduced salinities may occur in intertidal beaches due to fresh water run-off over the beach during low tide or else due to heavy rainfall. Such salinity changes are usually confined to the upper layers of the beach because the lower layers retain, through capillary action, elevated levels of salt water. Since fresh water is less dense than salt water, fresh water cannot penetrate below that point where marine water is held by capillary action. This means that only the uppermost layers are subjected to salinity changes. Organisms occupying these layers are usually adapted to tolerance considerable salinity changes.

Another environmental factor is that of wave action. Wave action acts both intertidally and subtidally, but is most dramatic intertidally. Breaking and moving waves have the effect of resuspending the upper layers of sand and, seasonally, of completely removing or depositing large amounts of sand. Whenever the beach is thus churned up, the whole array of internal space is rearranged, the organisms themselves move about, a process that has profound effects on organisms dependent upon staying in the space between grains. It means that for those organisms, the interstitial space are being constantly rearranged and they are in constant danger of being thrown out into the open water where the risk of predation is very much greater. Special adaptations for coping with this risk are known.

Light rarely penetrates below 5-15 mm deep into sand beaches. This means that microscopic plants should generally be restricted to the very surface layers. However, diatoms have been found at depths considerably below this, raising speculation as to how they are managing to survive. Most of the meioflora and microflora are attached to the surface of the sand grains and form an important food source for meiofaunal grazers.

Composition

The types of organisms constituting the psammon are drawn from a very broad range of invertebrate phyla, but some are represented by one or a few species, whereas other are abundant both in terms of individuals and species. Those invertebrate phyla that normally have small bodies are, of course, preadapted to live in the small spaces between sand grains and are represented by many species and individuals. The phylum Protozoa is represented abundantly by a large number of species of ciliates, which are very diverse in form and, interestingly enough, often larger than the metazoan animals found with them (Fig 6.1). The class Turbellaria, of the phylum Platyhelminthes, is also abundantly represented by numerous small, flat, and elongate worms (Fig 6.2). Most abundant of all, perhaps, are the ubiquitous roundworms, the nematodes.

Several inconspicuous invertebrate phyla are well represented in the psammon, probably because they are normally of small size. These include the phyla Gastrotricha (Fig 6.3),

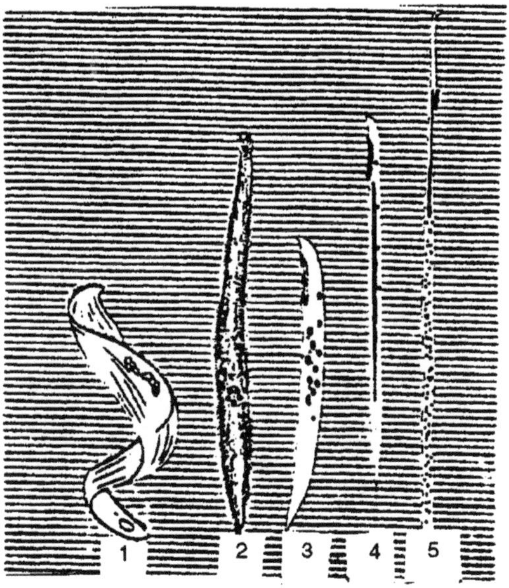

Fig. 6.1: Interestitial Ciliate Protozoa:
1. *Loxophyllum verniformae* 2. *Trachiloraphis remanei* 3. *Remanella faurei* 4. *Ramanella caudata* 5. *Geleia gigas*

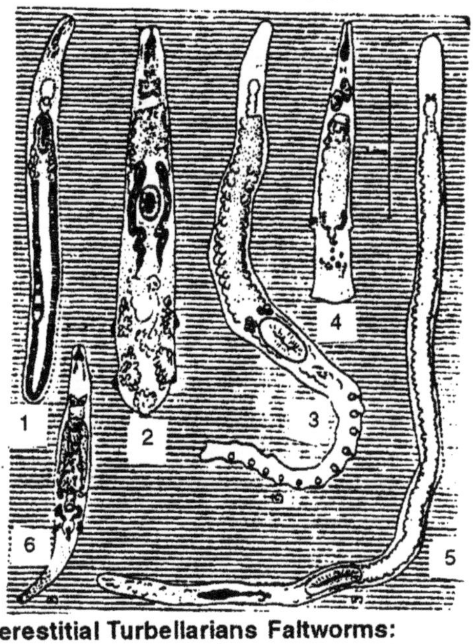

Fig. 6.2: Interestitial Turbellarians Faltworms:
1. Member of Macrostomidae 2. Calyptorhynchidae 3. *Ploystylophora sp.* 4. *Proschizorhychus sp.* 5. *Cicerina sp.* 6. *Nematoplana sp.*

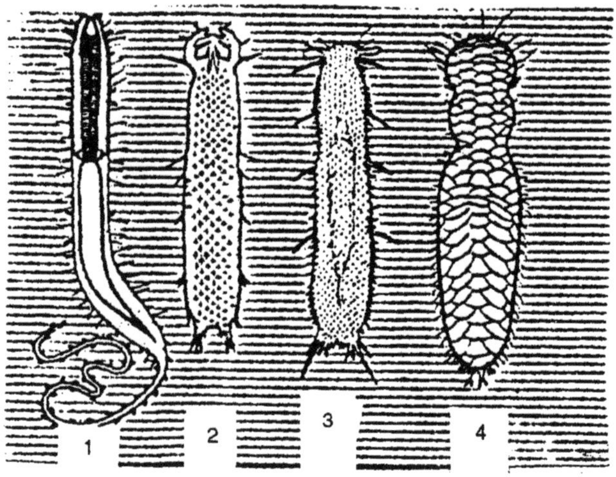

Fig. 6.3: Interestitial Gastrtrichs:

1. *Urodysis viviparus* 2. *Pseudostomella roscovita* 3. *Thaumastoderma heideri* 4. *Dipldasys ankeli*

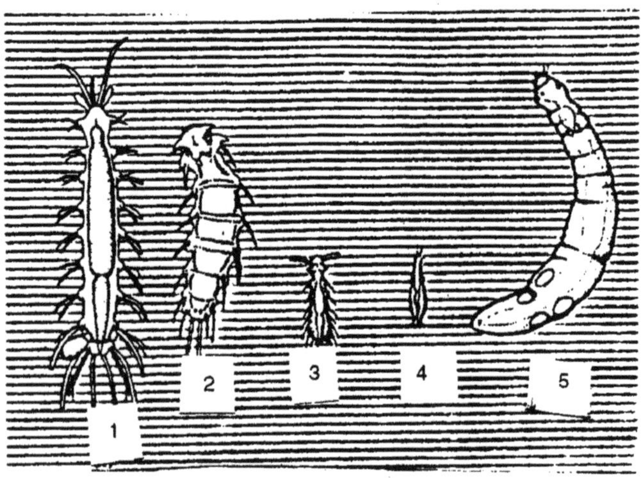

Fig. 6.4: Interestitial polychacte annelids:

1. *Mesonerilla intermedia* 2. *Paranerilla limnicola* 3. *Nerillidium simplex* 4. *Diurodrilus sp.* 5. *Trilobodrilus sp.*

Tardigrada and Rotifera. The Gastrotricha are particularly numerous in the thalassopsammon and Rotifera in the mesopsammon.

The worms of the phylum Annelida with their elongated shape are well suited to the interstitial environment and are abundant. Representatives in the meiofauna are from both the classes Oligochaeta and Polychaeta (Fig 6.4). Crustaceans are represented primarily by an abundance of harpacticoid copepods and Ostracods and a few other small groups such as Mystacocarida.

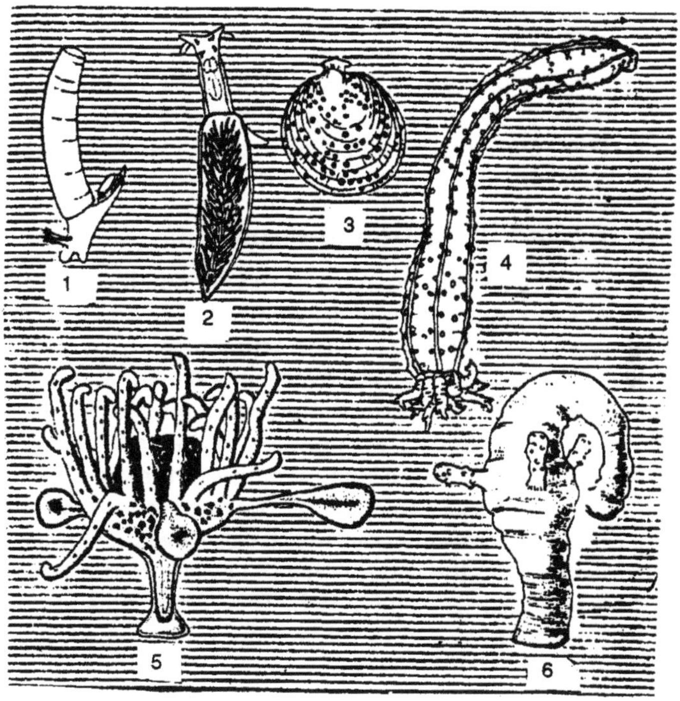

Fig. 6.5: Interestitial Mollusans, Cnidaria and Echinoderms:

1. *Caecum glabra (Gastropoda)* 2. *Hedylopsis brambeli (Gastropoda)* 3. *Gwynia capsula (Brachipoda)* 4. *Labidoplax buskii (Echinodermata)* 5. *Stylocoronella riedli (Cnidaria)* 6. *Psammohydra nanna (Cnidaria: hydrozoa)*

Phyla that normally consist of large organisms sessile forms are very poorly represented in the psammon. In the case of large-bodied phyla, this is certainly due to the body size, which does not allow them to fit into the small spaces. These groups are, therefore, represented only by a few rather aberrant forms. Examples of such phyla are Echinodermata and Cnidaria. Phyla of sessile or sedentary habit seem to have few representatives (Fig. 6.5), because the dynamic nature of substrate makes it difficult to exist if one is permanently fixed in one place. Moreover, most of these phyla are suspension feeders, depending upon filtering organisms out of the water. In the interstitial environment, there is relatively little water to filter and very few organisms in the water, as most are attached to or adhere to the sand grains. The sessile groups that are represented by a very few species include the phylum Bryozoa and the class Tunicata (Ascidiacea). The representative of the following phyla viz., Sipuncula, Photonida, Echiura, Pogonophora, Porifera, Ctenophora, Hemichordata, and Chaetognatha are absent in the intertidal community.

Sampling and Extraction of Psammon

Because interstitial organisms are very small and occupy a unique habitat, special techniques are needed to obtain samples. A great variety of sampling and extracting techniques have been employed by meiobenthologists over the years, and none is really standard. Most techniques are, however, simple and may be grouped into general classes. Because of the small size of the interstitial organisms, only a small sample of the sand substrate is required. This is most conveniently obtained by taking a core. Coring devices are usually hollow plastic or metal cylinders that may be pushed into the sand by hand, closed off at one or both ends, and removed from the beach. The enclosed sand column may then be extruded and the organisms are extracted. It is also possible to take a small volume or core from a large grab or dredge sample that may be taken for macrofauna. Majority of the meiofauna individuals will, however, be concentrated in the top 10-15 cm, so a depth of 25-30 cm is adequate to sample all but a few forms.

The diameter of the coring device is not critical, but those in common use range in size from 2-4 cm. Once the sample is in hand, it is necessary to extract the organisms. Interstitial animals are reluctant to leave the interstices and must be removed by special methods in order to be studied. A number of different extraction techniques have been used in the past and each gives somewhat different results. It should be noted that the fauna may be extracted alive or preserved. In most cases, it is preferable to extract the animals alive, because live animals are much more easily seen and also because groups like flatworms and gastrotrichs are virtually unidentifiable when preserved.

The simplest extraction is to scoop up a bucket of sand and let it stand until organisms are driven out. Another simple technique makes use of the response of the organisms to temperature. This is the Uhlig Seawater Ice Technique (Uhlig, 1968). Here, sand sample is placed into a cylinder that has fine mesh at the bottom. On top of the sand is placed sea the whole device is then suspended such that the bottom touches the water surface in a bowl. Over time the sea water ice melts, runs through the sand and drives out the animals into the bowl below.

Even more efficient is the method of elutriation. In this technique, a sand sample is introduced into a container with openings at the top and bottom. The sand is then constantly stirred up by water entering from below. As the sand stirs, the organisms are dislodged and carried over and out with the water through the upper opening. This water is filtered on a very fine mesh, which then catches the organisms. In this technique, only filtered water is used as the input, otherwise, the sample will be contaminated by plankton organisms. This technique can be used with either living or preserved material.

Adaptations

Most of the morphological adaptations observed in the interstitial fauna appear to be correlated most closely with a few aspects of the environment. The environment is a dynamic one in which the individual sand grains of the upper layers are constantly being resuspended and deposited, thus, constantly changing the location and amount of interstitial space.

Perhaps the most obvious adaptation to this environment in that of size. All interstitial organisms are very small. Further more, in this fauna, one finds the smallest representatives of most large phyla. For example, mollusks, echinoderms, and annelids are usually fairly large animals. All have representative in the psammon, but they are remarkable for their small size. The holothurian *Leptosynapta minuta* is 2 mm, the gastropod *Caecum glabrum* is 2 mm, the Polychaeta *Diurodrilus minimus* is but 350 mm, and the hydroid *Psammohydra nanna* is 1 mm. In the case of ciliate protozoans, however, the size is generally larger than that of ciliates living in other environments. At the same time, however, these larger ciliates are considerably flattened and elongated, which allows them to fit more readily in the interstices.

Another adaptation concerns the shape of the body. Most of the interstitial organisms have elongated or vermiform bodies, even in those organisms that are normally not vermiform. Coincident with small body size, there is a reduction in the complexity and number of the body organ systems in metazoan interstitial animals. For example, in the polychaetes, the complex pharyngeal apparatus found in macrofaunal polychaetes often disappears, as do the parapodial gills and even the kidneys (nephridia) in the interstitial hydroids, the number of tentacles is reduced from that observed in the forms not living interstitially.

These organisms are always at the risk of being crushed as sand grains banged into each other. As a response to that many psammon animals have developed various types of reinforcement to their body walls. Skeleton of spicules are found in ciliates such as *Remanella*, turbellarian flatworms such as *Acanthomacrostomum spiculiferum* and the molluscan, *Rhodope* and *Hedylopsis*, Gastrotrichs on the other hand have an armor of scales while the nematodes depend on a thick heavy cuticle. Those delicate animals lacking a protective armor seem to have developed great abilities to quickly extend and contract the body which also permits escape from crushing. Because the environment is so dynamic, we find that virtually all the inhabitants are free moving. This is true even for groups that are normally all sessile such as the Bryozoa and Tunicata.

Two more vital adaptations noted and both are in response to the necessity of the interstitial organisms to remain in the sand and not be thrown out during the resuspension process. Many species of various phyla such as flatworms and gastrotrichs have adhesive organs to fix them to the sand grains. Crustaceans and tardigrades effect the same ability using hooks or claws on their limbs. Statocyst organs that detect gravity and hence differentiate up and down are also common. Such sense organs enable an organism to quickly determine which way to move during resuspension and, thus, return to the sand rather than up into the open water.

Life History

We know very little about the life histories of many psammon. There are really two groups of interstitial organisms a temporary psammon which consists of the newly settled juveniles of the macrofauna organisms, and a permanent psammon, consisting of those organisms that spend their entire lives in the interstices of sand. Since the permanent interstitial fauns are so small in size, in number of gametes that they can produce at any one item is also limited. In contrast to the macrofauna, which often produce thousands or hundreds of eggs, the production of eggs by thalassopsammon is almost always below 100 per individual and usually ranges between one and ten.

When the number of eggs is that small, a species cannot afford to lose many or there will be no next generation. As a result, these animals have evolved a number of adaptations to ensure survival of the few propagules. First, there are adaptations that ensure that fertilization will occur. One way is to have copulation in which the sperm are directly transferred to the female. Such is the case with harpacticoid copepods. Another way is to pack all the sperm in a single unit called a spermatophore and attach it to the female where it will be available to fertilize the eggs as they issue. Such a situation occurs in certain of the polychaete worms and the Acochlidiacean mollusks. Finally some species are hermaphroditic having both male and female systems in the same individual and, thus, ensuring fertilization. This situation occurs in gastrotrichs some hydroids (*Orohydra*) and some polychaetes (*Protodrilus*).

Once fertilization has occurred it is important that the embryos be protected from loss. Usually benthic organisms put their embryos into the plankton in order to obtain maximum dispersal. However, the plankton is a notoriously dangerous place with the result that most of the embryos are destroyed via predation. As a consequence to compensate the benthic animals using this route must produce large numbers of larvae. Since the thalassopsammon cannot produce large numbers it follows that it would not be a good survival strategy to put larvae in the plankton. In fact this is what is observed. Very few interstitial animals have planktonic larvae and those that do are the ones that produce the largest numbers of eggs. Most interstitial organisms have brood protection. In such a case the eggs are kept with the female until the young have reached a sufficient size to be self-sustaining. That selection pressure for brood protection is high in this environment can be deduced by observing that brood protection occurs here in groups of invertebrates in which it is otherwise virtually unknown such as cnidarians (*Otohydra*) and gastrotrichs (*Urodasys viviparus*).

For those few that do not have brood protection additional safeguards are found. Eggs that are spawned are either sticky so as to quickly attach to sand or else are confined in cocoons that are sticky. In any case the development is usually direct, that is the egg hatches as juvenile and does not have a free-swimming larva or a larva is produced that is nonpelagic and remains in the interstices of the sand. Studies by Gerlach (1971) suggest that generation time of interstitial organisms range from a few days to over a year. The average number of generations per year is three.

Ecology

The composition of the interstitial fauna varies both vertically and horizontally in the substrate. The factors creating this zonation are the grain size differences and physicochemical factors particularly temperature, oxygen and salinity. Characteristic patterns of species assemblages are correlated with these physical factors. Most of the thalassopsammon seems confined to the uppermost strata usually the top 5 cm. In investigations in Europe, McIntyre (1969) reported between 80-90 per cent of the interstitial organisms have been found as deep as 70 cm in some beaches and there is often a definite vertical zonation.

There is a specialized faunal assemblage that is confined to the anoxic layers. Characteristic of this assemblage are ciliates of the orders Trichostomatida, Heterotrichida and Odontostomatida of which the latter seems to contain only anaerobic species and is confined to this area. Zooflagellates, certain nematodes, turbellarians, gnathostomulids, rotifers and gastrotichs are also present but it is not known whether these latter forms are permanent members of the community or transients from the upper oxidized zones. Vertical migration of the organisms also occurs. This migration is usually triggered by temperature or salinity changes and is especially prominent in beaches in the temperate zone.

There are also seasonal changes in abundance of interstitial organisms. Certain species are most abundant at one season than another. For example on Danish beaches Muus (1967) found that copepods were at a minimum in winter and peaked in numbers in the summer. Fenchel (1978) has noted that in temperate shallow water areas there is an annual change in the densities of the dominant groups with copepods succeeded by oligochaetes, nematodes, ostracods and turbellarians. This temporal succession appears to be regulated by primary productivity and predation. Trophic relationships among the interstitial fauna are not well known. It is known that certain flatfishes such as *Pleuroneces platessa, Platichthys flesus, Microchirus boscanion* and *Limanda limanda* in European waters actively prey upon meiofauna when they are young and recently metamorphosed. McIntyre (1969) reports that their stomachs have been found to contain numbers of harpacticoid copepods ostracods oligochaetes and nematodes. Other macrofaunal predators on thalassopsammon include shrimps larger polychaetes and hydroids. In addition to macrofaunal predators, there are interstitial predators, including certain nematodes, turbellarians, some tardigrades, and hydroids such as *Protohydra*. The latter is a particularly voracious feeder on copepods and nematodes.

In addition to predators, the interstitial fauna contains three other feeding types, herbivores, detritus and suspension feeders. While some turbellarians have been shown to be predators, others appear to be scavengers, feeding on dead meio and macrofaunal organisms. Harpacticoid copepods seem to be primarily herbivores feeding on the few attached diatoms or else feeding upon detritus. Ostracods also ingest diatoms. Detritus feeders include gastrotrichs some nematodes and archiannelid polychaetes.

A considerable degree of food specialization appears to exist among the thalassopsammon. Fenchel (1968) studied the interstitial ciliates and demonstrated a surprising degree of food specificity for various bacterial and algal taxa. In addition he showed that in certain habitats whole groups of species subdivided their food resource according to size. A similar food niche separation has been observed in nematodes.

Biodiversity and Distribution

One of the more remarkable features of the interstitial fauna is the species richness. As many as 70 species of psammon organisms in 50 cm^2 of sand and they regularly obtained 30-50 species of ciliate protozoan alone. Based on the limited studies available now it appears that there is no gradient in species richness in this fauna moving from polar regions to the tropics. This is in-direct contrast to the common distribution pattern of most macrofaunal organisms where lower latitudest end to have greater richness than higher latitudes. Zoogeographically, the psammon have a pattern in which the common genera are often cosmopolitan but the species are not. As a result several different zoogeographical regions may be recognized.

Chapter 7
Pearl Culture

Introduction

Pearls are nature's perfect creations with exquisite beauty and serene luster. They come in limitless array of shapes, sizes and shades. Unlike diamonds, topaz, rubies, emeralds, sapphires or other precious stones, pearls are not dug out from the ground. They are the only precious gems to be formed by the natural secretion of living organisms, a particular kind of molluscs, the oysters and mussels.

Natural pearls are the oldest gems known to mankind. The Vedas of India, the Bible and the Quran make several references of pearls as objects of adoration and worship. The natural pearls of the Gulf of Mannar and the Persian gulf enjoyed very good reputation in the world trade from the time immemorial. The pearls from the Gulf of Mannar are famous throughout the world as Orient Pearls.

Man has wondered how the oyster produces pearl and many imaginative minds propounded many interesting theories. Some believed that pearl was formed when a rain drop or a drop of dew fell into the shell when it came to the surface. Some others felt that it was the tears of angels that crystallized into pearls. The truth behind this mystery was unraveled in 1907 when a Japanese scientist Tokichi Nishikawa proposed an explanation "the pearl sac theory" for pearl production. Pearl secreting cells of the mantle migrate into the body of the oyster under the stimulus of a foreign body and by a series of cell division form a pearl sac around the foreign body. The pearl-sac in turn secretes the nacre which becomes deposited over the body forming a natural pearl in course of time. Natural pearls are rare in occurrence, small in sizes and generally irregular in shape.

The cultured pearl is produced by human interference by inserting a shell bead "nucleus" and a mantel piece into the gonad of the oyster and growing the oyster in the sea. Artificial pearls are those made of plastics, glass, etc. They are painted with pearl essence which is a mixture of enamel and silvery extracts of fish scales. The credit for making the first cultured pearls goes to the Chinese. As early as the 12th century, Chinese fishermen introduced miniature buddhas and similar objects inside oysters and removed them after they were covered with pearl forming nacre. But the technique was soon forgotten. Swedish botanist Carl Linnaeus re-discovered the process to produce cultured pearls in salt and fresh water (Fig 7.1).

The Central Marine Fisheries Research Institute (CMFRI) based at Kochi, Kerala has started a research project on pearl culture in 1972 at Tuticorin. An early break through was achieved and the first cultured pearl was introduced in 1973, heralding the development of pearl culture technology in India (Alagarswami, 1974).

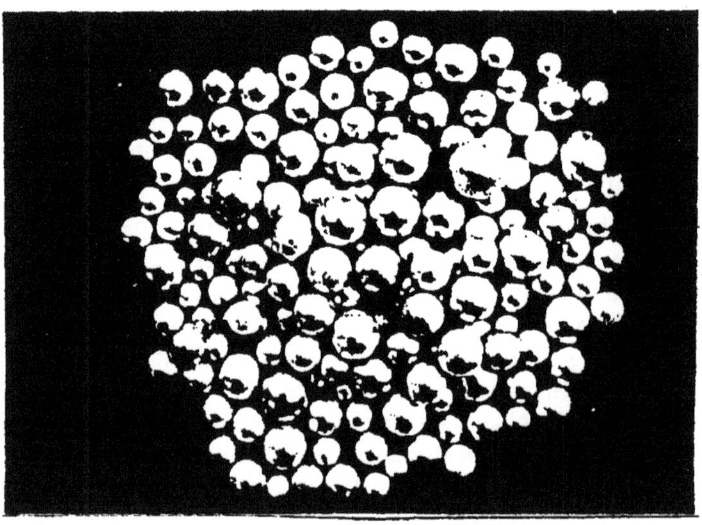

Fig. 7.1: A collection of cultured pearls.

2. Distribution of Pearl Oysters

The pearl oysters belonging to the genus *Pinctada* come under the family Pteriidae. They occur in almost all the seas of the tropical and sub-tropical belt. The pearl oysters occur in Persian Gulf (Bahrain, Kuwait, Dubai, Musket, and Bushira), Red sea (Farasan Islands, South of Sabia and Jidda, west of Mecca and Sudan), Philippines, Japan, China, Korea, Myanmar, Indonesia, Papua New Guinea, Fresh Polynesia, Cook Islands, Australia, Gulf of California, Mexico, Panama and Venezuela.

Distribution in Indian Waters

In Indian waters, they occur in 6 regions namely the Gulf of Mannar, Palk bay, off Trivandrum, the Andaman and Lakshwadeep Islands and the Gulf of Kutch. Among these the Gulf of Mannar and Gulf of Kutch are the most productive regions where pearl oysters occur in large numbers. Six species of pearl oysters occur in Indian waters. They are *Pinctada fucata (Gould), P. margaritifera (Linnaeus), P. chemnitzi (Philippi), P. sugillata (Reeve), P. anamioides* and *P. atropurpurea (Dunker)*. Of these *P. fucata* occurs extensively in the beds of Gulf of Mannar and to a less extent in the Gulf of Kutch. The black lip pearl oyster, *P. margaritifera*, occurs in stray numbers in some pockets of Andaman and Nicobar islands (Fig 7.2). Here they are fished more for the shells than for their pearls. *P. chemnitzi* are spat of *P. anamioides* has been observed on the ridges of the rocks and corals in some islands. In the south-west coast of India, at Vizhinjam, the spat *P. fucata* were collected from mussel culture ropes.

Morphology and Anatomy

A. **Classification**
   Class : Pelecypoda
   Order : Dysodonta
   Family : Pteriidae
   Genera : *Pinctada*.

B. **Maximum Size:** 77-92 mm in shell height and shell length. Commercial size: 55 to 60 mm in shell height and shell length, 30 g in total weight.

C. **Colour:** Outside of the shell dark brown or dark reddish brown with six radial bands of dark violet. Inside of shell having strong pearly lustre.

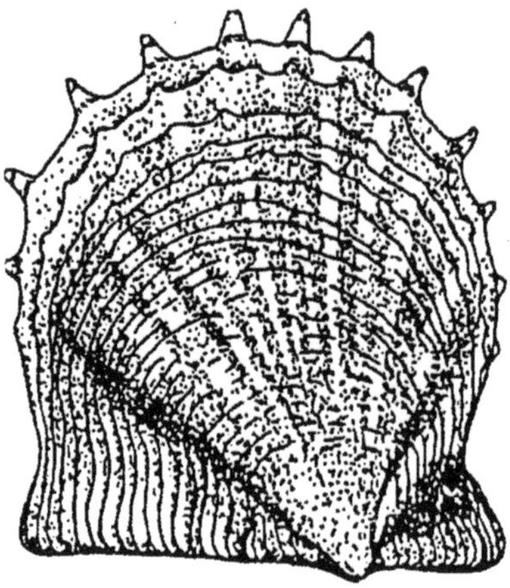

Pinctada fucata

Pinctada margaritifera

Pinctada chemnitzii

Mytilus edulis

Fig. 7.2: Pearl maker organisms

D. **External characters:** Hinges line straight umbo located at anterior part of the shell. Lower part of the shell almost round in shape. Surface of the shell covered with thin scaly leaves. Left shell is more concave than the right shell (Fig. 7.3).

Life Cycle (Fig 7.4)

Pearl oysters are sedentary animals found attached on corals, dead coral outcrops and sand grits by means of byssus threads. They inhabit at depths from 10 to 20 m. Like other bivalves, pearl oysters are filter feeders, feed mainly on phytoplankters. The analysis of stomach content of pearl oysters showed the presence

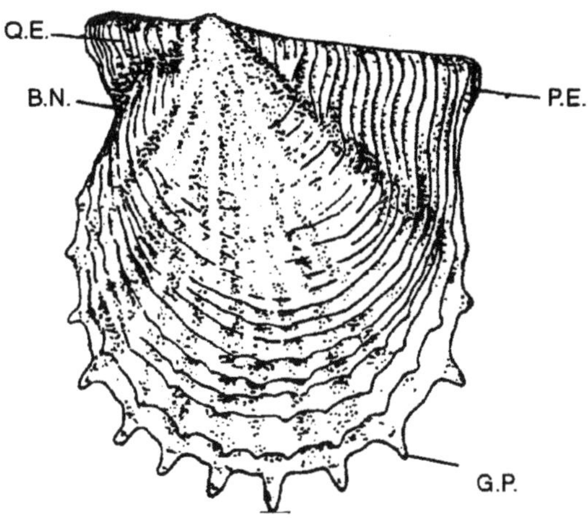

Fig. 7.3: External features of shell of the pearl oyster *Pinctada fucata*

A.E.	-	Anterior Ear	P.E.	-	Posterior Ear
B.N.	-	Byssal Notch	G.P.	-	Growth Process

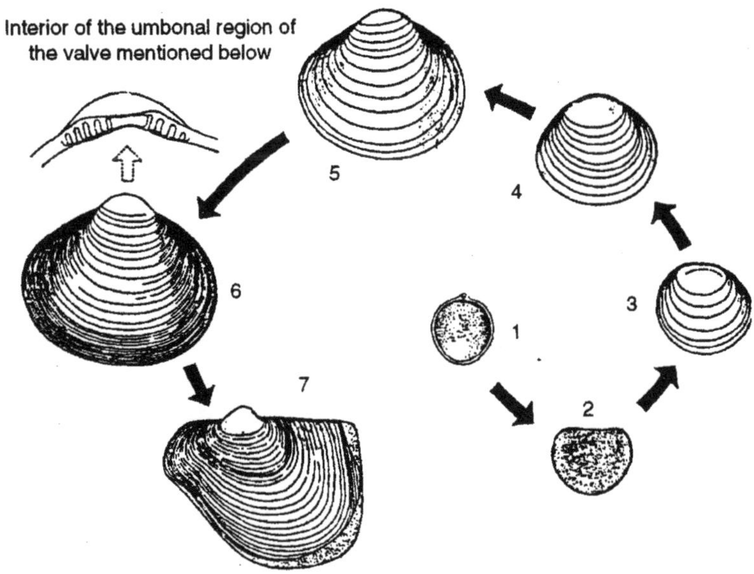

Fig. 7.4: Young stages of pearl oyster.

1. Just Fertilised Egg (41 x 53 u)
2. D Type Larva (63 x 80 u)
3. Umbo Stage (S.L. 115 u)
4. Umbo Stage (S.L. 135 u)
5. Umbo Stage (S.L. 225 u)
6. Full Grown Stage (S.L. 240 u)
7. Young (S.L. 1.1 mm)

of diatoms, flagellates, larvae of lamelibranchs, gastropods, nauplii of crustacenas, algal filaments and sand particles.

In pearl oysters, the sexes are separate and they do not exhibit sexual dimorphism. In Indian waters, 2 seasons of spawning, one in June-September and the other in November-December coinciding with the southwest and northeast monsoons observed in a year respectively. The matured oysters release their eggs and sperms in the water and the eggs get fertilized as soon as they come in contact with the sperms. The fertilized eggs pass through different larval stages and finally settle down into a spat. They grow fast during the first year and during subsequent years growth is slow.

Fertilization

Invariably males spawn first, which triggers the females to spawn in about 45 minutes. Soon after the eggs are released, they are fertilized by the sperms. The fertilized eggs measure 47.5 um in diameter and they settle at the bottom.

Early Development (Fig 7.5)

The first cell division takes place 45 minutes after fertilization. At two celled stage, a micromere and a macromere are developed. Macromere does not take part in further cell divisions whereas micromere divides repeatedly. Each micromere in the morula develops cilium. The morula swims and congregates at the surface. It exhibits phototrophism. By the reorientation of the cells, a blastocoel and a balstopore are formed. The stage is called blastula. This is reached 5 hours after fertilization. The cells convolute in and form dermal layers. Archenteron is formed. The gastrula stage is reached 7 hours after fertilization.

Gastrula transforms into trochophore by developing a long single flagellum and pre-oral tuft of cilia at the apical side and post-oral tuft of cilia at the rear side. The larvae swim with flagellum. The ectodermal cells secrete embryonic shell material and assume a `D' shape or veliger or straight hinge stage. At this stage, the flagellum and tufts of cilia disappear and new locomotory organ called velum is developed. The early embryonic development of the larva is completed by this stage. To reach this stage it takes 20 hours.

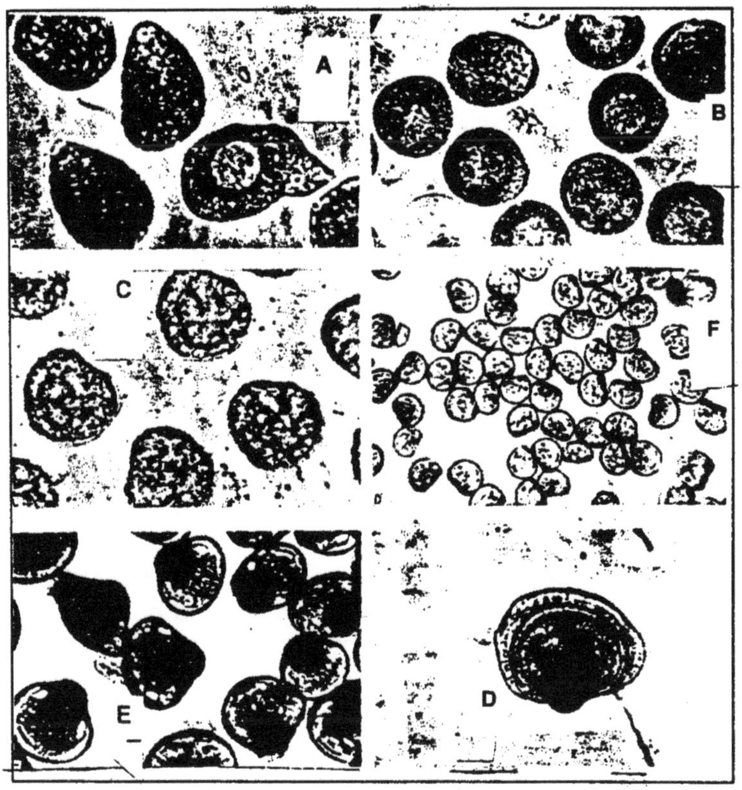

Fig. 7.5: Developmental stage of *Pinctada fucata*

A. Just Released Eggs B. Fertilised Eggs
C. Trochophore Larva D. Veliger Larva
E. Pediveliger Larva F. Plantigrade Stage

Umbo Stage

On 10-12th day, the veliger reaches umbo stage by the addition of further shell materials and measures 135x130 um. The straight hinge disappears and umbo region is formed. The shell valves are equal with mantle folds on the inner side. The larva swims with velum.

Eyed Stage

On 15th day umbo stage transforms into eyed stage. The eye spot is found at the base of the foot primordium. At this stage the larva measures 180x170 um. During the transitional stage from eyed to pediveliger, the larva has both velum and foot functional.

Pediveliger

The pediveliger stage is reached on 18th day. Pediveliger has a functional foot which helps in locomotion and in attaching to the substratum temporarily at the time of settlement. The velum disappears. The size of the pediveliger is 200x190 um.

Plantigrade Stage

The plantigrade stage is reached on 20th day by the growth of additional shell material all along the periphery. Byssal threads are produced for attachment. It measures 220x200 um.

Spat

The plantigrade stage metamorphoses into a spat by the formation of straight hinge line and anterior and posterior ears. Byssal notch is well developed. The size of spat is 300 um at the time of settlement.

Biology of Pearl Oyster

The pearl oyster's mantle is a thin fold of skin covering the soot body below the shell. It is formed of two lobes on either side. The two lobes are fused together dorsally below the hinge line, from this point they hang down on both sides of its body. The nacre or mother of pearl, found on the inner surface of the shell is formed by the secretion of the outer epithelial layer of the mantle. When some minute extraneous matter such as sand grain, silt particle, parasite larva etc., accidentally enters within the tissue of the oyster, it causes irrigation to the oyster. As a means of defence against the irritant, the oyster encloses the foreign particle in a sac of epithelial tissue. If the sac happens to be formed by the outer epithelial cells of the mantle, it secretes nacre which gets deposited around the particle. The sac is the pearl sac and the foreign particle will be the core material. The pearl sac keeps growing with the growth of the pearl inside. The secretion continues throughout the active life of the oyster leading to the formation of a natural pearl. The natural pearl generally takes the shape of the core material and is small in size and irregular in shape. The natural pearls are formed either within the mantle or in outer soft tissues of the oyster or between the mantle and the interior surface of the shell. Therefore, it is evident that the two pre-requisites for the formation of a pearl are the outer epithelium of the mantle lobe and a core substance or nucleus. Formation of a natural pearl is a matter of chance and, therefore, only a very small percentage of pearl oysters produce pearls in nature.

In the cultured pearl technique, the same process is manipulated in the oyster by man. A small piece of mantle from the donor oyster and a spherical shell bead, the nucleus, are implanted into the gonad of a recipient oyster in proper orientation by skillful surgery. It was found that cut pieces of the mantle epithelium would provide the pearl secreting cells and that processed shell beads would be accepted by the oyster as the foreign body. The surgery operation is called implanting or grafting the inner epithelium and connective tissue of the mantle disintegrate and become absorbed by the surrounding tissue. The outer epithelium of the mantle piece undergoes cells division and rearrange themselves over the shell bead nucleus forming the pearl sac. The epithelial cells of the pearl sac secrete and deposit nacre or mother of pearl over nucleus in the form of concentric micro-layers. When a large number of these layers have been deposited over a period of time, the pearl grows in size and attains its gem quality. Since the pearl is produced in the oyster by manipulation through surgery and by further cultivation of the seeded oyster, it is called the cultured pearl.

In cross section, a natural pearl will reveal a minute nucleus of extraneous origin surrounded by nacre. In a cultured pearl, the nucleus is large and the nacre forms only, one fifth to one third of the pearl in thickness.

Nacre (Fig 7.6)

The nacreous matter consists of thin alternate layers of aragonite and conchiolin deposited around the nucleus. The conchiolin is organic in nature and are made of mucopolysaccharides. It forms the binding layer for the aragonite crystals. The aragonite layers are 0.29 to 0.60 mm thick and are made of calcium carbonate in the form of highly laminate crystals.

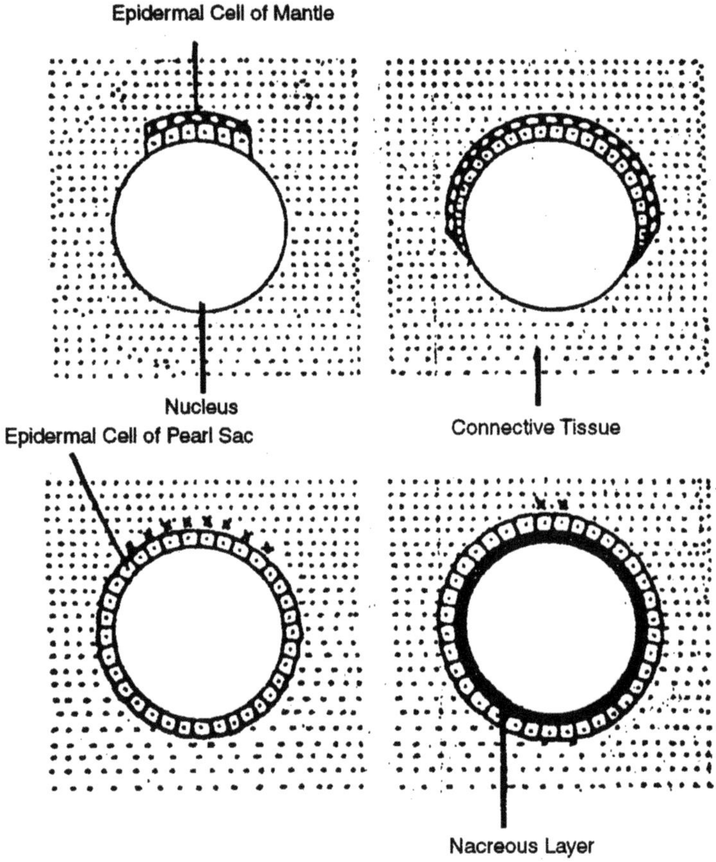

Fig. 7.6: Nacre secretion by the mantle layer encloses the nucleus to form a pearl.

In cultured pearls the nacre quality and the process of pearl formation are the same as in the formation of natural pearls. Cultured half pearls (blister pearl) are produced by affixing many nuclei on the inner surface of the shell valves. The outer epithelium of the mantle forms the pearl sac on the free surface of the nucleus and the half pearl is formed. Hydro-biological factors such as primary production, temperature, water current, trace metal content and inherited capabilities of the oyster constitute to the colour and quality of the pearl.

Oyster Farming

Selection of a suitable site for pearl oyster farming is very important. Due consideration should be given for the water inflow, primary productivity of the water, siltation, nature of sea bottom etc., to obtain good growth and production of quality pearls. Unusual fluctuation in the salinity and temperature, appearance of red tides, presence of hydrogen sulfide and industrial and domestic pollutants can affect the oysters and pearls. Sheltered bays and calm coastal waters where sufficient depth of water is available are suitable for pearl oyster farming.

In a pearl culture establishment, farming activities go round the year. Therefore, the establishment must have good laboratory for oyster surgery, pearl collection center, pearl grading and processing unit, farm stores, cage cleaning yard, mechanical workshop and boat for transport of workers, oyster and farm materials.

Pearl oyster farming can be divided into two phases, viz., mother oyster culture and post operative culture. Mother oyster culture refers to farming of oysters from the time they are brought to the farm till they are used for nucleus implantation. The post-operative culture refers to rearing of oysters from the day of implantation up to the day of harvest.

Oyster Sources

There are three ways to raise pearl oyster for mother oyster culture in the pearl culture farm. They are (1) collection of oysters from the natural beds (2) Hatchery production of seed and (3) collection of spat through spat collections.

In the Gulf of Manner, the pearl oysters occur in large numbers on the submerged rocky substrates known as paars. Oysters form these rocky beds are collected by skin and scuba divers. In the Gulf of Kutch, the pearl oysters are found sporadically on the intertidal reefs known as Khaddas and the collection is done by hand picking. There source from hatchery is more dependable than the sources from natural beds as the hatchery alone can ensure a sustained supply of pearl oysters for pearl culture throughout the year. The pearl oyster hatchery of CMFRI'has produced several million seeds of oysters over the last 13 years.

Hatchery Production of Spat

Success was achieved in the breeding and production of spat of *P. fucata* at the shellfish hatchery laboratory of CMFRI at Tuticorin in 1982 (Alagarswami *et al.*, 1983b). Ripe pearl oysters are held in groups of 25 in the room conditions of 25°C and fed with a mixture of micro-algae grown in outdoor tanks and spawning is induced by slowly raising and the water temperature to above 33°C. Spawning has also been induced by chemical methods (Alagarswami *et al.*, 1983a) such as transferring the oyster to sea water containing hydrogen peroxide (62.5% success), tris buffer (78% success) and injecting hydrogen peroxide at the base of the foot (48% success).

The larva is reared in tanks and the sea water (sand filtered) is changed on alternate days. *Isochrysis galbana* cultured in 100 litre tanks at 25°C under axenic conditions is given as food. The baby pearl oysters or spat settlement occurs in about 20 days with 20-30 per cent survival from the initial larval stock. The spat are fed with mixed microalgae and reach 3 mm size is about 2 months from spawning. The hatchery has capacity to produce one million spat per spawning run under favourable conditions.

Success was also achieved in the breeding and spat production of *P. margaritifera* (Alagarswami *et al.*, 1989). The oysters (or spat) obtained are cleaned and placed in cages and hung vertically in the sea from the raft/rack at desired depth.

Rearing Methods

Raft Culture

This method is one of the most suitable method of farming oysters in sheltered bay and open areas. A raft of the size 6 m x 5 m is found to be suitable. Rafts are usually constructed with teak, casuarina or eucalyptus poles of chosen length. The poles are lashed with coir ropes. Four floats are attached to the four corrners of the raft to give buoyancy. The oyster carrying cages are suspended from the raft at 5 m depth in the sea (Alagarswami, 1991).

Rack culture

This is possible only in shallow and calm seas of 2-4 m depth. In rack system a fixed structure, 121 teak or eucalyptus poles of chosen length are driven vertically into the sea bottom in 11 rows, 1 m apart. Cross and horizontal poles are arranged as per requirement on the top of the poles and lashed with coir ropes as a convenient height of 0.5m above the water level so that the rack thus erected remains always above the water. A total of 400 box-type cages holding oysters can be suspended from the wooden frames.

On Bottom Culture

On bottoms of coast areas culture of pearl oysters can be done in the places where the sea bottom is hard or rocky type. Oysters are placed in box-type cages and arranged on the sea bottom in rows.

On Shore Tank Culture

Pearl oysters can be successfully reared in a 50 litre capacity concrete tank filled with clean sea water. Mother oysters numbering 1250 can be successfully stocked and grown in the tank. The microalga *Chaetoceros calcitrans* cultured in the out door tanks is given as food for the oysters. This technology has greater advantages over the open sea culture system, as the growth of the oysters in the tanks is much faster, survival rate is high, predation and fouling are practically absent and ensures greater control over the environment.

Besides the above four methods, long lines and underwater platforms are used in some parts of the world. In the long line culture method, a series of spherical or cylindrical float are attached to strong ropes at uniform intervals. The ropes are marooned with anchors. The oyster cages are suspended from the strong rope of the line. Long lines are ideal for turbulent sea conditions. In another method of hanging, a hole is drilled near the hinge of pearl oyster. A small thread is put through the hole, which is then tied to a straw rope coated with tar. The straw ropes are hung from the raft. Underwater platforms are used in French Polynesia for farming black-lip pearl oysters in deep lagoons. Oysters in strings are suspended from these platforms (Alagarswami, 1991).

Rearing of Juveniles

Juvenile pearl oysters are reared in the cages. Net cages made of synthetic fabric of velon screen whose side are stretched in the form of a prism are used for rearing juveniles of the size of 3 to 20 mm. To prevent clogging by silt and other outgrowths, the velon screen net bags are replaced periodically. As the juveniles grow, number of juveniles in the cages are reduced to obtain fast growth.

Culture of Mother Oysters

Box cages, measuring 40x40x15 cm with a lid and covered by synthetic webbings are used to rear mother pearl oyster and nucleated oysters. The sizes of the mesh varies with the size of the oysters to be reared. The frames of the cages are made up of 6 mm mild steel rods, coated with anti corrosive paints or cool tar.

Environmental Conditions

The success of pearl oyster farming and pearl production depends largely on the environment in which the oyster are reared. The oyster being a filter feeder wholly depends on the environment for all its life functions such as osmoregulation, respiration, nutrition and reproduction. Many industries let out agro-chemical and industrial effluents into the sea which adversely affect the well being of the oyster population. Therefore, the set of environmental conditions have to be ideal and this should be ensured while selecting the site for the farm.

a. Topography

Sheltered bays are ideal locations for establishing pearl culture farms. They offer good protection to the culture structures such as rafts and cages from strong winds and waves. Shallow coastal waters where the sea is calm during most part of the year can also be considered as a suitable site. The Gulf of Mannar in India where pearl farming is practiced has been found to be moderately acceptable in the absence of better locations on the coast of mainland. The Andaman and Nicobar Islands have many areas which provide ideal conditions for pearl oyster farming. The lagoons of some islands of Lakshwadeep especially Bangaram, would seem to be favourable.

b. Hydrography

Depth is an important factor in rearing pearl oysters, the preferred depth is about 10 m in the open sea and 2-4 m in the sheltered bay. The minimum depth required is more than 5 m which is favourable for formation of high quality pearls of pinkish colour though the deposition of nacre is slow. Sea bottom should be either gravel or rocky. Muddy or sandy bottoms should be avoided. A high amount of silt in the farm water may choke the gills and affect the filtration efficiency of pearl oysters. A mild water current is necessary, not only as a source of oxygen rich water, but also to bring in fresh plankton on which the oysters feed, as well as for the removal of metabolic products and fecal matters from the farm. Repeated culture in the same ground often deteriorates the quality of the pearls. Accumulation of biological waste of the oysters and fouling organisms on the sea bottom often affect its physical and chemical state.

The environment factors such as temperature, salinity, hydrogen ion concentration and nutrient salts play a crucial role in the settlement, growth and reproductive pattern of oysters both in the farms and natural beds. Unlike in Japan, the variation in temperature and salinity is not much in the Gulf of Mannar. The lowest sea surface temperature is about 23.8°C and the highest 33.5°C, the monthly averages ranging from 25.9°C to 31.5°C. In the Gulf of Mannar and the Gulf of Kutch, oysters show a higher growth rate in winter months in the lowest range of temperature. Higher temperatures within the ambient range helps in faster growth of pearls. The salinity of the Gulf of Mannar normally varies from 30-35 ppm. If salinity falls below 15 ppm. and if such condition is progloned, it may lead to mortality. Pearl oysters seem to prefer high salinities but oysters raised in such waters produced pearls with a golden tint. This may happen during unusual heavy rain and heavy discharge of fresh water by rivers in the farm. The pH should be around 7-8. The rich nutrients discharged by rivers into the sea increases the productivity of water. The oysters probably derive their chief source of conchiolin from nitrogen substance of the plankton. The organic matter and calcium dissolved in the sea water are directly absorbed by the food consumption cells. In Tuticorin bay, it ranges from 316-485 mg/litre. The calcium passes through the mantle to be deposited on the surface of the shell or pearl in the process of their formation. The presence of trace elements in small quantities influences the colour of nacre. Gold and cream coloured pearls contain more copper and silver, whereas skin and pink coloured pearls contain more sodium and zinc (Matsue, 1958).

The farming area must be naturally sheltered against violent winds and waves. Tides and currents must be sufficient to change the water completely and frequently with oxygen and fresh plankton and carry away the excretory waste. If the water current is strong, the formation of pearl layer is fast but the quality of the pearl produced will be lower.

Biofouling, Boring and Predation of Pearl Oysters

During the course of the experiments in pearl culture since 1972 in the Gulf of Mannar, several problems were encountered in the farm due to biofouling, boring and predation of oyster stocks. Wada (1973) reported that animals and seaweeds settling on the oysters and baskets inhibited the growth not only of the oysters but also of the pearls.

Fouling Organisms

Barnacles, bryozoans and molluscs were the most numerous and are larger organisms in the foulding complex. Tunicates, decapod crustaceans, hydroids and anthozoans were the less significant ones. Biofouling organisms settle and grow on the shells and cages. Most of the fouling organisms are filter feeders and hence they are competitors for food and removal of them is a labour intensive process. Seasonal dominance is exhibited by the fouling organisms and they have to be removed periodically.

The barnacle *Balanus amphritrite* is one of the major fouling organisms. Heavy settlement of this causes physical obstruction to the opening and closing of the valves, if left uncleaned. The removal of these organisms sometimes, damage the shell margins and result in the recession of the shell growth.

Boring Organisms

These riddle through the shells making them weak and friable. They include the polychaetes, sponges, molluscs and isopods. *Polydora sp* causes simple and compound blisters on the inner side of the oyster shells. The sponge borers belonging to the species *Cliona* infests initially the umbo region and spreads to the surface of the valves. By heavy attack, the shells become extremely fragile and susceptible to further infestation and damage.

The oysters infested with boring polychaetes and sponges were treated using 1.0 per cent formalin which killed the borers.

Treatment with brine solutions having the salinity range between 42.3 per cent and 78 per cent for 7-21 hours is also proved effective to kill all polychaetes.

Predator Organisms

Besides the boring and fouling organisms the predators like *Murex* (molluscs), crabs and the puffer fish are a menace in the damaging the juveniles and mother oysters in the oyster farm. Large scale mortalities of pearl oysters have been reported from Japanese waters due to red tides and other causes (Chellam *et al.*, 1987 a,b,c).

Production of Cultured Pearls

After the oysters have grown in the farm and attained the size suitable for nuclear implantation, they are brought to the laboratory. The oysters which are collected from distant seas are first given a rest in shallow waters for a few weeks allowing them to recover from the stress of the journey. Their shells are cleaned of all fouling organisms and other encrustations and kept in sea water.

Pearl production can be divided into two phases:
- (i) The laboratory phase and
- (ii) Farm phase

The laboratory phase is very short, which includes selection of oysters, conditioning, graft tissue preparation, nucleus, surgery and convalescence. The farm phase include post-operative culture and pearl harvest.

Laboratory Phase

Selection of Oysters: The factors to be considered in the selection of oysters are their age, weight, stage of sexual maturity and overall health. Oysters of 25 g weight and above (1.5 to 2.0 years) are ideal and even smaller sizes are also considered for implanting smaller nuclei of 2 to 3 mm diameter. The gonad of the oyster should be in the spent resting stage. Nucleus implantation in these oysters yields good results since correct orientation of mantle piece with the nucleus is always ensured. Ripe gonads are not suitable, since during surgery the gametes tend to flow out and block the visibility of the implantation site so that proper orientation of the mantle piece and nucleus cannot be ensured. In addition, the oysters infected with polychetes, sponges and trematodes are to be avoided. In order to avoid any difficulty, the contents of the gonads, eggs or spermatozoa may sometimes be artificially removed from the mother shells prior to the operation.

In order to do this, mother shells are placed in a basket specially designed for this purpose at a very high density are hung in a deeper part of the sea than usual. Once a day basket is lifted up to the surface of the sea and the shells are checked to determine whether they have released their contents. They are stimulated to release the contents of gonads. Care must be taken not to weaken the mother shells during this treatment. After removal of the contents, treated shells should be kept under the favourable conditions for a few days to recuperate before the operation.

Conditioning

The selected oysters are conditioned for nucleus implantation. Preconditioning of oysters for surgery is an essential process. Healthy oysters, when taken outside the seawater close their valves tightly by contracting the adductor muscles. For surgery a gap of 1.0 to 1.5 cm between the valves is required. When the valves of the oysters are opened forcefully their adductor muscles get cut and the oysters die immediately. Hence the oysters are allowed to open their valves gently by themselves. This is done by using a chemical Menthol. The oysters are arranged in a plastic basin with their hinges pointing downwards. Seawater is slowly poured into the basin till the oysters get immersed in the water. Sprinkle a little amount of menthol powder over the water and cover it with a lid. In about 60-90 minutes the oysters get narcotized. The adductor muscle relaxes and the shell valves open. Immediately a small wooden peg is inserted in between the valves to keep them open and thus the oysters are conditioned. The conditioned oyster should be operated as early as possible, as the delay causes swelling of tissues, increased mucus secretion and mortality. For this reason conditioning is done in batches. The conditioned oysters are washed individually in seawater and arranged in empty plastic basin.

Surgical Instruments

A set of specially made surgical instruments are needed during tissue preparation and nucleus implantation. The instruments can be made to specification by any surgical instrument manufacturing company. The following are the necessary instruments: 1. Oyster stand, 2. shell speculum, 3. incising-cum-grating needles, 4. nucleus insertion needles, 5. graft cutting knife, 6. spatula, 7. needle hook, 8. forceps, 9. knife and 10. scissors. The other instruments needed for graft cutting wooden blocks, wooden pegs, camel hair brush, trays, rubber sponges, towels, glass beakers etc.

Preparation of Graft Tissue

As soon as the oysters are narcotized the preparation of graft tissue should commence. Preparation of graft tissue is an important step in surgery. Many graft tissue can be prepared from the mantle of a single oyster. A healthy oyster is taken from the stock of oysters. The donor oyster should not have been subjected to any narcotizing process. A sharp knife is inserted in between the 2 valves of the oyster up to the adductor muscle and the latter is cut vertically. The knife is pushed further down so as to cut the soft body into two. Care is taken not to injure the mantle tissue in any way. With a pair of scissors, a strip of 5 cm long and 0.5 cm wide mantle is cut and transferred on to a rectangular soft wooden block. The mucus of the mantle is removed gently with a blunt end of a scalpel. With the graft cutting knife, the thickened outer edge of the mantle is cut and removed. In the same manner, the inner muscular portion of the mantle on the opposite side also cut and removed. Now a long ribbon of 5 cm length is obtained from the pallial zone of the mantle. The ribbon is cut into pieces of 2 to 3 mm squares. Mantles from both the sides are used in the preparation of graft tissues. About 20-25 pieces can be cut from each ribbon. The graft tissues should be kept wet with sterilized filtered sea water all the time during preparation. Smearing a weak solution of Azumin will help to keep the cells alive for a longer duration. The pieces are sterilized with mercurochrome or an antibiotic solution to prevent the spread of disease to recipient pearl oysters. The place from where the mantle piece is cut will affect the quality and the colour of the pearls to be produced.

Precautions

1. Use only filleted, clean sterilized sea water for this operation.
2. All instruments should be washed in fresh water and sun dried.

3. Only clean and wet sponges are used for wiping the tissue.
4. Clean, smooth, moist wooden blocks or glass plates can be used.
5. The mantle of healthy oysters are most suitable. Shrunken mantles are difficult to handle.

Nucleus

A piece of graft when inserted into the gonad of an oysters results in the production of an irregular pearl. In order to get a larger and spherical pearl in a short period, a spherical shell bead known as nucleus is inserted along with the graft tissue. The beads are manufactured in Japan from the thick shells of fresh water mussel, pie toe, washboard, dove etc. The hardness and specific gravity of these nuclei are nearly identical with that of the deposited nacre. These shells are collected and exported to Japan, where they are cut, ground and processed into precisely spherical shell beads of different diameter generally 2-7 mm. In India, chank shells have been cut and processed into beads and used in experimental pearl production.

Surgery

The conditioned oyster is mounted on the oyster stand with the right valve upwards. The shell speculum is inserted in between the valves to regulate the gap. The mantle, gills and labial palps smoothly pushed aside to get a clear view of the gonad region. A sharp incision is made with incision-cum-grafting needle at the base of the foot of the oyster and a passage is cut subcutaneously through the gonad up to the predetermined site of implantation. In the case of single implantation, the site is close to the turn of the intestinal loop or within the gonad. In double implantation, a second site is chosen close to the hepatopancreas. In multiple implantation several other sites between the above two are selected. A piece of mantle is inserted through the passage with the incision-cum-grafting needle and the left at the site. Then the shell bead nucleus is inserted through the same passage and placed in such a way that it is in contact with the outer epithelium of the mantle piece. The incision is smoothened and the two margins of the incision come in contact. With this step, the nucleus implantation operation is complete.

Implanting of nuclei of 3-4 mm can done through the byssal groove. The byssal threads along with their base can be pulled out and a passage is made with the needle up to the gonad. The graft and the nucleus can be inserted through this passage. Care is taken not to make the passage deep into the muscle of the oyster. By this method of implantation, ejection of the nucleus is minimum.

Operations are conducted from spring to autumn when the water temperature is above 18°C. They should not be performed under water temperature conditions above 28°C or under conditions of low specific gravity of sea water, such as after a rainfall. (Alagarswami, 1987 a,b).

Convalescence

The operator oyster is then removed from the stand and the peg or shell speculum is removed from the valves. The oysters are then placed in a plastic tank, where seawater is allowed to flow gently. If no flow-through system is available sea water is to be changed frequently until the oyster fully recovers from the effect of narcotization. Within 30 minutes, the oysters slowly reopen their valves and commence their pumping and filtering activity. Some pearl culturists examine the pearl oysters individually after recuperation by fluoroscopy to check the condition of the inserted nucleus. Then the oysters are placed in baskets and hung in the one ton FRP tank containing fresh sea water. A mild flow of sea water is always good and this is ensured by aerating the sea water. The fecal matters are either washed away by mild current of kept away from the oyster. The operated oysters are kept in the laboratory for 2-3 days under observation. The dying oysters and the oysters with reject nuclei are removed from the basket. Within 2-3 days the incision wound heals completely.

Farm Phase

Post-operative Culture

After Convalescence, the oysters are placed in box cages, transferred to the farm and suspended from the raft/rack at 5m/3m depth in the sea. During the post-operation rearing period, the oyster density in culture cages or culture grounds should be kept at a minimum, over-crowding may cause adverse effects such as production of low quality pearls, slow formation of nacreous layer, etc. In a box cage, 50-75 oysters with size 40-45 mm (25 g wt) can be accommodated. The cages are numbered with plastic number plates. The oysters are reared in the farm without much disturbance. It is ideal if oysters are reared in areas of high phytoplankton production. Once in a month, the cages are lifted to remove the predators from the cages and the scrap off the epifauna from the outer shells and cages.

In another method of hanging, a hole is drilled near the hinge of a pearl oyster. A small thread is put though the hole, which is then tied to a straw rope coated with tar. The straw ropes are then hung from a raft.

In Indian waters, the deposition of nacre on nuclei is much faster than in sub-tropical and temperate waters. The duration of post-operative culture varies from 4-18 months depending on the size of nuclei inserted and the desired size of the pearls to be obtained. When a 3 mm nucleus is inserted in an oyster, it takes a minimum of 4 months for the pearl to attain maturity and it is 15-18 months for a 7 mm diameter nucleus.

Pearl Harvest

Harvesting of cultured pearls is usually carried out manually. The oysters are brought to the laboratory from the farm. With a sharp knife, the oyster's adductor muscle is cut and the pearl is squeezed out of the gonad region. In case the oysters need to be reused for a second time, the oyster's valves are gently opened without damaging the adductor muscle and the pearl is carefully removed with instruments. The oysters are then returned to the farm for recovery and after a certain length of time they can be operated for a second time to produce additional pearls. Pearl harvest is done usually during the cooler season of the year during which time the pearl coating is thin and fine.

Cleaning of Pearls

After harvest, the pearls are cleaned with distilled water and given a salt wash. The pearls are mixed with powdered salt in equal volume and placed in a tube with small amount of water. Then the pearls are taken out and cleaned with distilled water. The residual mucus on the surface of the pearl is removed by rubbing with salt to obtain good luster. The cleaned pearls are sieved to size (mm) using gauging sieves. They are then sorted by size, shape, colour, luster, surface quality and flaws.

Colour of Pearls

The colour of cultured pearls largely follows the colour of the nacre of the shell of the pearl oyster which produces the pearl and is genetically determined. Besides this, the physiological condition of the oyster, the nature of the culture ground, depth, light penetration, feed, water quality, minerals and trace elements in the sea water also determine the pearl colour to some extent. The Indian pearls show diversity in colours. In the Gulf of Mannar, the pearls produced are largely yellow in colour followed by white, ivory white, cream, grey silver and light pink. Indian preference is mainly for yellow, golden yellow and white colours. Europeans prefer pink coloured pearls followed by white and cream coloured ones. In the USA the preference is for rose, pink and cream coloured ones. In general, the demand for white cream coloured pearls is always good.

Other pearls colours produced are yellowish pink, green, green pink, blue, steel black and black. Pearls produced by *Pinctada maxima* are silver white, *P. margaritifera* black or steel gray, abalone green, and fresh water mussels produce pink coloured pearls. The site at which the nucleus has been implanted also determines the colour. In *P. fucata* while the pearls produced in the ventral region of the gonad are white or golden, whose place was in proximity to the hepatopancreas are grey or white.

Matusui 1958 found that the golden and cream coloured pearls contain more copper and silver, skin coloured and pink pearls contain more sodium and zinc and that gold pearl contains more metal elements than the green ones. The colour varies according to the amount of porphyrins and metalloporphyrins present in the pearls. The organic substances deposited at the beginning of pearl formation also would decide the colour. The good quality blue pearls are of this origin. Above all the granulate and laminar structure of nacre produces the iridescence of pearls.

Quality of Pearls

Gross production of pearls obtained during harvest comprises of finest pearls. Some of the pearls may be of outstanding colour and perfectly round shape, many are inferior, some are totally valueless as gems or jewels, in some cases the oyster would have only the nuclei or would not be present. Such composition is common to pearl culture anywhere in the world. The economical success of pearls culture depends on the percentage composition of the various categories.

Pearls are graded into 3 categories:

A–Grade: Flawless, one flaw, small flaws, small stain, pink, silver or light cream.

Further categorised into

A–1: Top pearls-perfectly round, pink flawless and lustrous, but may include those with small blemishes about the size of a pin point.

A–2: First class pearls with slightly larger pits and protruberances once treated will become indistinguishable from top pearls.

B–Grade: Fairly large flaws, stains, cream colour and irregular shape.

C–Grade: Trash pearls, wild shaped, bad coated, heavily pock-marked, clayey lumps, half good and half bad.

Pearl Processing

It is rather a common practice of the trade to improve the quality of cultured pearls through processing. The pearls that need processing are drilled for holes using specialised drilling machines. The drilled pearls are treated with dilute hydrogen peroxide which not only removes the organic impurities but also bleaches the pearls.

Quality Improvement of Pearls

For obtaining good quality pearls, the oysters should be grown at depths of 5-10 m. Strong sunlight on oysters must be avoided since sunlight can induce nacre secreting cells to produce calcite crystals to form prismatic layer over the nucleus resulting in poor quality of the pearl.

Quality is also genetically controlled. The cultures ground, the water quality, the feed and physiological condition of oyster also determines the quality to some extent. Graft tissue preparation is also an important factor. The tissues must be taken from the central portion of the mantle of a healthy oyster. Now a days in Japan, various chemicals and medicines are used to condition the oysters and make them grow healthy. Food dyes are also used in seeding operation to colour the graft tissue, activators are used to activate the growth of pearls sac and growth promoters are also used to accelerate growth (Daniel and Durairaj, 1993).

Modern Trends

Fresh Water Pearl Culture

The Indian pearl industry is all set to get a new shine. An indigenous technology for growing pearls in fresh water mussels. *Lamellidens corrianus* and *L. marginalis* is now available in the country. The technology, when adopted widely, can help to produce pearls in quantities which besides meeting the demands of local jewellery makers, it might also leave the country with surplus pearls for export.

Janaki Ram, senior scientist at CIFA, Bhubaneshwar, who was the first to develop this technology says, "it was a challenge for me as China and Japan, which are the only nations producing fresh water pearls, have kept the technology a closely guarded secret". He was confident that the experiments he has conducted and the technology he has developed will help and break the stronghold of China and Japan over the world pearl market.

Pearl culture is nothing new to India. For instance, CMFRI scientists have been at it since late 1970s. Sharma (1994) have perfected the technology of inducing pearls in many pearl oysters (*Pinctada fucata*) besides the techniques of breeding them. But Janaki Ram's achievement will move the pearl industry which for ages has been confined to the coastal towns, away from the seas. Now, farmers can venture into producing pearls in small ponds in their farms. If that seems impossible, Ajay Kumar Sonkar, an enterprising youth who learnt the techniques of fresh water pearl culture from Janaki Ram has not only repeated the experiments in his farms at Allahabad, at least 1500 km away from the seas, but using innovative methods has brought down the costs of investments too. He has demonstrated that pearl culturing is a profitable job. In just a year he has earned back most of the money he spent on pearl culture.

Sonkar is planning to sell this technology through the Biotechnology Consortium Limited, New Delhi. At CIFA, Janaki

Ram is also busy at finding an alternative to the imported nucleus. He is also planning to find ways of growing coloured pearls. He thinks it might be possible to feed the animal with specific chemicals, say, metal salts, which will make it to secrete coloured nacre. It would mean going a step further than nature itself.

Giant Size Pearl Production

Working in his backyard fresh water pond, Mr. Ajay Kumar Sonkar of Allahabad has perfected a new technique for creating giant sized cultured pearls. American pearl experts believe that Sonkar's technique could result in pearls as large as 24-25 mm in diameter. Sonkars break through rests on his ability to create nuclei of 22 mm which can be implanted into a pearl oyster to form a perfectly round pearl. As of today, the largest nuclei are about 18 mm in diameter which produce pearls of 20-21 mm diameter.

Sonkar presented a paper on his work on developing large sized pearls at an international conference held at Honolulu, Hawaii, USA in 1994. The paper dealt with the technique of making his own nuclei and using them to cultivate pearls from fresh water oysters, occurring in his back water pond. Sonkar has borrowed a few tricks from medicine to prevent rejection of nuclei by pearl oysters. He introduced steroids and antibiotics in the oysters and then inserted growth hormones to encourage their growth. He has now filed an application for a patent for his breakthrough invention (Joshi, 1997).

Sea Ranching of Pearl Oysters

Sea ranching of laboratory/captive reared organisms/ animals is a technique which aims at rebuilding the wild population from its destruction/catastrophe by man-made and natural causes. The main aim of sea ranching pearl oysters in their natural habitat is to revive the pearl oyster population from extinction and create new beds. The main difference between land-rancher and sea-rancher is that the land-rancher retains ownership of his animals whereas the sea rancher cannot possess the ownership of the animals since they are ranched in the sea. The ranched animals become part of the wealth of the sea which can be harvested by anybody (Victor, 1993).

The development of technology for mass production of pearl oyster spat under controlled condition in 1981 (Alagarswami *et al.*, 1983 a,b) has opened the possibility of sea ranching pearl oysters. An experimental sea ranching was commenced in December, 1985, in the pearl oyster beds of Tuticorin waters.

The wide spread mortality of young and old pearl oysters in the pearl banks may be due to certain physical causes such as shifting of sand, strong current, destruction by natural enemies etc. Factors like over fishing, over crowding of oyster and diseases may also be responsible for the depletion of pearl oyster population. The predatory fishes such as *Balistes sp* and *Serrance sp.* destroy the oyster population. Many consider the starfish, *Pentaceros lincki* as the greatest enemy of pearl oysters and suggest for their removal from the beds. Moray seals and octopus are also considered as the destructive agents of pearl oysters. Replenishment of natural stock on the beds is possible, if only mass settlement of spat take place every year.

For the conservation of pearl oyster in the natural bed, scientists suggested transplantation of young 'strides' or brood of oysters from unproductive paars to the places where better growing conditions prevail.

Mode of Ranching: The spat to be sea ranched are kept in the hatchery tank and spat are allowed to settle on synthetic materials like old fish nest, velon screen fabrics and tufts of monofilaments. These materials with the spat were placed in large rectangular cages (90 x 60 x 15 cm) covered with synthetic webbing. The cages were further enclosed with old fishnets. The spat could crawl out through the meshes for dispersal on the paar. The spat-filled cages were lowered and kept inside hollows or secured coastline projections with synthetic ropes to prevent drifting.

Pearl Trade

Black Pearls of French Polynesia

The Tuamotus Archipelago is made up of 2 mountainous islands and 76 atolls. Each atoll traps and holds a piece of ocean, a lagoon that acts like a giant soup tureen for plankton. The lagoons also protect and nourish *Pinctada margaritifera*, the black lipped pearl oyster. Here Robert Wan, known as the king of pearls, owns a company Tahiti Pearls, which is the largest producer of pearls in French Polynesia.

This black pearl oyster of French Polynesia cultured in Robert Wan's company is four times the size of Japanese Akoya pearl oyster, which produces most of the pearls in the world. Black pearls are rarer, thicker, and bigger. They are also richer in orient, the reflection of light beneath the surface of the pearl. Other colours such a silver grey, white and green (called as Peacocks) are also reared.

In the lagoons, the oysters act as barometers, registering the health of their environment. When the lagoon declines, so do the oysters, they require temperatures of about 75°F and clear, unpolluted water. Most Japanese oysters are maricultured, born in a hatchery where egg and sperms are artificially combined. But in the Tuamotus, the lagoon is the true parent of the oysters, which spawn naturally.

In Wan's farm, they use plastic garlands suspended in the lagoon to provide an anchorage for the drifting pinhead size larvae. In a few months each garland is choked with little oysters, which soon grow to the size of a silver dollar. At 6 months the oysters are placed in hanging baskets where they grow for another year and a half. Then the oysters are removed and wedged open, one by one. With surgical precision implantation of a piece of mantle on a nucleus (a bead carved from the shell of an American fresh water mussel) near the gonad is done. The mantle tissue forms a sac which secretes the nacre to form a pearl. After surgery the oysters are returned to the lagoon in 3 years the pearls will be ready to harvest.

After the May harvest, thousands of gleaming black pearls are got. These are graded for quality, then sold in the USA, Europe and Japan. In 1996 French Polynesia exported a million pearls, worth 140 million dollars. More than half came from Wan's operation. The rest came from 500 other farms throughout Polynesia. Wan's biggest pearl measures 21 mm got from his farm on Nengonengo (Doubilet, 1997).

Australia's Magnificent Pearls

In the global gem trade, one type, the Australian South Sea Pearl, outshines all. Plucked from giant oysters in tropical waters off northwestern Australia, they fuel an industry that has grown six-fold in the last decade (Raffaele, 1997). Here the great tropical bivalves called *Pinctada maxima*, nourished by nutrient-rich tides, produce large, lustrous pearls.

Pinctada maxima, known in Australia as the silver lip and in Papua New Guinea as the gold lip, is cultivated in a wide belt of warm water stretching from Burma to Australia and are called the South Sea pearls. Australian South Sea pearls have recently emerged as the Rolls-Royces of the pearling industry, rare, robust, enormous (10-20 mm in diameter compared with the 2-10 mm range of Japanese cultured pearls) and are expensive. North west Australia has proved to be a perfect place to raise a pearl oyster where there is little industrialization, few people, no fertilizer run-off from agriculture and tides as high as 33 feet.

With only a handful of farms, Australia is producing 60-70 per cent of the world large South sea pearls. But Australian oysters are not yet conceived on farms as Japanese akoyas are. Young oysters must still be found in the wild and gathered by divers on the flat sea bottom off Australia's north west coastline. Most of the harvest of about 3000 pearls are white, the kind most desired by jewelers but some are tinted gold, pink or blue.

Current Status of Marketing

The A grade pearls are sold at the rate of Rs. 150-200 per carat, 'B' grade Rs. 100-150 per carat, and 'C' grade Rs. 50-100 per carat. The first batch of free, spherical cultured pearls were produced at the Central Marine Fisheries Research Institute, Kochi (Alagarswami, 1974). In a recent study conducted by CMFRI at Valinokkam (Tamil Nadu, Southeast coast) by scaling up the operations, out of a total of 9414 oysters implanted (single implantation with 3-5 mm nuclei) mortality during one year post-operative culture was 2108 oysters (22.39 per cent). On harvest, the remaining 7306 oysters have yielded 1849 pearls (25.31 per cent). A total of 5457 oysters (74.89 per cent) did not contain pearls due to rejection of nuclei or non-deposition of nacre. This variation in the production of pearls is indicative of the range of production under variable conditions.

Based on the technology developed by CMFRI the Tamil Nadu Fisheries Development Corporation (TNFDC) is operating a pearl culture project at Krusadai since 1990 and has produced more than 3 kg of pearls and marketed them. Since the revival of the natural stock of pearl oyster in the Gulf of Mannar pearl banks is uncertain, to meet the requirements of the commercial venture, TNFDC has set up a hatchery in 1992.

Future Scope

In recent times, with the Indian cultured pearls available in the market, there is increasing awareness among the people about the technology and product. It will not be long before the massage "Pearls give the highest profit return of all marine products cultivated in coastal waters" (Wada, 1973) reaches the entrepreneurs and farmers, venturing into pearl culture.

At CMFRI, priority is given for short range projects on location testing of pearl culture at several places along the Indian coast. R and D efforts, already directed to take the technology to the grass root level as is being done at Valinokkam, will be strengthened. While the programmes on the upgradation of the technology in *P. fucata* would continue, pearl production in *P. margaritifera* would receive attention. Also plans are underway to initiate a project on *P. fucata* mantle epithelium culture, leading to in vitro production of pearls of desired colour.

The world production of marine pearls accounted for 78 tons at US $ 1042 million. Japan still holds the monopoly in the production of sea pearls. Even though India has the distinction of producing cultured pearls in 1973 itself, it could not produce cultured pearls for world trade. India has ample scope to develop and expand the cultured pearl industry in different locations, mainly in the Gulf of Mannar and Andaman and Nicobar Islands. There is good scope for privatisation in the field, if firms which have technical and marketing collaboration with foreign countries are willing to come forward. Pearl culture is long-term investment and huge profits can be made in a successful culture operation as there is still a great demand for pearls. The pearl industry is a biology based industry and is risk-prone as any other marine industry. India can also join with other leading pearl producing countries, if sincere and intensive efforts are put in the right direction.

Chapter 8
Planktonology

General Aspects

The aquatic systems like a pond, lake, river, estuary or the ocean contain a microscopic little world in it. The water of these lotic and lenthic systems also consists of inorganic and organic minerals in the form of suspended, dissolved or settled matter at the bottom. Decomposers such as bacteria and fungi and plants and animals both microscopic and macroscopic are encountered. All theses different components of the water body together constitute an ecosystem.

The life in the sea can be divided into three basic categories: the benthos, the nekton and the plankton. Benthos is the term given to the life on or in the sea bottom. Example are the starfish and sponges. Nekton includes the large swimming animals which can swim sufficiently to be able to control their movements irrespective of wind and currents. Example are whales, large squid and fish such as tuna, wahoo and sharks. Included in the plankton are all the animals and plants which live in the water but drift about at the mercy of wind and currents. The word plankton originates from Greek meaning wandering. Plankton is a collective term used to cover all those organisms whose power of movement are insufficient to prevent them being moved out by currents and waves, plankton means 'that which drifts'. Henson first coined the term. A single organism in the plankton is known as a 'plankter'.

The seas form some 70 per cent of the worlds surface and have an average depth of about 3800 m (12,500 ft). Plankton can be found in the three dimensions of the sea: the length, the breadth and the depth. Consider the volume of the seas is about 325,000,000 cubic miles. One cubic mile is equal to about 4,173,281,000 cubic metres. Each cubic metres have in it from about 400 (in open seas) to 4000 (coastal waters) animals above 1/4 mm up to 5 mm in size. This will give some idea of the vast numbers of living organisms which form the plankton. The plankton are largely grouped into phytoplankton (plants), zooplankton (animals) and the third group include all the microplankton. It is composed of both plants and animals below certain size of about less than 25 μ (Note 1 μ is = 1/1000 of a millimeter). Due to the wandering character the plankton are scattered throughout the pelagic zone of the oceans. It has been found that there exists a regular system of ocean currents throughout the world. The difference in the physical properties of water masses, as cold and dense water at the polar regions, warm waters of varying densities at the equatorial region, are sufficient to initiate and maintain deep water circulation.

Plankton Environment

The environment of marine plankton is the sea. Primarily sea water contained more than 46 to 50 essential and non-essential elements dissolved in it. Most of these elements are present in so small quantities that it is impossible to detect them by our routine spectrometric methods of analysis. Certain salts, especially sodium chloride are present in large quantities. Chemicals can be present in various states such as ions, colloidal state and particulate matter in suspension etc.

The basis of much analytic work on the sea is the fact that the major constituents of the sea water present in the same proportions over very wide areas of the sea. There are exceptions, of course such as those regions where land drainage, bringing down to the sea such compounds as calcium and magnesium salts, upsets the relationship and uniform composition of the constituents.

Salinity can be regarded as the total amount of dissolved salts in grams, contained in one litre of sea water. Salinity is measured indirectly by its chlorinity or by its conductivity.

The sea in areas of river mouth receive ingress of water and due to dilution, low salinity will be recorded. In an area which is under water at high tide but cut of from the sea at low tide, will have a high salinity due to evaporation by sun. Based on the capacity to tolerate the salinity the euryhaline and stenohaline organisms inhabit such waters. Other factors such as temperature, viscosity, pressure, upwelling, mass sinking, eddy conduction and stratification are important factors influence the plankton population in the sea water.

Classification

The plankton can be classified into different ways:

1. First of all the plankton can be grouped under two heads, 'euplankton' (true plankton) and the pseudoplankton. The latter often appears to the naked eye to be no different from the living animals and plants, but it consists of dead plankters and non-living debris.
2. Plankters may be classified on the basis of their nutritional requirements as 'phytoplankton' and 'zooplankton'. The phytoplankton proper consists of chlorophyll bearing plants, which are therefore capable of performing photosynthesis. The zooplankton consists of those plankters with a holozoic nutrition and it includes all of the planktonic animals.
3. According to the size plankton are classified as follows:

Megaplankton	>	10 mm	Only animals, e.g. large jelly fishes, euphasids, mysids, salps and corethra
Macroplankton	>	1-10 m	Larvae visible to naked eyes.
Mesoplankton	>	0.5-1 mm	Large diatoms, small zooplankton and many larvae
Microplankton	>	60 μm	Most animals and a large number of diatoms and dinoflagellates among the plant species
Nanoplankton	>	5 μm	e.g. Most plant species especially the Coccolithophores, the Silico-flagellates, diatoms and unicellular algae
Ultra plankton	>	5 μm	Bacteria and small flagellates

4. Again, plankters may be classified with regard to their environments. The limnoplankton are the plankton of

lakes, the rheoplankton are from the rivers, the haliplankton are the plankton of salt water, and the hypomyroplankton are the plankton of brackish water. The hypoplankton consists of plankters living near the bottom. The epiplankton is the plankton living in the euphotic zone, and thus includes most of the phytoplankters and many animals. The bathy plankton lives in the aphotic zone, and, therefore, contains no living phytoplankters proper.

5. Furthermore, the plankton can be classified according to the life history of the organisms involved. Animals whose life history is completed without ever leaving their planktonic life are called holoplankton. On the other hand, many important and characteristic plankters are only temporary members of the plankton community. These constitute the meroplankton (e.g. Balanus).

Distribution of Plankton

The wide distribution of many plankton species is one of the most striking features encountered in the study of planktonology. The wide distribution of phytoplankters and their frequent abundance, accounts for the great importance of the phytoplankton as a major basic food source in the food web of aquatic systems. It is generally stated that the production of phytoplankton is higher in high latitudes in the ocean that in temperate regions, and that the production in tropical seas is always higher.

The nutritive substances chiefly determine the abundance of plankton at any place. The foundation of any community is the proper assemblage of the inorganic substances in combination with sunlight and photosynthetic organisms. The upper illuminated waters are most richly populated and densities of population show some sort of stratification with respect to distance from land. But currents carry nutritive substances, algae and animal plankton, from place to place. Nutritive materials sink to the depths for temporary loss, and much of this is brought back into illuminated waters along shore or somewhere out at sea through currents and upwelling processes.

For various reasons, littoral and neritic waters are generally richest in plankton. Organic and inorganic nutritive materials drained from the land and enrich the littoral waters. The algae growing in the littoral regions after death undergo decomposition or broken up into fragments by the wave action and form detritus that is carried by the water as food for animals of benthos and zooplankton. Little portion of this gets mixed into the open sea. Along the coasts and away from them as well, abundant development of plankton may be expected where ascending currents brings up nutritive material. Where water descends from the surface into the depth, plankton in the upper depleted waters will be usually scarce. When warm and cold currents come together and mix, the rising temperature may promote the growth and reproduction of algae, and of their predators, utilize more rapidly the materials carried in the cold waters.

The physical conditions of temperature and viscosity have much to do with the geographic and depth distribution of particular kinds of plankton regions. There are species tolerant only of cold and others only of warm waters. Such animals as can endure only a narrow range in either the upper or lower part of temperature scale are called stenothermal. Others are eurythermal, with apparent indifference to changes of temperature.

There is great diversity in the composition of plankton, not only in different regions and at different depths, but also at different seasons, and at different hours at the same place and depth and probably even in different years. The conditions that lead to maxima and minima, as well as to minor fluctuations of abundance of any particular plant or animal, are complex indeed in their physical, chemical and biological aspects. Just one of these conditions is migratory habit of many plankters. The power of locomotion of such small organisms are slight compared with those of larger animals in respect of horizontal movements over broad areas. The locomotion of individuals may result locally in the formation of dense schools. Most notable are the vertical migrations, which are generally toward the surface when the light is dim and away from it when the light increases.

The plankton communities are always in the process of change. At any spot changes may occur in the kind of plants or animals that are present and in the relative and absolute number of the several species. There are seasonal changes as the prawn, crabs, worms, starfishes, molluscs and other benthic animals rise to constitute for a time prominent elements of the drifting populations. Among the biological conditions causing changes in plankton population, is the work of predators. Increase of algal density promotes the growth and reproduction of small animals plankters, which in turn are devoured by larger predators. Marine plankton live generally in a moving home, whether the movements are rhythmic and reversible tidal currents of harbors or the progressive currents, drifts and eddies of wide ocean.

Indicator Organisms

An indicator species is the one chosen to represent conditions in an ecosystems either because it is especially sensitive to changes in the environment or because as an indicator of ecosystem conditions, it is particularly easy to work. An indicator species might be used in management to monitor the effects of pollution or other degradation of an environment by comparing the present standing crop of the species to the standing crop during the previous disturbed conditions, or to some specific standard. The indicator species system was based on the law that any organism can only survive for long periods where its ecological expectation is met by the ecological supply to a substantial degree. The best indicators are the micro-organisms which directly utilize the substances present in the water. The presence of any species whose environmental limits are clearly understood could be used as an indicator. For most reliable use of indicator species the following pre-requisites are essential.

1. Exact identification of the species or varieties of organisms involved must be possible.
2. Their propagative and sterile distribution must be satisfactorily known.
3. The morphological stages of development in the life history must be known for the species or must be deducible from what is known of the group to which they belong. In this connection the time and depth at which spawning occurs and the duration of the various stages are important.
4. It would be known to what extent ecological adjustments can be made by migrations into deeper or

shallower water in the extended geographical distribution. For e.g. *Uaneus hyperboreus* and *Parthemisto oblivia*, arctic species, may live in deep cold water in lower latitudes and by vertical migrations they may enter deeper waters and they may also found in surface layers.

5. A clear distinction must be made between animals brought in from outside geographical areas and those of biological groups that develop locally when conditions stimulate those of the outside geographical area, specially with respect to temperature.
6. Associations of organisms transported from a common centre of distribution are of more value that occasional strays. Benthic species and communities have often been regarded as being the best indicators of organic pollution because of their constant presence, relatively long lives, sedentary habits and differing tolerances of stress.

a. Plankton as Indicators

Based on the knowledge of the distribution pattern of the plankton the quality of the environment can be assessed. It is commonly observed that the surface distributions of phytoplankton and of zooplankton are often patchy in character. It is observed that great swarms of zooplankton may consume the phytoplankton to such an extent that they create discontinuous distribution of plankton. The patchy distribution of phytoplankton may also be due to the differential rates of multiplication which might produce similar effects, the fast growing phytoplankton producing dense blooms where they are more abundant in water at a favourable time slower growing zooplankton might graze upon this, and by the time the population became heavy, the bloom of phytoplankton would have passed its peak because of exhaustion of surface nutrients. It seems probable that these mechanisms may play a role in producing the discontinuities at various times and places.

Fishes have specific taste for certain planktonic organisms. The specificity in feeding effects to a large extent related to the life and movement of various species of fishes. Therefore, the occurrence of a particular plankton species become the indicator to the availability of fish having specific preference for it.

In Indian waters, the appearance of sardines and mackerel has been correlated respectively with the swarms of diatoms, *Fragellaria oceanica* and Cladocerans, *Evadne tergestina* and *Penilia avirostris*. The source of protein, oil, vitamin-A etc. in a fish depends upon the richness of the plankton on which it feeds.

Hardy and co-workers devised a very simple plankton indicator for herring fishermen to aid them in locating areas where zooplankton especially Calanus are found in large numbers. It had been found that herrings are usually found where there are clouds of Calanus, and tend to avoid phytoplankton bloom. The indicator consisted of a cloth disc that was towed behind the fishing boat for a given length of time, and then examined. If it showed a reddish colour as a result of having retained many of the red copepods, the area was judged to be less favourable. Boats making use of the plankton indicator reported consistently higher catches than those which did not use plankton indicators.

At times this drifting life is so abundant that it colours the sea for many kilometres around. Such expressions as red water, yellow water and green water are used by fisherman who become expert to decide where to shoot their Mackerel or pilchard nets by slight differences in the tint of water that lands man would not notice. In every case the characteristic tinge given to the sea is known to be due to the presence of certain kinds of organisms in innumerable quantities. Darwin in his "Voyage of a Naturalist" noted such discoloration and mentions passing through two patches of reddish-coloured water. "Stinking Water" is another expression used by fisherman and in this case a characteristic odour is given to the sea also by the presence of hordes of certain organisms. A very large number of these small drifting creatures are capable of emitting a phosphorescent light. It is these that make the sea sparkle with little glowing points of fire when we observe the water on a dark light.

b. The Plankton as a Community

The assumption that plankton is scattered randomly throughout the sea is false. Similarly an assumption the plankton is scattered throughout any given system is also false. In a similar way to land fauna planktonic animals can be linked with the environment that is, the conditions in which they are living and also with other types of plankton animals which co-habit their particular system of water. Some species have a wide distribution throughout the greater part of the world, while others are very restricted.

The diatoms, dinoflagellates and small plants, with their chlorophyll and the sun's energy, synthesize simple compounds to make food in a form which can be consumed by animals. On these plants with their organic matter and food reserves, will feed the herbivores; these include many crustaceans and molluscans. Two the special features of the plankton community are the relatively high numbers of omnivorous animals and the filter feeders. Filter feeding is virtually impossible on land, but in the sea or in fresh water, where numerous tiny organisms are distributed throughout the three dimensions, it is an efficient means of getting food. It is perhaps developed to highest degree in the bottom living fauna. However, some larvae and crustaceans are considered as efficient filter feeders in the aquatic systems. Carnivores are numerous and constitute about 5 per cent of the number of the zooplankton. These include, chaetognaths, medusae and some crustaceans. Planktonic carnivores are usually larger than the herbivores on which they prey.

Although many plankton animals feed on dead materials if available, otherwise feed on live ones, because the dead animals sink to the bottom and are not available. Among the bottom dwelling forms there is high a percentage of scavengers. The colourless flagellates, yeasts, and bacteria form a large percentage of the nanoplankton.

c. *Temporary and Permanent Plankton*

These two divisions of the plankton are based on the life history of the animals and plants. The common rock-lobster of the tropics, *Panulirus* is quite obviously not a plankton animal but its larva the so called phyllosoma larva is well adapted to plaktonic life. This larva belong to the temporary class of plankton or meroplankton, since only part of their life cycle is spent in the plankton stage. Other examples of temporary plankton are larval stages of many starfish, worms, bottom living fish, crabs, and prawns. Other examples of this group which live on or in the bottom by day and ascend into plankton by night are some ostracods, some prawn larvae and species of Leptochela.

The permanent plankton or holoplankton consists of those animals which spend all their life in the plankton. These include pelagic copepods, chaetognaths, some medusae and pteropods. Shallow tropical waters support both large numbers and an extraordinary variety of temporary plankton whereas deeper oceanic waters contain relatively few.

Buoyancy Mechanisms

As stated earlier plankton is a drifting life as against swimming life. It happens that living matter is generally denser than water. It follows that any matter denser than water and drifting along in the water will sink to the bottom. This would be a problem to planktonic life, so various mechanisms are evolved to maintain them to remain suspended in the water. With the high salinity values density of the water also increases. There will be gradations of density from one to another type of water in the oceans. Hence the less saline water will have low density and in such waters, there will be great difficulties for the plankton organisms to remain in afloat. The temperature of the water is also important. There are different mechanisms by which the organism overcome the problems. One of the method is to make the body fluid of the organism less dense than the surrounding sea water. Two examples from tropical water, are *Noctiluca,* a dinoflagellate contains principally a solution of ammonium chloride within its cell wall. Now a solution of ammonium chloride which is isoosmotic with sea water is less dense (Sp. Gr. 1.01) than sea water (Sp. Gr. 1.025) and so gives buoyancy. While there is conclusive evidence that *Noctiluca* regulates its ammonium chloride level to some extent and the vertical movements are controlled with this.

The development of special gas filled sac is also seen in some animals, the swim bladder possessed by many fishes is an example. Many Siphonophores develop such a sac. During breeding period the *Tretomphalus,* a foraminiferan, develops a terminal large chamber filled with gas. An aid to the buoyancy is associated with the chemical process of the body, is excretion or exclusion of heavy sulfate ions. Soft bodied animals such as salps, heteropds and ctenophores are fairly permeable to water. When the animal excrete or exclude the heavy sulfate ions, they replace them isoosmotically by lighter chloride ions to avoid loss of water which would otherwise takes place. The density of the animals in relation to sea water is, therefore, less than it would be if they retained the sulfate ions.

The food stored in the form of oil droplets as in the case of copepods which undertake vertical migration, the oil droplets act as an hydrostatic organ. Another aspect is the body size and, thus, the body surface also important in maintaining the process of afloat.

Non-Seasonal Phytoplankton Blooms

In addition to the regular seasonal blooms of phytoplankton that are characteristic of most bodies of water, there are occasional blooms also found in some waters. Frequently the water is greatly discoloured and it may sometimes become slimy or even matted on the surface with phytoplankters. Such blooms may occur at any season of the year and either in fresh water or in marine environments. They may build up slowly to enormous numbers and decline equally slowly back to ordinary quantities, or they may increase in numbers extremely high and suddenly die out.

Gunter et al. (1963) observed the progress of the phytoplankton bloom on the west coast of Florida in 1946-47. First the water became turbid or slightly changed in appearance from a clear green or bluish-green to a more opaque green. Then it became yellowish green and later greenish yellow. At the next stage and possibly only in spots a few yards or a few hundred yards across, the water became a bright saffron yellow in colour. Gunter and Smith observed a patch of this water. The water colour was very striking and peculiar in contrast to this green waters of the bay.

The lakes in the Madison area of Wisconsin with a higher degree of fertilization and pollution were subject to unusually frequent, non-seasonal blooms of blue-green algae and other plankters, and the blooms often were very obnoxious to the near-by residents. The blooms were most frequent in the lakes receiving high concentrations of inorganic nitrogen and phosphorous. The ratio between inorganic nitrogen and phosphorous was found to be of importance.

The most conspicuous biological effect of excessive phytoplankton bloom is the extensive animal mortality. Such mortality may be due to suffocation or to the direct or indirect production of biotoxins by blue green algae. Suffocation may be caused, for e.g. during the hours of darkness when photosynthesis is not taking place, and when the metabolism of the large number of phytoplankters is using up the available oxygen, or it may be caused by death and decay or large quantities of the phytoplankters. There may be an actual direct production of poisons by the phytoplankters themselves.

A survey conducted on algal poisoning of domesticated and wild animals and birds revealed that not only ingestion of the algae themselves, but also the drinking of filtered water in which they had been growing is very toxic to such animals as mice. Further more, the liver of animals which has died of algal poisoning was very toxic to dogs etc. That time failed to isolate a specific toxins, but was able to show that it was a water soluble substance, which is heat stable and able to resist ordinary domestic water purification methods. Details about some of the aquatic biotoxins is explained in Chapter 11.

Plankton Collection and Preservation

There is no single piece of apparatus which will collect the whole range of plankton satisfactorily. Based on the desired type of plankton material the suitable plankton net should be selected. Collection of plankton is a problem common to all areas. Thus, most of the information which refers to plankton collecting in colder waters will also be applicable to the warm waters of the tropics.

In general sampling equipments can be divided into three types. Sampling with a bottle, with a pump and with a net. Bottle samples are used normally for very small organisms. The bottles are lowered into water open and when they have reached the required depth, a messenger, that is a weight that can be slipped on to the wire is sent sliding down the wire. It hits the top of the bottle, the force of the impact releasing a catch, which allows the top and bottom lids to spring closed. This method has an advantage of sampling accurate volume of water (1-8 litres) at particular depth. After sampling the organisms can be separated by centrifuging or by pouring the water through a disc of very fine nylon or filter paper.

The pump method of plankton sampling is simple. It is an open ended inlet hosepipe, which can be lowered into the water, a pump to which the inlet hosepipe can be attached, and a small, flexible outlet hosepipe. The diameter of the inlet hosepipe is usually between 5-10 cm. Small organisms cannot be retained by the pump method.

Numerous types of plankton nets with various provisions for adjustments to depth and speed and size of apertures are available in the market. The base shape of the net is cone. There is a cone of nylon mesh which ends in another short length of nylon. The net bucket in which plankton are collected, is tied to the short end of nylon. Net buckets of many design have been made, but for most occasions a simple cylinder, closed at one end and lippped at the other is quite useful.

The purpose of net is to filter the water. That is the water entering the mouth of the net and pass out through the mesh apertures. Any thing larger than mesh size will be trapped by the nylon around the mesh and so captured. The size of the individual apertures or holes, in the nylon mesh used for plankton nets varies from 15 micron to 1 mm and still larger if required. An aperture size of 150 to 175 micron will be suitable for sampling plankton from coastal tropical waters. For collection of small early stages of plankton animals and plants an aperture size of 30-50 mm is required. The smallest mesh size can be used for organism such as small plants, flagellates and tintinids.

Only by studying the living organisms, it is possible to be fully aware of the significance of the adaptations. Any important features of the plankton can be studied by observing a living species. The nanoplankton organisms are so delicate that the only really satisfactory way of examining them is while living. For late observation or electron microscopy or when huge samples are collected under such conditions preservation is needed. Generally osmic vapour (Osmium tetroxide) is used. For such structures as flagella, Lugols solution is used. Fixing and preserving such minute organisms is a specialised field which needs more experience.

High Speed Plankton Nets

One of the problems at plankton sampling is catching the larger animals say about 10-50 mm. Usually the plankton net is too slow moving; these animals can be usually swim out of the way. Nets which have been designed to catch those animals are the so called high speed plankton nets. The general principle of these nets is to reduce the area of the mouth in relation to the area of the filtering surface. These high speed nets are the type generally used to catch the larger planktonic animals. Other special forms of plankton nets have been devised for catching plankton swimming above the sea bed. Essentially they are sledges on which a net is mounted. Some have an opening and closing device in order to restrict the sample to the plankton which is actually near the bottom. Others are without this device. A bottom plankton net is very useful if one wishes to catch mysids, amphipods and similar animals.

Many fishes large and small, feed selectively on planktonic animals and concentrate them in the stomach. Animals which are rare in the plankton sample are common in the stomach of fish. When the pelagic fishes are removed immediately their stomach may be cut open and the plankton collected may be collected in a preservative immediately.

There is no doubt that the best fixative and preservative for general use is formalin. This is a saturated solution of the gas formaldehyde in water, giving the so called 40 per cent solution. If the 40 per cent solution is diluted 8 times it will give a 5 per cent solution, commonly used in the storing of plankton samples. The formalin is conditioned by adding marble chips to neutralize free acid and Rochell salt (sodium potassium tartarate) which removes free iron molecules respectively. By doing so the plankton organisms will not undergo decaying the discoloration, stay a long period of time without much decay.

Causes of Phytoplankton Blooms

In cases of general blooms of phytoplankters the most frequent cause is an increase of available plant nutrients. This may be due to various reasons. The trade winds may stir up the bottom deposits, liberating accumulated nutrients into the water, and in this way lead to increased phytoplankton production. An increase of nutrients sometimes also comes about by an upwelling of water from the aphotic zone, where lack of photosynthesis has allowed an accumulation over the years. Such upwelling may occur in conjunction with the high offshore winds, which drive surface water away from the shore and, thus, force deeper water move up to the surface.

Heavy rains and floods frequently carry rich supplies of nutrients from the land into the water and, thus, pave the way for a blooming of the phytoplankton. The artificial use of fertilizers in ponds can increase the quantity of phytoplankton and that this can favourably influence the productivity of fish in the pond. Often a phytoplankton bloom consists of a great growth of a single species rather than a heterogeneous bloom of a number of different species together.

Plankton as Indicators of Pollution

The assessment of the effect of pollution on the plankton of any region is possible only if a thorough knowledge is available of spatial and temporal changes of plankton before the area gets polluted. Benthic species and communities have often been regarded as being the best indicators of organic pollution because of their constant presence, relatively long lives, sedentary habits, and differing tolerances of stress. Those forms indicative of pollution have certain characteristics such as tolerances of low oxygen levels, high biotic potential and small size.

Certain algal forms grow in the special type of polluted waters and these species are characteristic features for the particular environment. *Microcystis aeruginosa* is the best single indicator of organic pollution. *Euglena* and *Oscillotoria* could also be reliable indicators of eutrophication. Palmer (1969) in his valuable review on algae as biological indicators of pollution found certain algae tolerant to relatively raw sewage or organic waste. Hosetti (1987) in his studies on the role of algae in the sewage purification revealed that *Chlorella, Microcystis* and *Scenedesmus* may be considered as indicators of organic pollution.

Saprobic System

In the course of self-purification of a lotic water body would result in a series of zones of decreasingly severe conditions succeeding one another at the downstream, and each containing characteristic animals and plants.

The zones are defined as follows

(a) *Polysaprobic*: Zone of gross pollution with organic matter of high molecular weight, very little or no dissolved oxygen and the formation of sulfides is evident.

(b) *Mesosaprobic*: With simpler organic molecules and steadily increasing oxygen content. This zone is divided into an upper zone with many bacteria and often fungi and with more types of animals but few algae and a lower zone where mineralization has proceeded further and the conditions are suitable for many algae, and tolerant animals and some rooted plants may also occur.

(c) *Oligosaprobic*: The zone of recovery where mineralization is complete, the oxygen content is back to normal and a wide range of organisms occur.

The order of dominance of flora in the polluted and unpolluted stations of rivers Cauvery and Kapila of Karnataka state are as follows:

Planktonic algae :Unpolluted

River Cauvery

Chlorophyceae > Bacillariophyceae > Cyanophyceae > Euglenophyceae.

River Kapila

Chlorophyceae > Cyanophyceae > Bacillariophyceae > Euglenophyceae

These were called positive indicators.

Polluted zones

River Cauvery

Cyano > Chloro > Bacillario > Eugleno

River Kapila

Cyano > Chloro > Bacillario > Eugleno

These are called as negative indicators.

Diatoms as Indicators

Notably diatoms have been employed as indicators of water quality by a number of investigators. Schoeman (1976) classified the entire group of diatoms under major heads as oxygen indicators and as varying nitrogen level indicators.

Nygaard's Trophic Level Indices

Nygaard in 1949 established five idices to get meaningful evaluation of the extent of pollution viz., Chlorophycean, Cyanophycean, diatoms, euglenophycean and compound index. These five indices represent ranges of values established by Nygaard for Danish lake.

In applying the indices, each of the major groups (Cyanophyceae, Desmidaceae, Chlorococcales, Diatoms and Euglenineae) of algae are determined up to the species level for both investigated ponds for all the months. The ratio of these groups are calculated and used to express the degree of eutrophication. It is determined and compared with physical and chemical analysis.

One of the remarkable pollutions causing tremendous biological and economic damage is the oil pollution. At low levels of oil spill, physiological changes occur prior to changes in population numbers and community structure. It is reported that oil affects the cells membrane. Therefore, membrane mediated processes, such as respiration and excretion rates should be good indicators of an organism response to oil. These rates are also good indicators of an organisms metabolic state, especially when combined as O:N ratio.

Nygaard's Trophic Level Indices

Index	*Calculation*		*Oligotrophic*	*Eutrophic*
Myxophycean	Mysophyceae	Desmideae	0.0-0.4	0.1-3.0
Chlorophycean	Chlorococcales	Desmideae	0.0-0.7	0.2-9.0
Diatom	Centric-diatoms	Pennate Diatoms	0.0-0.3	0.0-1.75
Euglenophy-phycean	Euglenophyta Chlorococcoales	Myxophyceae	0.0-0.2	0.0-1.0
Compound	Centric-diatoms	Euglenophyta Desmidae	0.01-1.0	1.2-2.5

The occurrence of oil spills in the major oceans of the world and oil spill identifications are well known problems. Investigations on oil spills occurring in the Madras area, caused during transport of oil from oil tankers to the Refinery Dock, and oil spills from ocean liners, tugs, mechanized fishing vessels and petroleum effluents and samples of Madras Refineries are undertaken during 1975-76. Inside the entrance of Madras Harbor and adjoining areas. The sea area off Santhome was not polluted during the period June, 1975. The oil spills have been mapped and compared to with the occurrence of nueston plankton in the oil contaminated and non polluted areas. The investigations have revealed the following points.

1. Nueston-plankton biomass was very low in the oil polluted zones, whereas in the adjoining waters off Santhome it was 4.5-6.5 times higher.
2. The nueston-plankton in the oil polluted zones constituted mainly of certain hardy species of crustaceans and coelenterates i.e., copepods, amphipods, and mysis larvae. In the non-polluted zones, in addition to the hardy species, molluscan larvae, decapod larvae, copepod nauplii, polychaete larvae, echinoderm larvae, adult copepods, cladocerans and fish eggs and larvae occurred in higher population density in the non-polluted zones. Green water: Phenomenon in the Arabian Sea of Mangalore.

Seasonal Algal Blooms

Incidences of algal blooms and discoloration of water in the Indian seas have been reported at irregular intervals over the last eight decades by a number of workers. Most of these blooms have been attributed to either the blue-green alga *Trichodesmum erythraeum* or the dinoflagellate *Noctiluca milaris*. The few instances of green water phenomenon in the Indian seas were observed when dense swarms of *Noctiluca* occurred along with innumerable green flagellates inside them. However, in one instance, green discoloration of the water due to the swarming of only a green flagellate, Subramanyam (1954) was observed over the period 1949 to 1953.

A study on the distribution of phytoplankton in the coastal waters (West coast of India) of Mangalore was studied by Ramesh *et al.* (1992). According to them the succession of phytoplankton blooms was influenced by the ingress of large quantities of freshwater from Netravathi-Gurupur estuary into the ocean. The frequency of occurrence and abundance of diatoms and dinoflagellates were high during post monsoon periods. The sequence of changes in the environmental parameters brought by the formation of euhaline community consisting of *Cosinodiscus* sp., *Chaetoceros* sp., and *Ceratium* sp., during premonsoon ; a euryhaline community of *Ceratium* sp., *Pleurosigma* sp., and *Cosinodiscus* sp. during the monsoon and the one consisting of *Chaetoceros, Asterionella* and *Ceratium* during latter half of post monsoon season were observed.

The plankton population of Kali estuary, Karwar followed a bimodel cycle of yearly abundance. One during the premonsoon and the other in the post monsoon period. The phytoplankton exhibited greater population in the February represented by 54,175 counts/m^3 and least number in the August with 1,229 counts/m^3. October was the peak for zooplankton population with a mean of 21,900 counts/m^3. The second peak of zooplankton was depicted in February with 19,451 counts/m^3. The phytoplankton consisted by diatoms such as *Thalassosira, Nitzschia, Cosinodiscus, Asterionella, Rhizisolenia* and dinoflagellates such as *Noctiluca* and *Ceratium*. The zooplankton include protozoa, larvae of polychaetes, chaetognathus and pteropods.

Indications of an unusual bloom of *Noctiluca miliaris* in the near shore waters off Mangalore were first noticed during the later half of October, 1986, at Thannirbhavi north of Mangalore. During the second half of January, 1987 the inshore waters up to 20 m depth contour off the Nethravathi Gurpur estuary at Mangalore acquired deep greenish discolouration and soup-like consistency. The green waters were highly phosphorescent during dark nights because of *Noctiluca*. The phenomenon of green water, which could be termed green tide, lasted from the last week of January to the end of February, while the *Noctiluca* bloom continues till the end of May. Samples collected at different centres during February revealed the occurrence of green tide along the Dakshina Kannada coast with high intensity from Someshwar to Surathkal coast. Detailed observations made off Mangalore and Thannirbhavi showed patches between 18 to 20 m depth contours. The biomass of *Noctiluca* reached very high (8.2 x 109/m^3) levels in the coastal waters of Mangalore. Significant differences in surface water temperature and dissolved oxygen content of regions were noticed with varying densities of the bloom Chlorophyll and Phaeopigments exhibited very high values in the green water regions. The study of various hydrological parameters recorded before and during the green tide could not give any clue to the unprecedented blooms of *N. milaris* and *P. noctilucae,* since no drastic changes could be seen in these parameters in the region as a whole. However, some distinct differences could be noticed in respect of some parameters (temperature, dissolved oxygen and phosphate) between regions of higher and lower densities of the bloom particularly during the height of the green tide. The distinctly higher temperature (up to 32.0°C) observed could be attributed to the absorption of most of Solar radiation by the thick soup like surface waters.

The phosphate content of water at all the Mangalore stations showed a distinct increase during the peak period of the bloom. The decomposition of the dead *Noctiluca* and the green flagellate was evidently responsible for the observed high phosphate content. It was also reported that the green *Noctiluca* are usually rich in brackish waters along the coast. In case of study on off Mangalore, they were found in distinctly higher numbers in the more saline waters of the sea. These were also reported in relatively higher concentrations from the coastal water of Karwar (Neelakanthan *et al.* 1988).

There have been conflicting opinions about the effect of Noctiluca blooms on the marine fauna, especially the fishes while some have considered *Noctiluca* blooms harmless, others have attributed large scale fish mortalities to these swarms. According to Bhimachar and George (1958) the shelling fishes avoid areas of red water or green water, thereby causing abrupt set backs in the commercial fisheries. They did not observed any severe fish mortality. The avoidance of *Noctiluca* patches by fishes could be either because of oxygen deficiency or changed ecological conditions brought about by the dying and decaying Noctiluca and the flagellate. The red tide occurrences are on the increase due to discharge of large amounts and varieties industrial effluents. However, this could only be one of several causative factors.

Plankton as Indicators of Water Current

Perhaps the most successful use of indicators in the marine environment relates to that of plankton characterizing particular water masses and their movements. Thus, a species or community of plankton organisms, plants or animals typical of a water mass of a given hydrogrpahic nature, may be used to define the movements of this water mass. The presence of oceanic water can, for instance, be detected in coastal areas by the presence of certain oceanic plankton organisms which, therefore, give an additional indication of the origin of the water to those chemicals and physical characteristics usually made use by hydrographers.

For example, *Rhinocalanus nastis* is found in the water off the coast of Green land, this means not only that the copepod is being carried to its death in a unfavourable environment but also it is direct evidence that water of tropical origin has reached the shores of this island likewise the presence of salps, or the copepods. *Microstella rosea* in Briscayne Bay, Florida would indicate that oceanic water has entered the bay, for these are oceanic and not found as a regular thing in such inshore localities. It is reported that the channel water along the coastal areas usually populated by a plankton community comprising of a Chaetognath, *Sagitta setosa*, as a dominant member (Kusum *et al.*, 1988).

A great deal of research needs to be carried out on the normal habitats of plankters in order to detect the currents through the examination of plankton content of water. It has been said that the plankton can tell us more about the currents than the most refined methods of hydrography, and this probably would be true if we knew enough about the plankters themselves, for example, in the work reported by Johnson (1939) a single zoea of *Emerita* was found in water which hydrogrpahic analysis showed to be of oceanic origin. However, it seems very likely that this larva was indicative of at least a minimum of admixture of coastal water with the oceanic water at the station where the larva was found. The only possible explanation would be that the larva migrated from a mass of coastal water into the oceanic when the two masses were continuous.

The plankton is for their economical importance have been worked on extensively in the modern biological world. Several systems are employed for the biological monitoring of water quality. The suitability of an index for a particular situation and its usefulness is the interpretation of the results depend on the discretion of a trained biologist in the field. Though previously the plankton study was regarded be uniform and disinteresting now has been considered to be of great interest. The diverse habitat of plankton is extensively worked on by the various scientists. The literature known about plankton is far less compared to the facts to be known. Thus, plankton study is one of the dynamic fields of research in marine biology.

Some Drift Organisms and Protozoa

This group include both plants and animals, because some of these were plants at some stage and animals at the other stage depending upon the environmental conditions and their life cycle. *Euglena* is about 30 μ long and 12 μ wide possess a flagellum, while *Chlamydomonas* is about 12 μ long and 5 μ wide also with two flagella. These organisms normally contain small green bodies, the chloroplasts, other forms may be colourless. Some species may have four or more flagella. When observed while swimming three or more different shapes of the same species are exhibited within few seconds.

The dinoflagellates have two flagella and cell wall. Some are photosynthetic but some such as *Noctiluca* are carnivores. Some dinoflagellates are referred as animals. A large number have the cell wall composed of a series of plates and called as armored plates. The other being naked. While alive they are brown but can vary in the colour from greenish brown to yellowish green. Some are distinctly reddish in colour. They occur in enormous numbers in the tropical waters. The so called red tide is caused by the dinoflagellates such as *Gymnedinium* and *Gonyaulax*. Their metabolic by-products are toxic to fish and man. These are referred as biotoxins. *Gymnodinium* is about 50 μ in diameter *Polykrikos* is a colonial alga 100-500 μ long found occasionally.

The armored forms include the genus *Ceratium*, size range from 200-500 μ. This is an ubiquitous form seen in chains more than a dozen. *Peridinium* is another common form with world wide in distribution. It is about 200-250 μ in diameter. *Pyrocystis* is characteristic component of tropical plankton often occur in large numbers. It looks like tiny soap bubbles.

Diatoms are small single celled plants with the cell wall containing silica. The diatom cell wall is known as frustule. The cells can occur singly ranging from abut 5 μ to over 1 mm in size or they occur in chains which may be 50 mm long or more. Many tropical forms of diatoms are also seen in the cold water.

Cosinodiscus is a diatom, about 1 mm length. The species of *Rhizosolenia* and *Bacillaria* are very common in the sea waters. *Hemidiscus hardmanianus* is a large diatom and characteristically a tropical diatom moderately common in the sea. Viviparous planktonic plants and drift animals are printed in Plates 1 to 13.

Trichodesmidium

A pelagic bluegreen alga more or less restricted to tropical waters. These appear as a filamentous tiny bundles. It discolours the water over a wide range of water surface to yellowish or distinctly red. These bundles may be 1 mm long being found in the brackish waters and presence of this alga is a nuisance.

Globigerina

It is a pelagic protozoa belongs to the group foraminifera. These are found both in tropical and temperate waters. But these protozoa are characteristic of tropical sea waters. The organism about 1 mm long and its body is covered by calcerous shell or test with long number of fine projections.

Radiolaria

It is another group of pelagic protozoa which also possess skeleton made up of silica. These are constant features of tropical waters. These include both symmetrical and non-symmetrical forms. While counting in a tray they look like tiny stars about 200 μ in diameter but colonial forms are also recorded (e.g. Acanthometra).

Other Protozoa

Coccolithophorids are tiny organisms possess a calcerous test as we find in Coccolithus. These found in small numbers. Silicoflagellates are also tiny organisms having a skeleton of silica which is commonly star shaped. These appear commonly as food items in the gut of copepods. Tintinids are ciliate protozoa live in a bell shaped jelly like structure. These being very tiny organisms cannot be taken into usual plankton nets.

Tretomphalus

These are foraminifera usually live in the bottom freely or attached to some substrate. During breeding the organism develops a gas chamber and float on the surface. At this stage it is found in the shallow coastal waters and can be sampled.

Coelenterates and Worms

These can be up to 2 metres in length but majority will be less than 1 cm. These constitute an important part of phytoplankton community and all are extensively carnivores. The size and shape of these animals vary a great deal but medusae are easily identified because of their jelly like consistency and tentacles. Alive medusae are transparent and while preserved ones are opaque in colour. Some species like *Liriope* and *Solmundella bitentaculata* are permanently planktonic. These can be identified by naked eye. Another common form is Aglaura which is bell shaped 1-2 mm long and the body wall is iridescent. Two more common forms of medusoid organisms, are Obelia and Bougainvillea. Obelia is 200-300 μ in diameter and at time occur in large numbers. The edge of the bell of Bougainvillea attached with a ring of tentacles of 2-10 μ or more in diameter. Many times Aurelia appears as a mass of slippery slimy medusae in large numbers. All these medusae have the power to string but majority are not toxic to man. The sting of medusae, Carybdea produces a powerful toxin and it also occurs in swarms.

Siphonophores

These are colonial organisms and the colonies are broken into pieces when sampled. It becomes very difficult to identify the organism as the organisms will not be intact. The other siphonophores include the Portuguese man of war or Physalia, Velella and Porpita. The large colonies of siphonophores are *Aglama stephanomia* and *Halistema*. All these are conspicuous organisms sting when they are touched. The colonies of siphonophores, Dipheys and Lensia are small and their length varies between 1-5 mm long.

Ctenophores

This group include relatively large sized predatory animals. The important feature of this group is presence of ciliary plates called as comb plates which are always 8 in numbers. These produce bioluminescence in the dark. The organism, Pleurobrachia is commonly identified in the sea. These look like a ball of jelly and 1-5 mm long.

Mnemiopsis

These are large size organisms (5-10 cm) brownish in colour and these can be easily identified under water.

Beroe

It is not usually taken into plankton samples, but young individuals 2-5 mm long are collected as plankton.

Pilidium Larvae

The nemertines constitute a group of worm like animals which are mostly marine. The pilidium belongs the sub-division heteronemertines and is a free living larva.

Polyclads

These are group of flatworms (Platyhelminths) and some are planktonic and rarely encountered in the plankton samples. Their size is 1-2 mm and the adults are up to 1 cm in length. These look like flattened leaves in appearance.

Trochophore Larvae

These stages are characteristic of some worms and molluscs. Generally these are small 200-500 μ long but some time larger ones are also found. The ciliary bands remarkably found on the body help to identify them.

Polychaete Larvae (Post Trochophore Larvae)

Their size varies from 750 μ to 3 mm. Possession of setae is characteristic of the organisms. These give the impression of the presence of needles emerging from the body. The number and variety of the polychaete larvae is great. The members of the family Spionidae, Chaetoperidae and Nereidae produce these larval stages.

Some polychaete worm larvae like Terebellids have a pelagic larval stage, live in tube like jelly structure which is transparent and smooth to touch. The polychaete adult form the permanent plankton. One of the common plankton is Sagitella. These are small and the size is 1-2 mm long.

The alciopids are also distinctive having two relatively large eyes. Another form is about 1-3 mm long is *Lopadorhynchus*. Some other polychaetes are large in size come to surface for breeding and keep on moving from place to place.

Tomopteris

It is best adapted to planktonic way of life and is a good indicator for the presence of oceanic water. In the nature it is transparent and difficult to identify whereas the preserved one is opaque can easily be identified. Their size varies from 2-5 mm in length. Tremopterisa is a rare plankton.

Crustacean Plankton

This group usually outnumbers the remainder of all the plankton organisms.

Cladocera

This is small easily recognizable group, the size of the individuals vary between 0.5 to 1 mm long. Two types of cladocera were found in the plankton samples. The first has a single species, *Penilia avirostris*, while alive it is transparent and difficult to locate and identify. But under preserved conditions it is opaque. An important feature of the cladocerans is to produce parthenogenetic eggs which hatch into only females and in some species there are no males at all. The other type of cladocerans include, *Evadne tergestina*. These are very common in the water, 0.5 to 0.7 mm long and are identified by the presence of round shaped back with pigments cells along the sides. These appear throughout the year.

Ostracods

This is large group of small animals. The majority of the ostracods live in the bottoms at day time and join the plankton during the night hours. These are abundant in the tropical shallow water and found in very few numbers in the cold waters. The body is completely enclosed in a bivalve shell which is hinged along the back. The shell is calcerous in nature. These also reproduce through parthenogenesis. *Pyrocypris* show 8 young ones developing inside the body. These are 1-2 mm long and the eyes are visible. The organisms *Pyrocypris* and *Euconchoccia* are commonly found in the oceans and *Codonocera* come to surface only in the night and are caught in the plankton nets.

Copepoda

These are most important group of tiny animals too small to be taken into plankton nets generally used but a mesh aperture size 60 μ will help in collecting them. Many time these are confused with the nauplii of others. The genera *Eucalanus, Rhinocalanus* and *Euchaeta* are fairly large up to about 0.5 mm long and are distinctly pear shaped.

Monstrilloid Copepods

A large number of these copepods are parasitic having free living life only as a nauplius. The adults 1-5 mm long found in fairly large numbers.

Harpactoid Copepods

This is large group of copepods and some individuals are completely planktonic. The general shape of the body is tapering towards the posterior end e.g. *Macrostella gracilis* – body length is 1 to 1.5 mm and long setae. These are more common in offshore waters. *Microsotella* is smaller form about 0.5 to 0.8 mm long fairly common at times and is pink in colour. Another pink type is *Clytemnestra*. These are found in shallow marine waters, the eyes are large and conspicuous. *Euterpina* is another bottom living commonly taken as a shallow water plankton organism.

Cyclopoid Copepods

These are small bright coloured when alive or freshly collected. These are found both in temperate and tropical waters. The examples for this includes, *Olithona, Corycaeus, Copilia* and *Suphirina*.

Calanoid Copepods

These are found in large numbers in tropical waters. The body is characterised by two divisions, the cephalothorax and the abdomen is separated by conspicuous constriction. The size varies from 0.4 to 17 mm long. The examples include, *Paracalanus, Acrocalanus, Acartia, Eucalanus, Rhinclanus, Undinula vulgaris, Euchaeta concinna, Centraphages furcatus, C. dorsispinatus, Temara, Candecia, Pontella, Laboidocera, Tortanus, Haloptilus* and *Pleuronema xiphias*.

Cirripedia Nauplii and Cyprids

Adults cirripedes live either as sedentary or parasitic life. The sedentary life is represented by the barnacles. The parasitic life is shown by the Rhizicephala. Both fixed and parasitic forms have a planktonic nauplii larvae which develop into cypris larvae.

Mysids

These constitute a group of small prawn like animals with biramous thoracic appendages and a short carapace. These are collected in the night samples. These vary in size from about 2-10 mm or more in length. These can be observed with naked eyes. These have a pair of stastocyst in the tail and also live in ponds and mangrove swamps.

Cryptoniscid Larva

These are parasitic organisms having free swimming larval life. These are small 0.5 to 1.0 mm long rarely taken into plankton nets. The amphipods include gammarids and hyperiids. These are bottom dwelling nocturnal organisms the size varies from 2-6 mm long e.g. *Synopia ultramarina* and *Hyperiids* have the size range 2-7 mm long, e.g. *Tulbergella* and *Rhabdosoma*.

Large Crustacea

Euphausiids are the off shore oceanic plankton animals the size of the adult varies from 10-20 mm long. These are caught by a high speed plankton net with a mesh size of 2 mm. The larval stage are more common, the eggs usually hatch into nauplii. These nauplii range in the size 200-500 μ and develops into further stage calytropis and furcilla larvae and into adult. The adult euphausiids are easily recognized as the anterior legs possess gills, e.g. *Thysanopoda tricuspidata, Nematoscelis, Stylochiron, Pseudosuphasis flatirons*. The true prawn, penaeids and caridea and sergestidae include *Acetes, Sergestes* and *Lucifer*. The larval stages of these true prawn are commonly caught in the samples of oceanic waters.

Some Mollusks

A great and number and variety of shells are to be seen in the any coastal beaches. Most of their larvae are pelagic in character. Gastropods have trochophore, veliger and echinospira larval stages. Apart from the above plankton described other gastropods belonging to the groups, Pterotrachids, Atlantids and Cephalopods, Echinoderms, Minor phyla and smaller vertebrates constitute the plankton community.

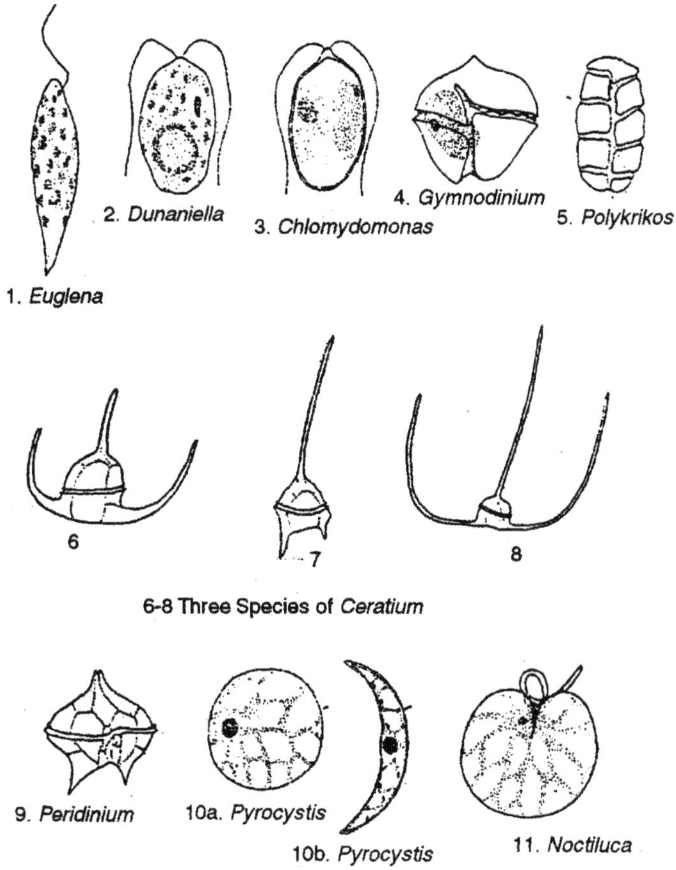
1. *Euglena*
2. *Dunaniella*
3. *Chlomydomonas*
4. *Gymnodinium*
5. *Polykrikos*
6
7
8
6-8 Three Species of *Ceratium*
9. *Peridinium*
10a. *Pyrocystis*
10b. *Pyrocystis*
11. *Noctiluca*

Plate 1

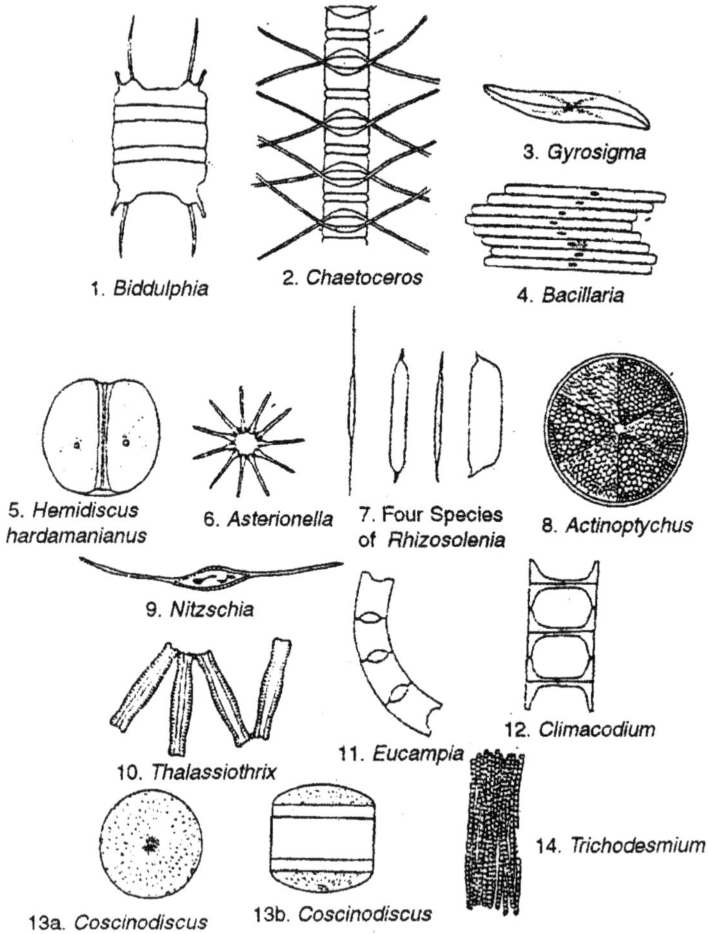
1. Biddulphia
2. Chaetoceros
3. Gyrosigma
4. Bacillaria
5. Hemidiscus hardamanianus
6. Asterionella
7. Four Species of Rhizosolenia
8. Actinoptychus
9. Nitzschia
10. Thalassiothrix
11. Eucampia
12. Climacodium
13a. Coscinodiscus
13b. Coscinodiscus
14. Trichodesmium

Plate 2

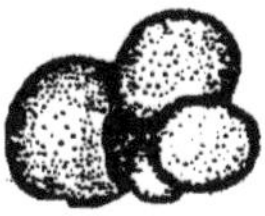
1. *Globigerina*

2. *Acanthometra*

3. *Coccolithus*

4. *Dictyocha*

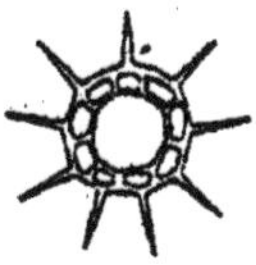
5. *Dictyocha*

6. *Tintinnopsis*

7. *Rhabdonclla*

8. *Favella*

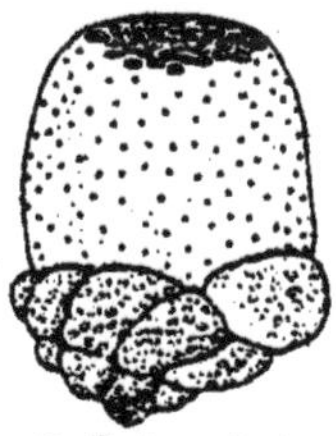
9. *Tretomphalus*

Plate 3

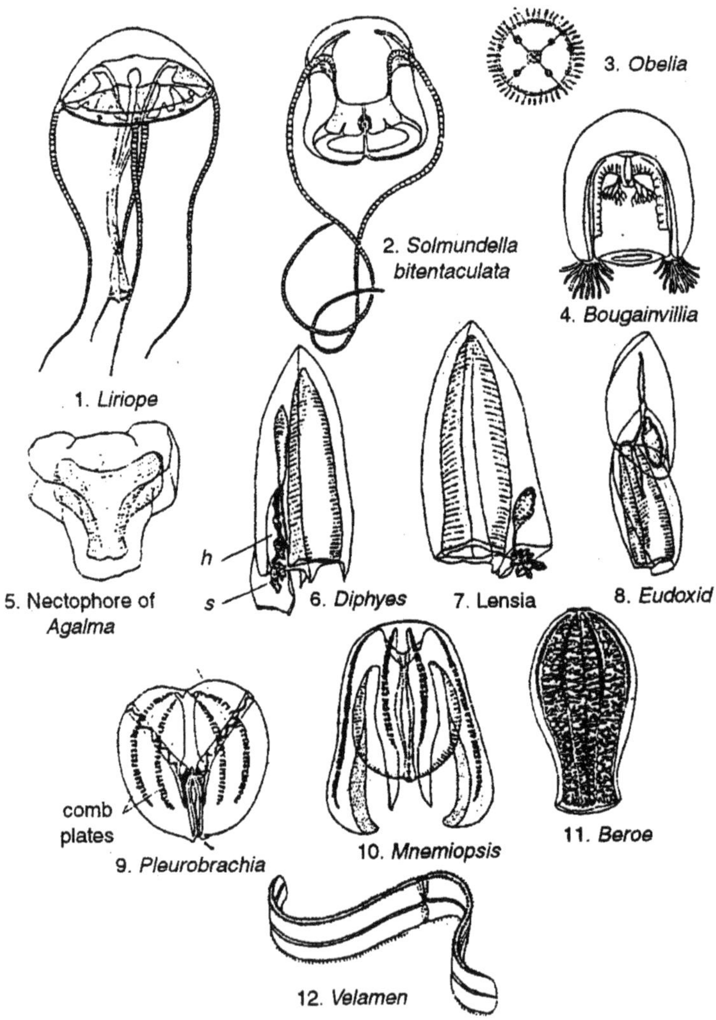
3. Obelia
2. Solmundella bitentaculata
4. Bougainvillia
1. Liriope
h
s
5. Nectophore of Agalma
6. Diphyes
7. Lensia
8. Eudoxid
comb plates
9. Pleurobrachia
10. Mnemiopsis
11. Beroe
12. Velamen

Plate 4

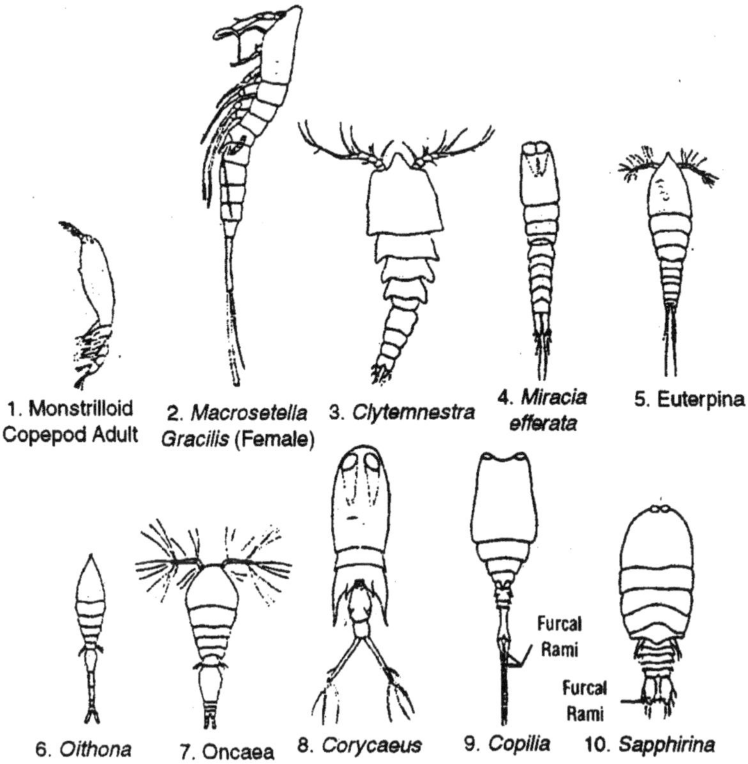
1. Monstrilloid Copepod Adult
2. *Macrosetella Gracilis* (Female)
3. *Clytemnestra*
4. *Miracia efferata*
5. Euterpina
Furcal Rami
Furcal Rami
6. *Oithona*
7. Oncaea
8. *Corycaeus*
9. *Copilia*
10. *Sapphirina*

Plate 5

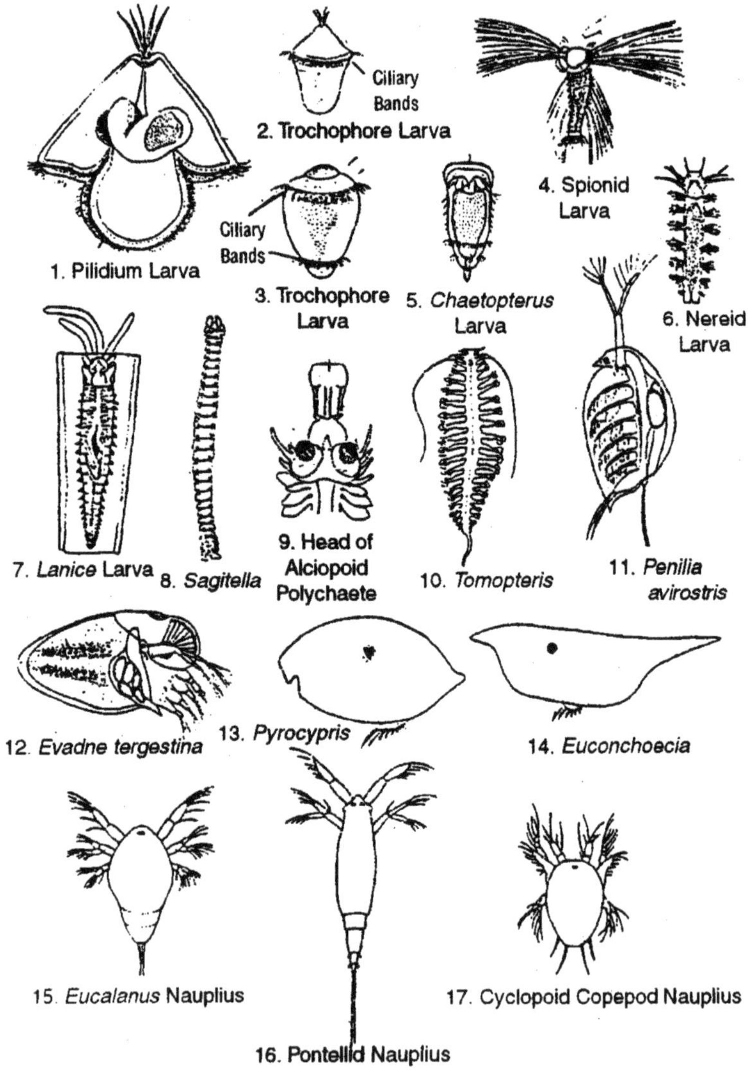

Ciliary
Bands
2. Trochophore Larva
4. Spionid
Larva
Ciliary
Bands
1. Pilidium Larva
3. Trochophore
Larva
5. *Chaetopterus*
Larva
6. Nereid
Larva
7. *Lanice* Larva
8. *Sagitella*
9. Head of
Alciopoid
Polychaete
10. *Tomopteris*
11. *Penilia*
avirostris
12. *Evadne tergestina*
13. *Pyrocypris*
14. *Euconchoecia*
15. *Eucalanus* Nauplius
16. Pontellid Nauplius
17. Cyclopoid Copepod Nauplius

Plate 6

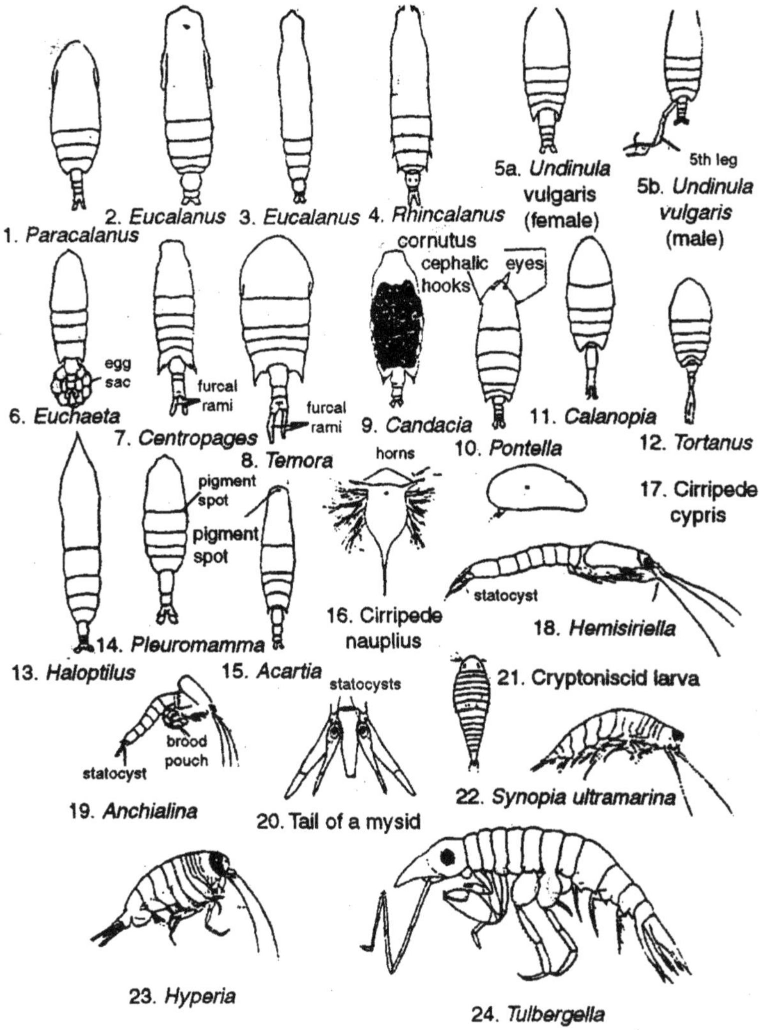
1. Paracalanus
2. Eucalanus
3. Eucalanus
4. Rhincalanus cornutus
5a. Undinula vulgaris (female)
5th leg
5b. Undinula vulgaris (male)
egg sac
6. Euchaeta
furcal rami
7. Centropages
furcal rami
8. Temora
9. Candacia
cephalic hooks
eyes
10. Pontella
11. Calanopia
12. Tortanus
13. Haloptilus
pigment spot
pigment spot
14. Pleuromamma
15. Acartia
horns
16. Cirripede nauplius
17. Cirripede cypris
statocyst
18. Hemisiriella
brood pouch
statocyst
19. Anchialina
statocysts
20. Tail of a mysid
21. Cryptoniscid larva
22. Synopia ultramarina
23. Hyperia
24. Tulbergella

Plate 7

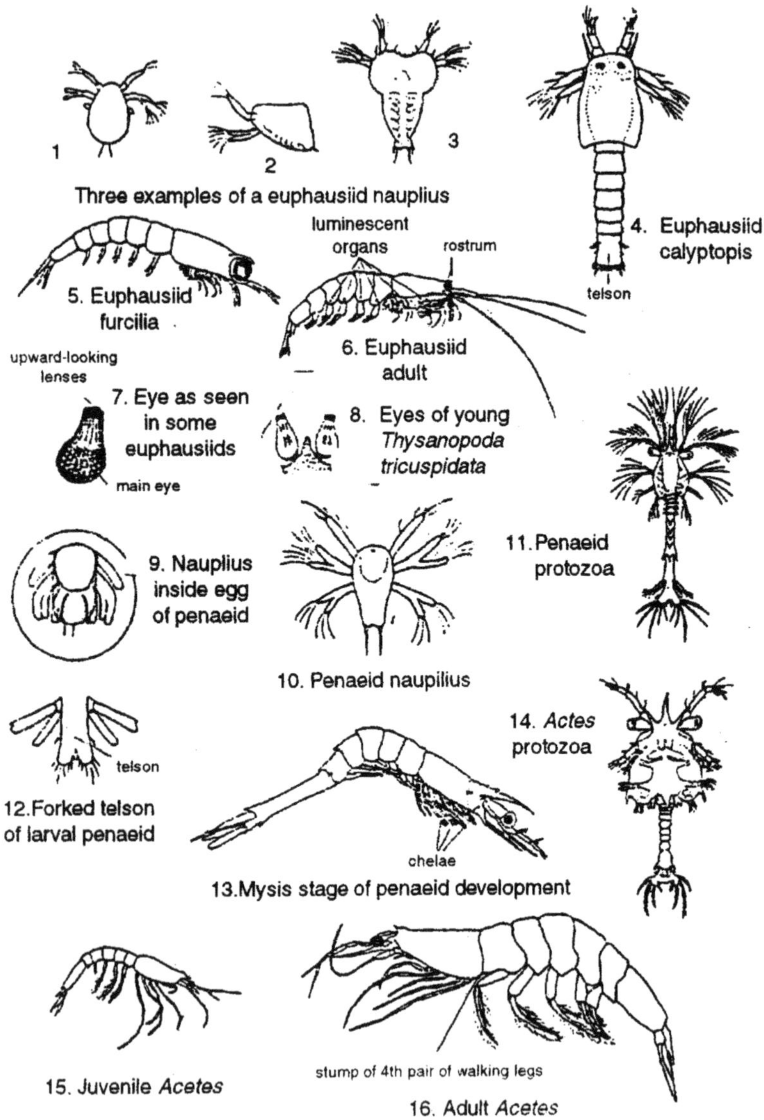
1
2
3
Three examples of a euphausiid nauplius
4. Euphausiid calyptopis
telson
luminescent organs
rostrum
5. Euphausiid furcilia
6. Euphausiid adult
upward-looking lenses
7. Eye as seen in some euphausiids
main eye
8. Eyes of young *Thysanopoda tricuspidata*
9. Nauplius inside egg of penaeid
10. Penaeid naupilius
11. Penaeid protozoa
14. *Actes* protozoa
telson
12. Forked telson of larval penaeid
chelae
13. Mysis stage of penaeid development
15. Juvenile *Acetes*
stump of 4th pair of walking legs
16. Adult *Acetes*

Plate 8

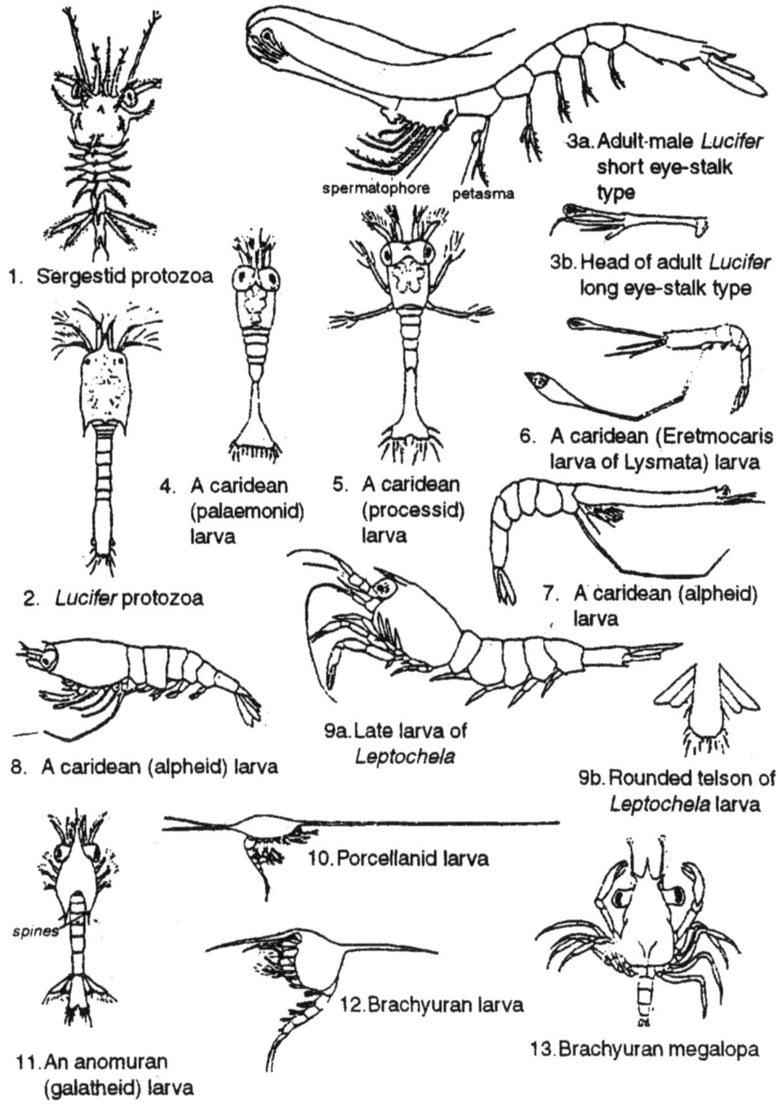
1. Sergestid protozoa
spermatophore
petasma
3a. Adult-male Lucifer short eye-stalk type
3b. Head of adult Lucifer long eye-stalk type
4. A caridean (palaemonid) larva
5. A caridean (processid) larva
6. A caridean (Eretmocaris larva of Lysmata) larva
2. Lucifer protozoa
7. A caridean (alpheid) larva
8. A caridean (alpheid) larva
9a. Late larva of Leptochela
9b. Rounded telson of Leptochela larva
10. Porcellanid larva
spines
12. Brachyuran larva
13. Brachyuran megalopa
11. An anomuran (galatheid) larva

Plate 9

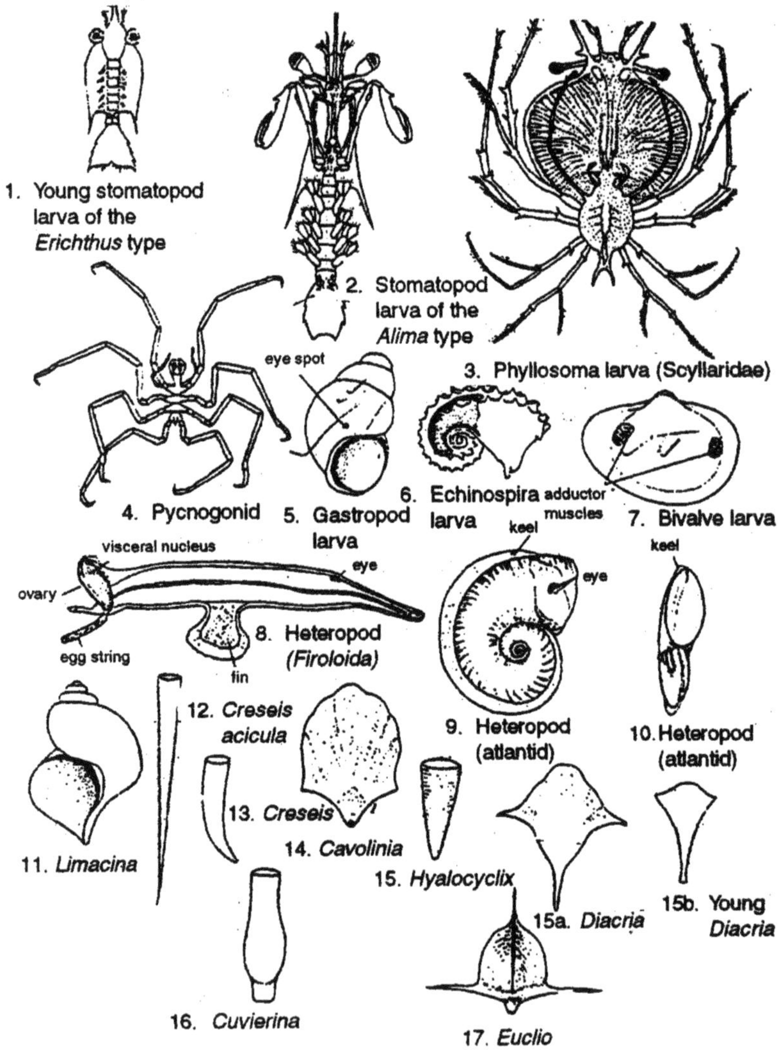
1. Young stomatopod larva of the *Erichthus* type
2. Stomatopod larva of the *Alima* type
3. Phyllosoma larva (Scyllaridae)
eye spot
4. Pycnogonid
5. Gastropod larva
6. Echinospira larva
adductor muscles
7. Bivalve larva
visceral nucleus
ovary
egg string
eye
fin
8. Heteropod (*Firoloida*)
keel
eye
9. Heteropod (atlantid)
keel
10. Heteropod (atlantid)
11. *Limacina*
12. *Creseis acicula*
13. *Creseis*
14. *Cavolinia*
15. *Hyalocyclix*
15a. *Diacria*
15b. Young *Diacria*
16. *Cuvierina*
17. *Euclio*

Plate 10

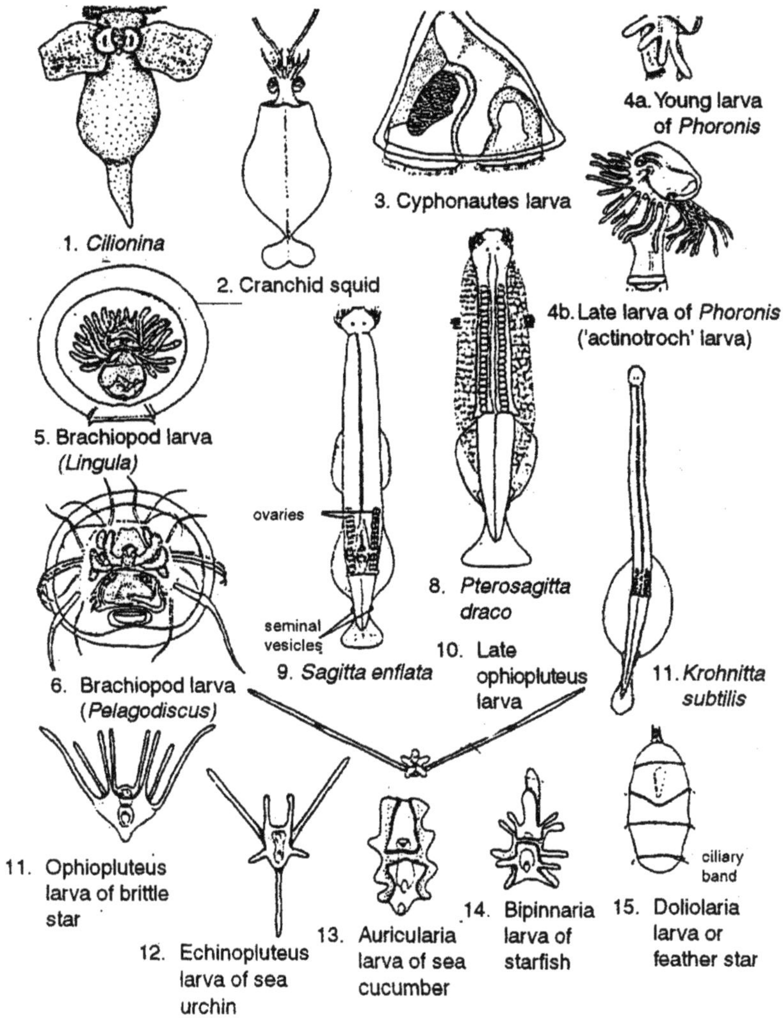

1. *Cilionina*
2. Cranchid squid
3. Cyphonautes larva
4a. Young larva of *Phoronis*
4b. Late larva of *Phoronis* ('actinotroch' larva)
5. Brachiopod larva (*Lingula*)
6. Brachiopod larva (*Pelagodiscus*)
ovaries
seminal vesicles
9. *Sagitta enflata*
8. *Pterosagitta draco*
10. Late ophiopluteus larva
11. *Krohnitta subtilis*
11. Ophiopluteus larva of brittle star
12. Echinopluteus larva of sea urchin
13. Auricularia larva of sea cucumber
14. Bipinnaria larva of starfish
ciliary band
15. Doliolaria larva or feather star

Plate 11

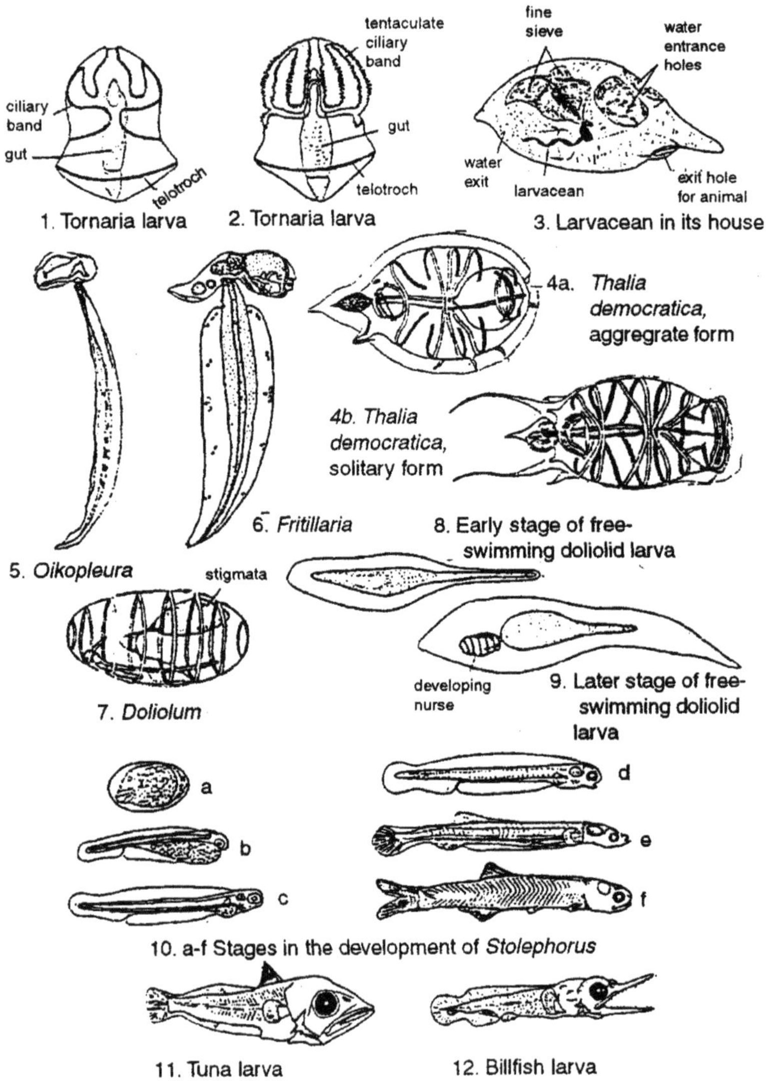
tentaculate
ciliary
band
ciliary
band
gut
telotroch
gut
telotroch
1. Tornaria larva
2. Tornaria larva
fine
sieve
water
entrance
holes
water
exit
larvacean
exit hole
for animal
3. Larvacean in its house
4a. Thalia democratica, aggregrate form
4b. Thalia democratica, solitary form
6. Fritillaria
5. Oikopleura
8. Early stage of free-swimming doliolid larva
stigmata
7. Doliolum
developing
nurse
9. Later stage of free-swimming doliolid larva
a
b
c
d
e
f
10. a-f Stages in the development of Stolephorus
11. Tuna larva
12. Billfish larva

Plate 12

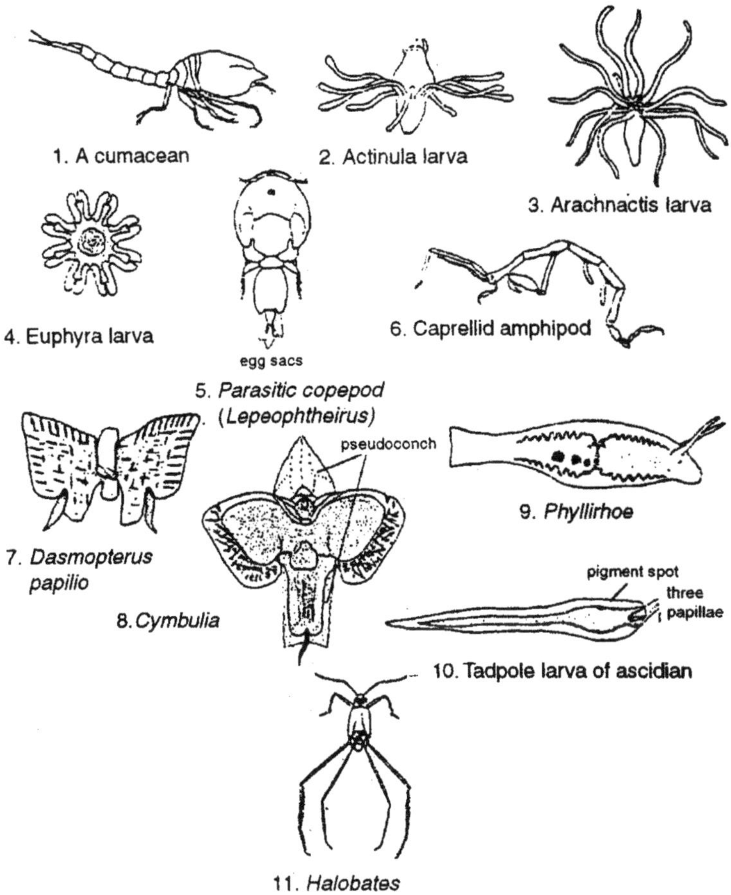
1. A cumacean
2. Actinula larva
3. Arachnactis larva
4. Euphyra larva
egg sacs
5. *Parasitic copepod*
(*Lepeophtheirus*)
6. Caprellid amphipod
7. *Dasmopterus papilio*
pseudoconch
8. *Cymbulia*
9. *Phyllirhoe*
pigment spot
three papillae
10. Tadpole larva of ascidian
11. *Halobates*

Plate 13

Chapter 9

Water and Pollution Abatement

Introduction

Water is the prime necessity of life. 70 percent of Man's body is composed of water. Protoplasm of plants contains 80-90 percent water. It has became an essential commodity for the development of industries and agriculture. General survey revealed that about 15 crores cubic kilometres of water is found on the average on the layers of earth. But inspite of such abundance only 0.03 per cent of world's total water runs through rives, streams, lakes and ponds, which is available to man for his various activities. Or if all the available water is filled in a vessel, of six litres, out of this only three spoonful of water is available for man.

Pollution of these water sources became a severe problem with the industrialization of nations, coupled with rapid acceleration in the growth of human population. Water requirements have ever since been increasing due to the pressure of population, urbanization, industrialization and demands for higher standards of living.

The water pollution is significant only when it influences biological system directly or indirectly. In India only about 14 per cent of people get clean drinking water through tap systems and rest quench their thirst from raw or polluted lakes, tanks, rivers and wells due to which more than 3 millions people are affected or die of enteric diseases annually.

Strictly speaking, pollution is any departure from purity. However, the term purity in this context is viewed differently, because water is never found in the pure state. A more reasonable definition might be that the water is polluted when any outside influence is sufficiently adequate to render the water unacceptable for its best possible usage. Another definition for pollution would be, the introduction into the environment, any physical phenomenon, biological agent of chemical substance in any quantity, that has an adverse effect on the receiving system. The substance introduced into the water system, which was not present in its natural state and causes adverse effects on the biotic community is a water pollutant. Or it may be defined as an unwanted or undesirable change in the water quality. Most recent and widely accepted definition for water pollution is an undesirable change in the physical, chemical and biological parameters which affects its ultimate use, it causes harm to man and his associated species and industrial and other anthropogenic utility.

Classification of Water Pollutants

Any pollutants polluting the aquatic systems could be derived from natural or due to anthropogenic (cultural) activities. But there are no hard and fast rules applied in categorizing the pollutants. Different authors employ different modes to explain them. We shall see two main and simple ways in which the water pollutants can be classified. According to Richard J. Schmitz, the following 10 categories can be considered as important pollutants of the water system.

(1) Oxygen Demanding Wastes

Oxygen being a basic requirement of almost all plant and animal life, its presence in sufficient quantities is important to support and sustain life. All organic matter undergo biodegradation and exhibit a biochemical oxygen demand (BOD) or a chemical oxygen demand (COD) as oxygen is required for their degradation. The typical sources include sewage from domestic and animal source, industrial wastes from tanning operation, food processing, papers mills, natural decaying vegetation and decay of dead plants and animals. This type of water pollution becomes a problem when the oxygen required for biodegradation or chemical oxidation is greater than the available oxygen in the ecosystem. Natural systems are capable to accommodate self purification through decomposition by employing a number of re-oxygenation processes. However, anthropogenic pollution often overwhelms the system.

(2) Disease Causing Agents

There are various types of pathogens entering the aquatic system due to contamination of water by human or animal excreta. Utilization of such wastes lead to the spread of various diseases such as typhoid, cholera, dysentery, hepatitis etc. Pathogens spread to water mainly sporadically and once entered they cannot survive for long. But as a result of man's continuous activities change water systems in such a way that fosters disease. This poses a great danger to human health and the aquatic life. Details about their diseases will be discussed later.

(3) Thermal Pollution

The discharge of hot water from industrial and economical operations may artificially heat up the aquatic environment. Organisms may become physiologically stressed from the exposure to heated waters or can be directly killed if escape is not possible. If artificial heat loading is combined with natural seasonal warming of an aquatic system the impact is quite severe. Besides causing outright death natural life processes like growth rate, respiration, reproduction and the distribution of the species is also hindered. It can also kill the aquatic species through asphyxiation because thermal pollution mainly depletes the oxygen content.

(4) Sediments and Debris

Soil, sands and silt washed from the surrounding lands naturally follow a progression of turning wetlands to dry lands. Man's artificial manipulation of the environment adjacent to wetlands and water bodies from activities like clearing and development activities may release weathered sediments through run-offs. These sediments directly kill the aquatic life. The siltation of rivers with ground bottoms directly destroys the spawning areas of salmon fish. As the sediments are terrestrial it can also include debris discarded by man such as plastic bags into the aquatic system.

(5) Plant Nutrients

Here, the primary focus is on fertilizer chemicals such as nitrates and phosphates, which are most important for plant growth. Too much fertilization encourages the over abundance of plant life and may result in an environmental damage called eutrophication. This process is normal until man accelerates it through sewage treatment plant effluents, use of fertilizers and a variety of point and non-point sources creating significant problems.

(6) *Petroleum Hydrocarbons*

Much oil seeps from the earth as a natural material, however, man's activities pose greater threat to the environment. Direct smothering, acute and chronic toxic effects loss of habitat, etc. can also result. Dairy waste water discharges, leachates from recreational boating, bilge cleaning, underground tank leaks all contribute to pollution.

(7) *Inorganic Chemicals and Minerals*

This type of pollution is generally associated with various types of mining and industrial processing operations. The most common operations include process of that can release acids and heavy metals into the sediments from leaching of raw or processed wastes or through direct discharges and leachates from wastes piles and stock piles of mining operations. Acid rain and run-off not only threaten aquatic life directly by lowering the pH but also indirectly through the above mentioned process and increase the heavy metal burden on the receiving water bodies.

(8) *Synthetic Organic Compounds*

The accidental and intentional discharges of these synthetic potentially hazardous materials have created many cases of serious pollutions. As pure products, by-products and waste products they may have acute or chronic effects on the aquatic ecosystem. These include plastics, plasticizers, solvents, detergents, pesticides, PCB, pharmaceuticals, paints and many other synthetically created compounds. Since they are synthetic the compounds may be resistant to biological degradation and hence accumulate in the environment. The potential for hazardous effects on humans is a primary concern especially from low level chronic exposures in drinking water supply.

(9) *Radioactive Substances*

Radioactive elements can be highly injurious to aquatic organisms by affecting biological processes or can directly be lethal. The sources of such materials include the mining and processing of ores, nuclear weapons manufacturing, testing, handling and disposal of nuclear fuels, the use of nuclear materials for medical, industrial and research applications and the legal and illegal discharges of radioactive wastes.

(10) Biological Pollution

The introduction of a non-indigenous species into an established environment by man, very often has lead to disastrous consequences. The introduction may be intentional or accidental. In other circumstances activity of man has enabled species to invade a habitat whose geography or other conditions have been precluded. The classic case of an aquatic species expanding its range through man's activity is the sea Lamprey. To facilitate shipping the Wetland Canal was constructed to bypass Niagara falls and associated rapids. This enabled the Lamprey to penetrate the lakes beyond lake Ontario. The result was almost total distraction of fishing industry followed by 40 years of eradication efforts at high cost and only limited success.

In western ghats range of Karnataka, seed of Eupetorium was sprayed in the buffer zones, in order to increase the green vegetation. Of course this act has resulted in the increase in the green belt and provided more suitable habitat for understory birds and other wild animals. But in contrast to these benefits the Eupetorium could exterminate all the fodder yielding grasses from areas which lead to the non-availability of food for grazer and the local villagers complained regarding this.

We have just seen the 10 different categories into which water pollution and their pollutants have been divided. The same can be understood better with a simple classification of these pollutants based on their nature i.e. physical, chemical and biological. This is most widely accepted by many authors (Table 9.1).

Physical

The physical characteristics which render water unacceptable for use are temperature, turbidity water colour and suspended floating matter.

Temperature

Thermal pollution is the subject of debate and extensive research has been carried out on the effects of discharging heat coolant water into the aquatic system. A major problem of excessive temperature is oxygen depletion. At 20°C the maximum solubility for oxygen is 9.2 mg/l and the same at 40°C is reduced to 6.6 mg/l. It has been found that at 28°C though species like catfishes and carps could survive while Atlantic salmon required less than 14.2°C during spawning and egg development. Moreover, rapid changes in temperature creates lethal shock effects on respiration and heart beat in fishes.

Table 9.1: Classification of water pollutants

Occurrence	*Nature*	*Examples*
Physical	Temperature	Wasteheat from industries, Thermal plume
	Turbidity Colour	Dye and Pigments Industry effluents
	Suspended and floating matter	Silt, sand, metal, rubber, wood pieces, paper, foam, scum, carcass and sewage.
Chemical	Inorganic	Nitrate, phosphates, chlorides, fluorides, salts etc.
	Organic	Pesticides, detergents, waste oils, tar and Plastics
Biological	Pathogenic	Bacteria, virus, nematodes, worms, protozoa and fungal spores
	Nuisance organisms	Slime producing organisms, molluscans, nematodes and biofouling agents.

Turbidity and water colour

These are of serious impact because they can block the sunlight from reaching the algae and other aquatic vegetation. The result is the retardation of photosynthesis and hence indirectly governs the euphotic zone. Turbidity also has a drastic effect on clams (*Mercenaria*). Tests conducted on animals with 3 g of non-toxic sediment added to 1 L of liquid, all test animals died. Turbidity also stems from masses of free swimming organisms. The dinoflagellates causing red tides are good examples.

Suspended and floating matter

This is often caused by the uncontrolled disposal of sewage, silt, sand and metal pieces, rubber, wood chips, paper, foam and scum. It is mainly due to high human activities near the water bodies.

Chemical

The changes brought about in the chemical characteristics through the domestic and industrial effluents causes the chemical pollution. There are innumerable types of chemicals which can cause pollution.

Inorganic

It was observed that all dead organic matter gets converted into nitrogenous organic matter and finally to nitrate by bacterial activity in polluted waters. It was also pointed out that non-polluted tropical waters are deficient in nitrogen (Ganpathi, 1972). Atmospheric Nitrogen can be fixed by certain bacteria while are mostly added by anthropogenic activities.

Phosphorus is found in the natural water is an extremely small quantity and any amount exceeding 0.5 mg/l have been regarded as indicators of pollution (Atkins and Harris, 1924). High phosphate content is attributed to drainage of unfiltered sewage and synthetic detergents.

Normally uncontaminated waters contain low amounts of chlorides (Hutchinson, 1957). However, the concentration is increased during summer due to decomposition of organic matter. Similarly fluorides are also less 0.5 mg/l in surface water and increase with the pollution. The most devastating among the inorganic chemicals are toxic elements namely heavy metals found normally in minute traces in surface waters. These are cadmium, mercury, lead, arsenic etc. Heavy metals produce toxic effects and endanger the aquatic fauna and flora of which fishes are the most sensitive group.

Organic

Synthetic detergents are toxic water pollutants. Concentrations as low as 1 ppm are toxic to fish. Detergents contain much phosphorus, which is one of the most important fertilizer when added to water it stimulates algal and aquatic vegetation leading to eutrophication.

Several groups of organochlorine compounds are of interest in water pollution. They are mainly discharges of agricultural land and industrial effluents. Because of their low solubility in water they have a tendency to be absorbed on solid surfaces and only traces of these compounds are found in raw and treated waters. Plastics also pose greater danger to aquatic life since these are non bio-degradable. They are left afloat in waters since they are dumped by humans. These are frequently mistaken for prey by turtles and birds leading to fatal consequences. Plastic either obstruct the digestive tract or cause ulcers.

Biological

The biological agents cause water pollution are pathogenic organisms while others are nuisance organisms.

Pathogens

The contamination of water by pathogenic bacteria, viruses, parasites, kind of algae and aquatic vegetation cause additional undesirable results. These organisms are found in water source by conveyance of water from the course to the consumer.

Nuisance organisms

Five kind of nuisance organisms have their greatest impact on aesthetic value and must be kept out of water.

1. Nuisance organisms normally found in drinking water and those are biological slime. These choke the treatment plants and distribution systems.
2. Molluscans choke the water mains in the thermal power plants.
3. Nematodes also become pollutants in possible concentrations.
4. Odour producing algae.
5. Filter clogging algae.

The above mentioned biological pollutants have the following adverse impacts:

(a) State and form objectionable deposits.
(b) Form mats or aggregations of scum and foam etc.
(c) Produce objectionable odour, taste, colour and turbidity.
(d) Toxic or produce undesirable physical response in human, fish, other animals and plants.
(e) Remove oxygen excessively stimulating growth of sludge worms and other creatures.

Sources of Pollution

To address pollution problems it should first be understood how pollution moves through and affects the environment. We need to identify the source and the pathway that enables the pollution to move and the end point of pollution or the sink. The sources of pollution are those from where the pollutants are released and to which extent they are released is very important. This plays a very significant role in solving all pollution problems because of the careless discharges done without giving a thought to its consequences. Fully understanding the source better defines the nature of contaminants and thereby affects the pathway available to it. Hence for the purpose of control and regulation it is convenient to distinguish between point and non-point sources of pollutions from human activities (Fig 9.1).

Point Source

Point source is the one which discharges the pollutants or any effluents such as sewage/waste water through pipes, ditches and sewage treatments plants, electrical power plants, coal mines, oil tankers and off shore oil wells. So far most water pollution control programmes are given importance to reduce the discharge to surface water from industrial and municipal point sources because they are easy to identify.

Non-point Sources

In contrast non-point source of water pollution are widely scattered. The sources that discharges the pollutants over a large area and whose actual source is difficult to identify is termed as non-point source. The run off into surface water seepage, into the ground water from crop lands and live stock feed lands, logged forests, construction areas, roadway and natural process like stoms, siltation etc., are good examples for non-point sources of pollution.

Generally the sources are categorised into domestic and industrial wastes. Though both go hand in hand it has been seen that the pollution caused by domestic sewage is posing several problems to mankind. Domestic sewage contributes about 86-90 per cent of the total water pollution and only the rest is conferred to industrial and natural processes.

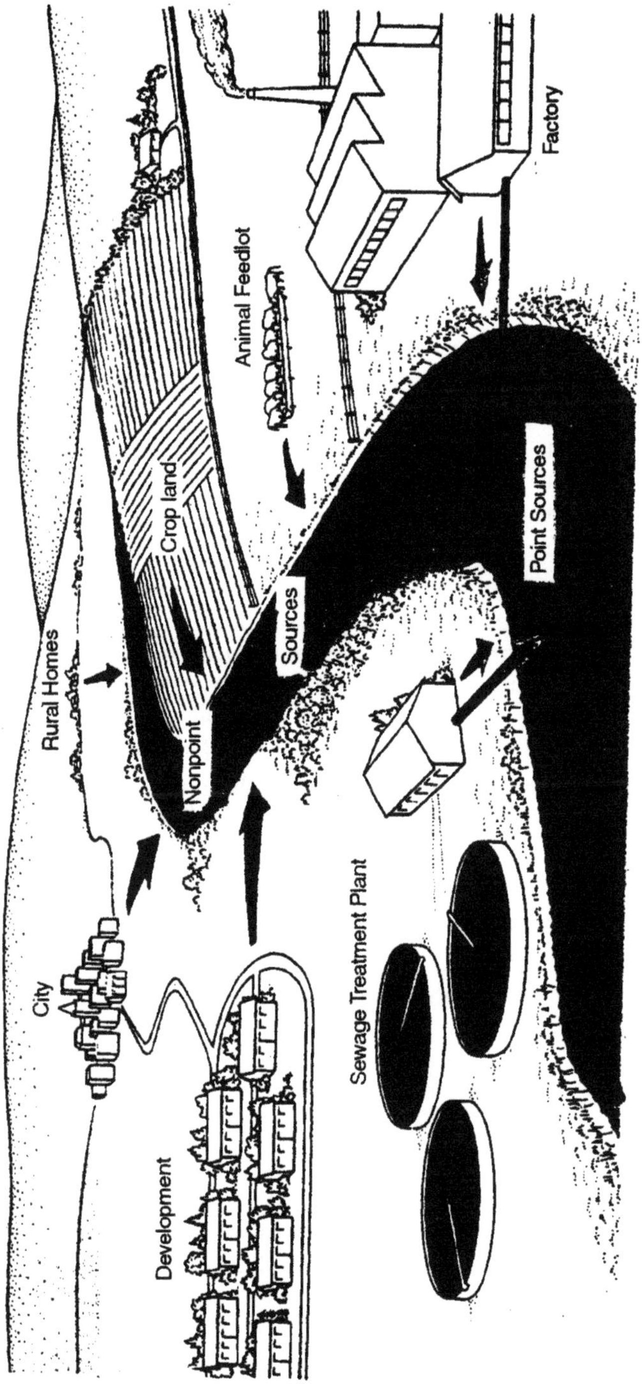

Fig. 9.1: Differentiation between point and non–point sources of water pollution in a typical river

Pollution of Rivers and Streams of India

India is extremely rich in the water resources. The various sources of water can be broadly classified into the following categories:

1. Underground water resources such as springs, infiltration glaciers, well etc.
2. Surface water sources such as rivers, streams, ponds, lakes, oceanic coastlines etc.

Now we shall study only the pollution problems related to rivers and streams of India. Broadly speaking rivers in India are of four kinds. They are:

1. The Himalayan rivers.
2. The rivers of Central India and the deccan.
3. The coastal rivers.
4. The rivers flowing into the interior drainage.

The Himalayan rivers are generally snowfed and continue to flow throughout the year. In their mountain courses they pass through deep gorges. The rivers of Central India and the Deccan are generally rainfed and their volume of water fluctuates considerably throughout the year. Coastal streams especially in the west have limited catchment areas. Most of the streams of interior drainage basin are of ephemeral character. The drain towards the individual basins such as salt lakes and have not outlet to the sea.

Based on the size of their catchment area the rivers are classified into four categories.

(a) Rivers of 20,000 sq. km catchment area and above constitute the major group and are 14 in number.
(b) Rivers between 2,000 and 20,000 sq. km catchment area constitute medium group and are 44 in number.
(c) Rivers with a catchment area below 2000 sq. km belonging to minor group and are 55 in number.
(d) The desert rivers

But unfortunately the fast pace of urbanization in India is creating a formidable problem of insanitation and pollution of aerial, terrestrial, aquatic and land lives of the country. In India there are 5,000 big and middle sized industries. In Mumbai, Calcutta and Delhi the industries pollute only 13, 11 and 10 per cent of the water and the rest of pollution is due to domestic water.

In Delhi about 20 crores litres of sewage water i.e. more than 5 kilometers of untreated waste of houses and markets pours in Yamuna without treatment. Down stream villages on the bank of the river are worst affected by this pollution. According to survey reports 60 per cent of children are infected by the TB germs. In Pali district 74 thousand people of 98 villages have deformities in the back bone due to excess of fluoride in water. More than 11,000 people in Eastern U.P. are suffering due to Cancer and other diseases caused by the toxicity of aluminium, chromium and steel factories.

The municipalities in big cities unaware of health hazards of water pollution or care little about the ecological balance in their respective territories. According to medical experts 2/3 of all Indian illnesses are water borne. It is not surprising that many cities in India are known for highest incidence of water borne diseases such as hookworm, helminths, round worm and protozoal amoebiasis. While small cities top with malaria, dysentery, hepatitis, typhoid and eye diseases.

Every day the effluents from large number of big and small industrial units are being swept into various rivers without treatment. Piles of metals, plastics, pesticides, urea, nitrogen, ammonia, chlorine and other toxic materials are liberally let out into the big rivers of the country.

The extent of pollution depends upon the volume of effluents discharged per day and their characteristics. A rough estimate of discharged waste water can be made from the volume of water required per unit of product and the total production in a particular factory. An estimated use of water in a few industries (billion litres/day) is shown in Table 9.2 and 9.3.

The upper permissible limit of effluent to be discharged into the river water has been proposed by Tebbutt (1978) as follows:

BOD – 20 mg/l, PV value – 15 mg/l, pH 5-9, suspended solids 30 mg/l, cyanamide 0.1, toxic metals below 0.5 mg/l phenols 0.5 mg/l sulphide 1 mg/l. There are many other agencies proposed the standards. Those are Indian standard Institute, Ministry of Works and Housing, WHO, NEERI, Central and state pollution control boards. Some of the characteristics of selected Industrial effluent and sewage in India are shown in (Table 9.4).

Table 9.2: An estimates water consumption by industries by the end of 2010 (Billion litres per day)

Industries	*1968-69*	*1978-79*	*1988-89*	*2009-2010*
Food products	0.12	0.22	0.46	1.37
Textiles	0.3	0.52	1.21	3.45
Mines and metals	5.25	13.42	32.5	106.00
Chemicals	0.41	1.61	4.8	14.2
Petroleum	0.45	1.19	3.54	10.2
Machinery	0.42	1.08	2.88	9.8

Table 9.3: An approximate estimate of water required in various industries

Name of Industry	*Volume of water required*
Dairy	6-10 1 per litre of milk processed
Sugar	1.5 to 4 1 per cane crushed
Distillery	20 1 per litre of alcohol
Cotton textile	10 1 per meter of cloth
Vicose rayon	1600 1 per kg fiber
Pulp and Paper	270-450 1 per kg paper
Tannery	40-50 1 per kg hide
Integrated steel	20-25 1 per kg steel
Urea	6-8 1 per kg urea

In India near the cities the surface water resources such as rivers, ponds and lakes where effluents are discharged from municipal and industrial works are highly polluted. Rapid increase of industries including textiles, rayon, paper, antibiotics, synthetic drugs, oil refineries, photo films, insecticides and several others discharge large quantities of objectionable wastes into the main source of water. Consequently water of all industrial centres viz. Mumbai, Bangalore, Kalyani, Pune, Calcutta, Durgapur, Kanpur, Delhi and others have been severely polluted as discharge of untreated water into the main source is rather a rule than an exception. A few general conclusions were drawn based on studies of such water.

Table 9.4: Characteristics of Selected Industrial Effluent and Sewage of India

	Industry	*pH*	*Suspended solids (SS) (mg/l)*	*BOD (20°C) (mg/l)*	*COD (mg/l)*	*Miscellaneous constituents*	*Pollutional aspects*
1.	Sugar Factory	7.8-8.0	1500-1832	650-820	60-98		Large volume, high pH, SS, colour
2.	Straw board	7.5-12.9	3000	2000	5000	–	High pH, SS and BOD
3.	Tannery	9.5	3200	7000	–	Chromium 15-20 mg/l	High BOD, SS, Cr and colour
4.	Cotton textiles	8.0-11.0	30-50	200-600	–	Detergents, dyes, chromium³ mg/l	Alkali, BOD, dyes and varying chemical quality
5.	Distillery	4.3	4000	29000	65000	Oil and grease	High BOD, SS, grease and ready putrescibility and low pH
6.	Steel mill coke ovens	–	–	630	–	Grease, oil and tar NH_3/N 1000 mg/l, phenols 1300 mg/l ONS	Highly toxic due to phenols Cyanides and ammonia
7.	Steel (finished)	–	310	280	–	Phenol 97.5 mg/l, NH_3/N 440 mg/l	– do –
8.	Refinery	–	–	200	–	180 mg/l and sulphides, oil 30 mg/l, phenol 30 mg/l	Mineral oil and phenols
9.	Fertilizer, Ammonia and Urea	8.0	3700	30	330	NH_3/N 510 mg/l arsenic	Toxic due to free ammonia and promotes eutrophication
10.	Dairy	8.0	690	816	1340	Oil and grease	High BOD, SS, grease and ready putrescibility
11.	Sewage	7-8	200	350	500	Ammonia Albuminoid	

Source: Course Manual on Industrial Wastes Treatment, NEERI, 1975.

1. Quality of water in all respects were deteriorated.
2. Maximum pollution was seen during the monsoons and minimum during winter (January and February).
3. Effect of effluents was variable with distance and nature of the effluent.
4. Coliform content abruptly increased even up to 100 fold in the river water during rainy season.
5. Turbidity and suspended solids increased during monsoon.

Table 9.5 gives a rough idea of the extent of pollution of a few selected Indian rivers.

Zonation of the Polluted Waters

When an Organic loading is placed in a river there is a mixing and acclimation period. This is followed by decomposition over time and distance during which the process of natural purification takes place. The environment downstream of rivers can be divided into zones as depicted in the Figures 9.2 to 9.4. Along with oxygen sag levels there is a corresponding alteration in the biotic and abiotic features of the ecosystem. Metcalf (1978) recognized the saprobic system of a river into 10 different types (Table 9.6).

Effects of Water Pollution

The organic matter play a major role in the water pollution problems of major rivers. Organic pollution results when large quantities of organic compounds which act as substrates for micro-organisms are released into the water courses. They origin from domestic sewage, urban run-off, industrial (trade) effluents and from farm wastes. The origin and their composition of effluents are extremely diverse and we can expect a similar diversity in the effects they have on the receiving system.

(a) Water Borne Diseases

The principal source of water borne diseases is through contamination of water supplies by untreated or partially treated sewage. Other sources include animal wastes contaminating directly or through run-off from feed lots and grazing areas. Certain industries that involve animal products viz. meat packing, tanning and other industries also are potential sources of bacterial contamination of water.

Table 9.5: Showing the Details of the Major Rivers in India

Sl. no.	Name of the River	*Origin*	*Tributaries*	*Catchment Area (in Sq Kms.) / Riverline length (in Kms.)*	Important Towns	Sources of Pollution	*Extent of Pollution*	*Sea/River to which the river joins*
1	2	3	4	5	6	7	8	9
1	Ganga	Gomukh confluence of two rivers Alakananda & Bhagirathi	Alakananda, Bhairathi, Betwa, Kosi, Chambal, Karnali, Yamuna, Ramganga, Kali and Ghaghra	861404 / 9500	Agra Bhagalpur Buxar Haridwar Kanpur Varanasi Hajipur	Effluents from the industries such as tannery, sugar, textiles, diesel locomotives, Heavy electrcials, domestic sewage agricultural runoff	River water is polluted at Haridwar, Agra, Kanpur. Water can be used for bathing & agriculture. For drinking water has to be treated.	Bay of Bengal
2	Indus	Tibet Behind the Great Mountain wall of Himalayas	Jhelum, Sutlaj, Chenab Ravi, Beas	966000 / 5600	Ladak Chitral Mirpur Kapurthala Kyelang	Domestic Sewage recreational tourism and agricultural run–off	Indicates slight organic pollution at Kapurthala can be used for drinking after treatment, high value of BOD kills aquatic life. Water can be used for agricultural purposes.	——

Contd.....

1	2	3	4	5	6	7	8	9
3	Yamuna	Bandar Punch in the Jamnotri Glacier	Tons, Giri, Asan, Chambal, Sind, Betwa and Ken		Delhi Agra Mathura Allahabad Etawaha Kanpur Karual	Effluents from industries such as textiles, paper mills, sugar, recreational sewage, domestic sewage etc.	At Delhi river water contains some toxic substances like As, Cr, Pb, SO which cannot be removed easily by ordinary treatment unfil for drinking organically polluted.	Ganga at Allahabad
4	Brahmputra	Tsang Po in Tibet	Amouchu, Raidak, Sankosh, Manas, Bareli, Subanseri, Dibang	$\frac{580000}{2900}$	Gowhati Dibrugarh	Petroleum industry, Agricultural run–off floods & domestic sewage	Organically polluted at Gowhati. Water is unfit for drinkng.	Bay of Bengal
5	Suvarnarekha	Southwest of Ranchi	——	$\frac{28000}{480}$	Singhbhum Mayurbhanj Midnapur	Industrial effluents from copper & Iron ore industries agricultural run–off	Water has found to contain some toxic elements at down stream of Mayurbhanj.	Bay of Bengal
6	Sabarmati	Mewar hills Gujrat	Sabar, Hathmathi		Ahmedabad Dharoidam Dehagam	Effluents from industries such as chemical dye stuffs, cement, glass, textiles, woollen & domestic sewage.	Water is unfit for drinking & bathing without proper treatment at Ahmedabad.	Bay of Bengal

Contd......

1	2	3	4	5	6	7	8	9
7	Narmada	Amarkantak at Reva	Jawa, Chotwjawa, Kundi	98420 / 1310	Jabalpur Indore Hugli, Hosangabad Khandwa Khargone	Effluents from textiles, glassware & carpet making industries & also agricultural run–off.	Cannot be used for bathing drinking & even for agricultural purposes at Hosangabad.	Arbian Sea
8	Tapti	Satpura plateau	Purna	65145 / 1450	Betul Berar Mandri Surat	Industrial effluents from cotton & silk industries domestic sewage.	Water can not be used for any purpose without treatment at Surat.	Arbian Sea
9	Godavari	Nasik Maharashtra	Purna, Maner, Parnhita, Indravathi, Tal, Sabari, Darna	393389 / 1500	Nasik Nevasa Nanded Paithan Dharmasagar Rajamundry Dhalegaon	Effluents from paper mills textile industries chemicals and agricultural run–off.	Water is unfit for liverstock consumption & also for agricultural purposes.	Bay of Bengal
10	Krishna	Mahabaleswar	Koyna, Ghataprabha, Malaprabha, Bhima, Tunghadra	260000 / 1300	Belgaum Bijapur Vijayawada Sholapur Raichur Ahmedanagar	Effluents from cement, paper, iron & steel, sugar, textle industries and also agrilcultural run–off.	Water is highly polluted at Vijaywada water cannot be used for any purpose without any treatment.	Bay of Bengal

Contd......

1	2	3	4	5	6	7	8	9
					Karnool Bellary Wazirabad Guntoor Shahabad			
11	Mahanadi	Sihawa	Mand, Hasdo, Sheonath, Telong	116000/886	Sambalpur Bolangir Kalahandi Raipur Raigarh Cuttack	Effluents from industries such as cement, jute, paper, textile, vegetable oil, leather, municipal sewage & agricultural run–off.	Water is not suitable for human consumption & bathing without any preliminary treatment.	Bay of Bengal
12	Cauvery	Talakaveri	Shimsha, Laxmanathirtha, Hemavathi, Kabini, Suvarnavathi, Lokapavani, Bhavani, Noyil, Amaravathi, Arkavathi	72765/802	O Srinaga-patnam Mysore Mettur Trirchirapalli T.N. Pura Thanjavur Karaikal Kaveripatnam Coimbatore	Effluents from industries such as paper mills tannery & sugar agricultural run–off, domestic sewage.	Water is not suitable for drinking & bathing at Karaikal. Water has to be treated properly for human consumption.	Bay of Bengal

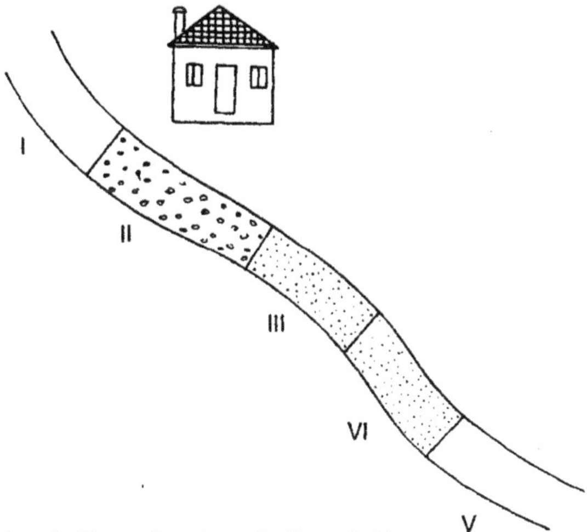

Fig. 9.2:: Simple Zonation in a Polluted River

I. Clear Unpolluted Zone
II. Septic or Gravely Polluted Polysaprobic Zone
III. Zone Under Self Purification or Meso Saprobic Zone
IV. Oligosaprobic Zone
V. Clear Zone

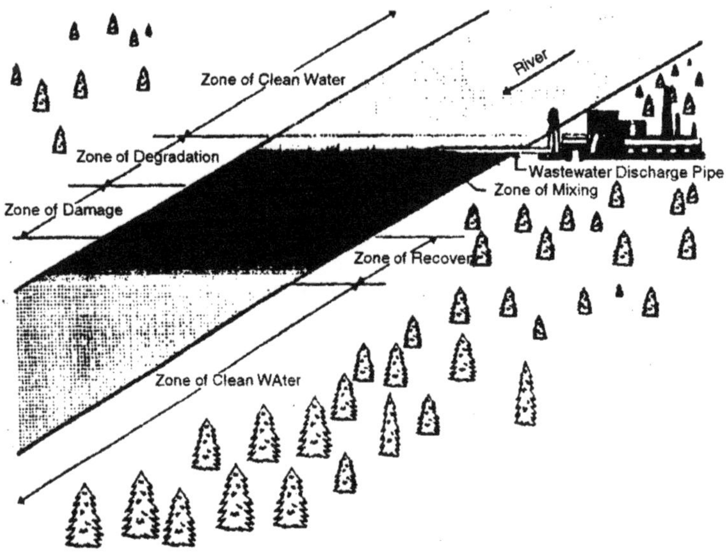

Fig. 9.3: Zonation and suprobic system in polluted a natural stream.

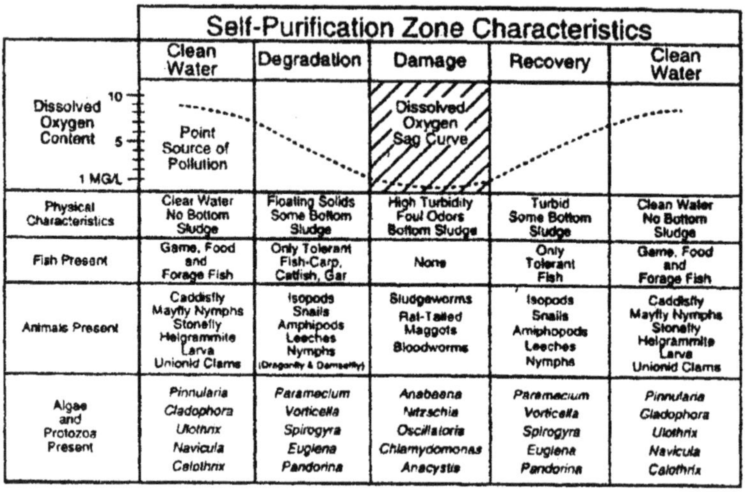

Fig. 9.4: Oxygen sag and self purification zones.

Table 9.6: Classification of Water Quality according to the Saprobic System (after Metcalfe, 1989)

Name	*Rating*	*Description of status with reference to organic pollution*
Katharobic	-1	Drinking water, purest water
Xenosaprobic	0	Pure water, unpolluted, biologically poor zone
Oligosaprobic	1	Clean, healthy, not adversely affected by pollution, game fish zone.
Beta-mesosaprobic	2	Mild to moderate pollution, lower recovery zone, fertile zone.
Alpha-mesosaprobic	3	Distinctly polluted, upper recovery zone, active decomposition, intermediate bacterial decomposition zone.
Polysaprobic	4	Heavily polluted degradation, active bacterial decomposition zone.
Isosaprobic	5	Active decomposition, degradation, sewage.
Metasaprobic	6	Septic, decaying sewage, H_2S zone.
Hypersaprobic	7	Strong decomposition, putrefaction, industrial liquors.
Ultrasaprobic	8	Abiotic lag-phase prior to degradation,.

Pathogenic bacteria in water are the source or vector of some of the most devastating epidemic disease in the world such as cholera, typhoid, dysentery, paratyphoid, tularemia etc. These water borne diseases posed tremendous problem in the developing countries of Africa, Asia and South America. Developed countries eliminated these problems through construction of wastewater collection systems, sewage treatment plants, the use of chlorination and implementation of strict regulations. Fortunately intestinal bacteria survive only for a few days in natural waters. Approximately 90 per cent of the bacterial population will die off within two days. However, the remaining 10 per cent decline much more slowly.

Some of the common pathogens found in potable water causing diseases in humans is listed in Table 9.7 (Classification by Geldreich (1972). Also the non-pathogenic bacteria are used as indicators of water pollution. They are the natural inhabitants of intestinal tracts of warm blooded animals such as *E. coli* and *Streptococcus* sp. Both are excreted in large number in feces averaging about 50 million per gram. A human may discharge as many as 100-140 billion bacteria per day which is almost fourteen times the human population of the world ! Though most of the strains of *E. coli* are non-pathogenic some cause severe infections as *E. coli* 0157: H7. The same strain was believed to be the cause of 2 deaths and illness more than 200 in 1993 from foods served at a fast food restaurant in Riverside California.

(b) Oxygen Sag Curves

When the organic load is discharged into a river it is gradually eliminated due to the activities of micro–organisms by the methods very similar to those occurring in sewage treatment process. Self purification requires sufficient concentrations of oxygen and complex organic molecules are broken down into simple inorganic molecule. Dilution, sedimentation and sunlight also play a part in the process. Esser (1978) has shown that attached micro-organisms in streams play a greater role than suspended organisms in self purification.

The deoxygenation of river caused by organic waste is generally a slow process so that point of maximum deoxygenation depends on a number of factors.

Table 9.7: Water borne diseases of man

Pathogen	*Disease*	*Mode of Entry*	*Symptoms*
Bacteria			
1. *Salmonella typhi*	Typhoid	Drinking water	Continued fever inflammation and ulcer in the intestine. Diarrohea.
2. *Shigella sonnei*	Shigellosis	Drinking water	Acute Diarrohea
3. *Leptospira sp.*	Leptospirosis	Drinking water skin abrasion	Infections to kidney & CNS
4. *Escherichia coli*	Gastroenteritis Diarrohea	Drinking water	Diarrohea, fever vomiting in children & urinary infections
5. *Fransciella tularensis*	Tularemia	Skin abrasion mucus membrane	Chills, fever general weakness
6. *Vibrio cholerae*	Cholera	Drinking water	Acute intestinal disease
Virus			
1. *Hepatitis A Virus*	HAV	Drinking water	fever, malaise fatigue, loss of appetite, jaundice
2. *Polio virus*	Poliomyletis	Occasionally drinking water	Paralysis, fatal
3. *Riovirus & Rotavirus*	Gastro-enteritis	Drinking water	Diarrohea and abdominal discomfort
4. *ECHO virus*	Aseptic meningitis	Drinking water	Diarrohea and abdominal discomfort
Protozoa			
1. *Entoamoeba histolytica*	Amoebic dysentery	Drinking water	Slight diarrhoea, constipation, problems in liver, lungs and brain.
2. *Negleria fowleri*	Amoebic meningoce-phalitis	Swimming in contaminated ponds/pools	Slight diarrhoea, constipation, problems in liver, lungs and brain.

(1) The dilution that occurs when the effluent mixed with the stream.

(2) The BOD of the discharge and of receiving water.

(3) Nature of organic material.

(4) Total organic load in the river.

(5) Temperature.

(6) Extent to which the re-aeration occurs from the atmosphere.

(7) Dissolved oxygen levels in the stream.
(8) Types of bacteria in the effluent.

Figure 9.5 illustrates the generalized curves known as oxygen sag curves which are obtained when dissolved oxygen is plotted against the time of flow at downstream. The curves show slight, moderate and gross pollution. A poor quality effluent contains a high concentration of ammonia and the eventual oxidation of this to nitrite and nitrate may cause further sag in the oxygen curve.

If the oxygen demand of a sewage effluent is studied over a number of days, it is found that oxidation proceeds quite rapidly to begin with but then slows down over a period of 15-20 days. However, there are often two further stages in oxidation which may account for a large proportion of total oxygen consumption. The oxygen demand in the first 20 days is due to the oxidation of organic matter (carbonaceous BOD) while later demand involves oxidation of ammonia to nitrite and then to nitrate.

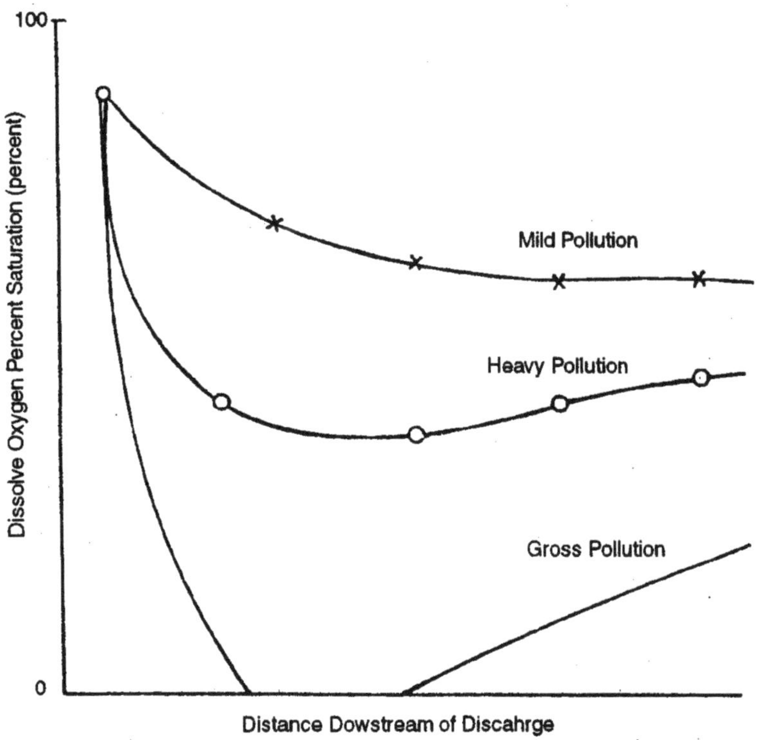

Fig. 9.5: Oxygen sag after discharge of sewage into a river

(c) *Eutrophication*

The pollutants do not have to be toxic themselves to result in the death of a species, but these may lead to destruction of an ecosystem or general degradation of water quality. Their presence is sufficient to set into play a chain of events that can have the same effect. Sometimes the presence of substances required for life in excess quantity receive the label pollutant when they cause an undesirable effect.

The addition of nutrients to an aquatic ecosystem is an example for the last type of pollutant mentioned above. This addition can be seen as an enrichment of the lake that leads to other slow processes collectively referred to as the natural aging of the lake. The enrichment process is generally irreversible and is known as "Eutrophication". Eutrophication will be defined "as an increase in the rate of income of nutrients" (Edmondson, 1974). The income of nutrients may be considered as artificial or cultural eutrophication if the increase is due to human activities or natural as in natural calamities like a forest fire. A number of compounds and elements act as limiting factors but it is only the excess of nitrogen and phosphorus which gives rise to algae nuisances. One of the above two usually act as the limiting factor.

Nitrogen and Phosphorus

Both are discharged by man as a product of wastewater treatment plants, city sewage, waste from animal forms, fertilizers etc. In addition many natural resources such as lightening and volcanoes are also responsible for the process. Leguminous plants also fix nitrogen. Wetland birds contribute significant amounts of nitrogenous waste to water bodies resulting in green-tides of Long Islands in 1950's. Since all these are the rich in nitrogen and nitrate sources, phosphorus is commonly referred as the limiting factor. Only traces of phosphorus exist naturally in the environment. Unfortunately man supplies a significant quantity of phosphorus particularly from treatment plant discharges and specifically from detergents. In the US, it is estimated that 80 per cent of nitrogen and 75 per cent of phosphorus is added to waters through anthropogenic activity. Given that sunlight and nutrients are plenty in such places algae proliferate which are eaten by primary consumers. When cultural eutrophication comes into play, abundant nutrients become available and the food chains and food webs are established.

Algal blooms can occur for long periods of time and extend over large areas having significant impact on the economy of ecosystem and even human activities. The aquatic system became turbid and thick mats of algae are formed. Oxygen levels decrease due to the elevated BOD as decay accelerates. Shallow waters become choked by the rooted plants. The water quantity and conditions deteriorate, when conditions become extremely eutrophic varying degrees of anoxia can develop and most species are eliminated. During this process many nutrients are locked in the benthic sediments so that even if major inputs of nutrients are cut off a lake can remain in eutrophic state for long periods of time drawing on these reserves.

It is easy to see in how many ways eutrophic lakes are significant examples of pollution problem. Not only there is a loss of food fish, but also the loss of a water supply and recreation facility, unpleasant odour and loss of aesthetic values are reported.

Effects of Eutrophication

(1) Species diversity decreases and dominant biota changes.
(2) Plant and animal biomass increases.
(3) Turbidity increases.
(4) Rate of sedimentation increases and shortens the life span of water system.
(5) Anoxic conditions may develop.

Problems to Man Due to Eutrophication

(1) Treatment of potable water may be difficult and the supply may have an unacceptable taste or odour.
(2) The water may be injurious to health.
(3) Increased vegetation may impede water flow and navigation.
(4) The amenity value of the water may decrease.
(5) Commercially important species may disappear.

Therefor it is apparent that the control of eutrophication is both ecologically and economically important. It is impossible to resolve natural eutrophication but is quite possible to address the problem of cultural eutrohpication. The productivity of freshwaters can be controlled by diverting the nutrients and limiting the quantities of nitrogen and phosphorus. The nuisance algae can also be physically removed. Sometimes increasing the circulation during stagnant periods also helps. This may add dissolved oxygen and lower the temperature of the surface waters to shift the conditions optimal for blue green algae. However, the most effective but least desirable is the use of chemicals, the most common chemical is the algicide, copper sulfate which has been used in controlling algae in the reservoirs but these kill more than just the algae (fish and benthic organisms). Other chemical strategies include addition of aluminum iron and calcium based coagulants to directly remove phosphate. Chemicals are also used to shift the pH to make it less favourable for offensive algae. In few cases different species of carp fishes have been used to eat excess algae or other plant life.

(d) *Quality of the Water System*

Organic pollution affects the organisms living in the stream by depleting the available oxygen in the water and thereby causing asphyxiation, by increasing the turbidity of the water and reduce the light available to photosynthetic organisms and by settling out on the bottom of the stream thus altering the characteristics of the substratum.

The general effects of fairly severe organic pollution are illustrated by Haynes (1960). The top part of initial one week carbonaceous BOD is responsible for the oxygen sag curve together with a massive increase of BOD, salts and suspended solids at the point of discharge followed by nitrogenous BOD causing oxygen sag. Later all these decline as the self purification occurs at the downstream reaches. The peak of ammonia is replaced by a peak of nitrates as nitrification proceeds and both are gradually diluted as the polluting load travels downstream (Fig 9.6).

There may be a large increase in the number of bacteria immediately below the outfall and these gradually decline as their substrate is depleted. There is also a large increase in the sewage fungus which disappears later as the stream re-oxygenates. The protozoa are chiefly predators on bacteria and increase in response to bacterial increase and subsequently decline as bacterial numbers fall. Algae especially cladophora increase in numbers as recovery begins, light conditions improve and nutrients are released from the oxidizing organic matter. They decrease as nutrients are used up or diluted.

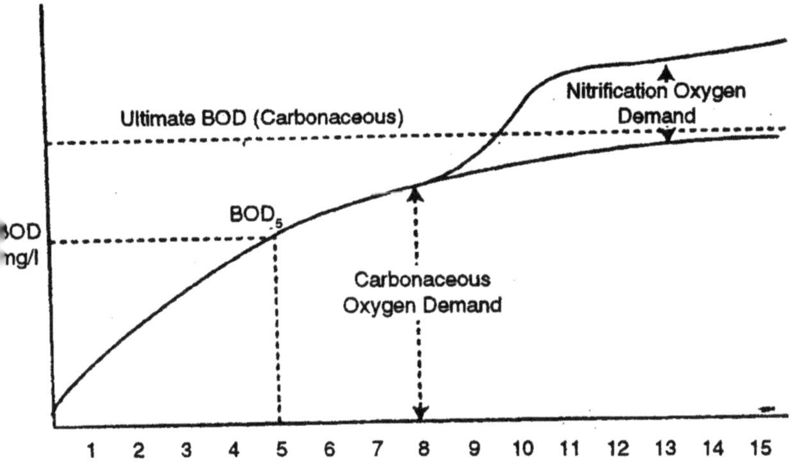

Fig. 9.6: Carbonaceous and nitrogenous BOD exerted by the sewage pollution in a river

The stream animals are eliminated at the point of discharge in the river due to their inability to tolerate the lowered levels of oxygen stress. Sludge worms may be the only macroinvertebrates present immediately below the discharge and in some cases of severe pollution these too may be absent. As conditions improve blood worms larvae of the midges, chironomids become abundant and as further amelioration occurs large populations of the isopod *Asellus sp.* build up. As the stream gradually reoxygenates the clean water fauna increases in the number and diversifies. All these processes occurs until the entire organic matter completely utilised and the water has regained its original clarity.

Case Studies

Pollution of River Tungabhadra

Tungabhadra is a major river of Karnataka state. This river is formed by the confluence of two tributary rivers, Tunga and Bhadra in Shimoga District. The river irrigates a total of about 4,80,000 hectares of land in Raichur and Bellary districts of Karnataka and Rayalseema regions of Andhra Pradesh. Several large and small manufacturing units have come into existence on the banks of this river during the past two decades and has resulted in severe pollution of the once clean river.

The water quality of River Tungabhadra was tested during February 1984 to January 1985 by Dr. B.B. Hosetti and H.S. Patil. The study was restricted to the pollution survey over a reach of 50 kilometres of the river. The river is polluted by M/s Harihar Pulp and Polyfibre Factory (HPF) and M/s Gwalior Rayon Silk Manufacturing Company (GRASIM), Harihar, wastes causing change in colour, pH and nutrients. Samples collections was made at seven points along the River Tungabhadra upstream and downstream of the two industries (Fig 9.7).

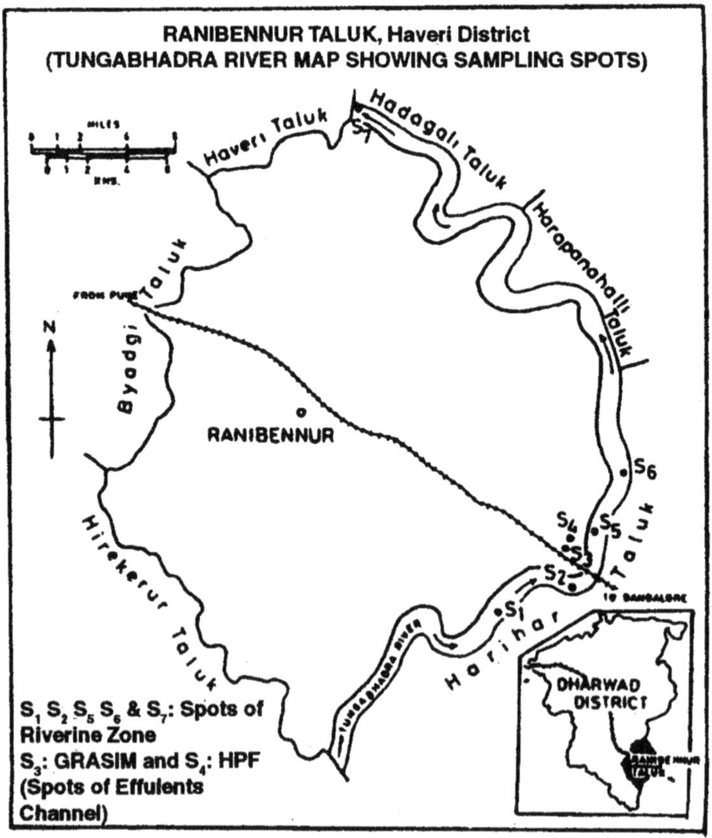

Fig. 9.7: Sampling stations on river Tungabhadra near Harihar town in Haveri District, Karnataka State.

Sampling Stations

- S_1 Unpolluted zone of River Tungabhadra.
- S_2 Below the National Highway bridge (agents of disturbance include washings, vehicles and bathing). This station is 4 km. downstream of S_1.
- S_3 Effluents discharge through a channel from M/s GRASIM. The rate of discharge into the channel is 20.4 m^3/day.
- S_4 Effluents discharge channel of M/s HPF. The rate of effluents discharge is 33.0 m^3/day.

Table 9.8: Characteristics of GRASIM and HPF wastewaters (Average Values ± SE).

	S_3 GRASIM			S_4 HPF		
	Min.	*Max.*	*Ave*[@] *± SE*	*Min.*	*Max.*	*Ave*[@] *± SE*
pH	5.1	6.9	5.72 ± 0.14	7.4	8.6	8.00 ± 0.12
Conductivity*	1400.0	2576.0	1900.90 ± 140.00	504.0	1736.0	906.27 ± 97.20
DO**	0.0	5.2	1.74 ± 0.61	0.0	6.48	2.83 ± 0.7
BOD**	110.0	236.0	181.64 ± 12.36	48.0	260.0	144.58 ± 20.31
NH_3-N**	1.8	3.6	2.43 ± 0.15	1.2	2.6	1.82 ± 0.09
SO_4**	2.0	24.0	15.27 ± 1.85	5.0	28.0	18.46 ± 1.9
PO_4**	1.56	5.2	3.28 ± 0.49	0.47	4.1	2.06 ± 0.31
Catalase***	9.99	15.82	12.87 ± 0.55	8.33	17.49	13.6 ± 1.02

* µ mhos

** mgL^{-1}

*** µ moles of H_2O_2 decomposed per minute per 100 ml of sample

@ Average values of eleven months. (Because during October 1984, effluents discharging channel was dried due to labour strike).

Source: Hosetti and Patil 1988

S_5 200 m downstream of confluence of effluents into the river.

S_6 The first village, Nadiraharalahalli about 2.6 kms. downstream M/s GRASIM and HPF effluents discharge on the left bank.

S_7 Mylar a holy place of Hindus, about 44 kms downstream of effluents discharge points on right bank of the river.

The minimum, maximum and average values of pH, electrical conductivity, BOD, DO, Sulphate and NH_3N and catalase activity, industrial effluents and in the river samples are shown in table 9.8 and 9.9.

Discussion

The colour problem was observed for about 12 kms from the point of confluence of the two industrial effluents into the river. The river was clean and unpolluted at S_1 became coffee coloured due to the disposal of HPF wastes. The colour decreased with dilution and finally regained its original colour at the downstream reaches at a distance of 12 km. Increased colouration also increased the turbidity and appearance of sewage fungi at the downstream seem to cause the growth slime forming bacteria.

Due to the disposal of highly turbid waste by the two industrial units sludge was noticed within a reach of 10-15 km during summer. This also resulted in the decline of DO levels and regular fish kills were observed in the summer months. Optimal levels of DO were recorded at S_1, a sudden decline was evidenced after discharge of HPF and GRASIM effluents at S_5. This was attributed to the wastewater microbes. The value of DO gradually increased at downstream stations and attained super saturation level of 8.0 mg/l at S_7. This may be due to the dilution, diffusion of atmospheric oxygen in the fast flowing water. The BOD was within acceptable limits at S_1 but raised to 166 mg/l at S_5 and then decreased gradually to 63 mg/l at S_7. This reduction in organic load is attributed to the self purification process occurring in the river. The pH of the river water at S_1 (8.0) was increased at S_2 due to the mixing of city waste with slight alkaline in pH. However, the effluents from the two industries vary in pH, one being acidic and the other was alkaline.

Increase in the pH from 7.8 at S_5 to 8.0 at S_7 indicates self purification capability of the river. Changes in ionic concentrations

Table 9.9: Changes in the characteristics of river water (Average values ± SE).

	S_1			S_2			S_5			S_6			S_7		
	Min.	*Max.*	*Ave ± SE*	*Min.*	*Max.*	*Ave ± SE*	*Min.*	*Max.*	*Ave ± SE*	*Min.*	*Max.*	*Ave ± SE*	*Min.*	*Max.*	*Ave ± SE*
pH	7.1	8.5	8.0 ± 0.12	7.2	8.7	8.12 ± 0.13	7.2	8.2	7.79 ± 0.08	7.0	8.4	7.84 ± 0.12	7.12	8.8	7.98 ± 0.15
Conductivity*	25.9	150.2	61.12 ± 12.26	26.62	257.6	84.34 ± 24.71	514.8	1624.0	1073.24 ± 122.73	117.6	504.0	267.73 ± 38.36	31.36	313.6	128.62 ± 28.67
DO**	7.2	10.9	8.67 ± 0.47	6.1	8.2	7.25 ± 0.17	0.0	7.5	2.82 ± 0.75	6.4	8.2	7.19 ± 0.18	7.25	10.8	7.98 ± 0.37
BOD**	2.8	32.0	15.64 ± 2.54	54.0	108.0	74.33 ± 5.41	32.0	166.0	87.58 ± 14.7	31.0	98.0	56.42 ± 6.49	5.8	63.0	28.29 ± 6.02
NH_3-N**	0.03	0.25	0.13 ± 0.01	0.4	1.3	1.0 ± 0.07	0.4	4.0	1.29 ± 0.28	0.5	2.2	1.4 ± 0.16	0.3	1.1	0.72 ± 0.08
SO_4**	0.05	1.1	0.67 ± 0.1	0.07	9.5	3.79 ± 0.77	3.2	17.4	11.37 ± 1.4	1.5	5.6	3.52 ± 0.39	0.2	1.6	0.95 ± 0.11
PO_4**	0.08	0.35	0.18 ± 0.02	0.13	2.0	0.58 ± 0.17	0.21	2.4	1.21 ± 0.23	0.5	2.2	0.95 ± 0.14	0.21	0.9	0.42 ± 0.05
Catalase***	1.66	3.33	2.2 ± 0.16	5.33	9.16	6.85 ± 0.33	4.16	13.32	8.25 ± 0.99	3.33	8.33	5.89 ± 0.54	1.66	7.49	3.67 ± 0.54

* μ mhos
** mgL^{-1}
*** μ moles of H_2O_2 decomposed per minute per 100 ml of sample

Source: Hosetti and Patil, 1988

reflected in conductivity measurements. The initial decrease at S_5 is due to the industrial effluents which was brought back to 130 mhos at S_7. High levels of SO_2 in the effluents was due to the utilization of large quantities of sulfuric acid in the pulping process. The nutrients viz. sulphate, phosphate and NH_3N are utilised by algae. Both PO_4 and SO_4 are reduced as purification increased during flow of the river.

Catalase, a hydrolytic enzyme was normally low in unpolluted waters. It was 2.2 units in the unpolluted zone at S_1 and increased to 6.8 units at S_2. Further it declined gradually reaching a normal level of 3.7 units at S_7. Presently as per the advice of Karnataka State Pollution Control Board the HPF industry has established its own wastewater treatment plants and are treating the effluents by adjusting the pH and by aerobic lagooning process. The final effluents are disposed into the river Tungbhadra as explained above. The treatment is not complete. The colour is also not removed. Sometimes sewage fungi also appear and the fish catch was reduced. Due to power failure and negligence the raw effluents sometimes are directly fed into the river leading to large scale fish kills.

Pollution in the River Ganga

A pollution survey recently conducted by the Central Water Pollution Control Board (CWPCB) has revealed that there are six major cities on the 2500 km course of the Ganga from Gangotri to the sea, where it is severely polluted. The most heavily polluted areas are Varanasi and Kannauj and comparatively less pollution witnessed at Allahabad and Patna. Studies conducted by West Bengal Water Pollution Control Board (WBWPCB) have revealed alarmingly high bacterial counts in the waters of Ganga between Uluberia in Howarah district and Diamond Harbour. This water is now considered unsafe even for irrigation.

Probably the biggest source of pollution in Ganga are the industrial effluents and next serious source is the raw untreated sewage discharged from drains into the river. Another significant source is the dumping of dead bodies, animal carcasses and corpses into the holy river. Washing of dirty things and human excreta along the river banks also contributed to the pollution. Another major source of pollution is the large number of dairy farms in urban areas. These discharge wastes into the river.

The crowded fairs along the Ganga contribute significantly to its pollution. During the bathing festivals, about 40-60 lakhs pilgrims will take bath each day (Kumbha Mela in 1986). The bacterial counts of water rise sharply with the arrival of pilgrims.

Experiments conducted by the Central Pollution Control Board at Ardh Kumbha Mela at Prayag in 1981 showed that a 72 hour pre-chlorination programme kept the bacterial counts in check, and this treatment is now recommended for such occasions. Another suggestion is to stabilise the waste treatment systems at initial low costs, introduce fisheries between municipal outlets of big cities and the Ganga river and installation of electric Crematoria.

Conclusion

For evolving policies for the best use of water resources, it is important to make an assessment of their magnitude, distribution and scope of utilization. It is also necessary to have an idea of the present and future demands of water for different uses. The pollution of water is more complex and of more important due to the following reasons.

1. Slower rate of regeneration and self purification.
2. Sources of contaminants being more diverse.
3. Contaminants of water being more sensitive to propagation of life.

It is necessary to develop a suitable technology to protect at least the drinking water not only against biodegradable organic and biological polluting agents, but also against non-biodegradable wastes and toxic inorganic compounds. Hence attempts should be made to study and examine the system as a whole and not in isolation and then develop a workable model to preserve and maintain water in its best quality.

B. Drinking Water Purification

Introduction

The vital element required for human survival next to air is the water. It is the nature's free gift to human race. It is available in various forms such as water, snow, hail, in rivers, lakes, streams etc. It is so important that the development of any city of the world has taken place near a source of water supply. Water is required for the manufacture of all essential commodities, recreation, industrial activities etc. The UN general assembly has declared the period 1981-1990 as the International Drinking Water Supply and Sanitation Decade (IDWSSD). For any water supply scheme the following points are considered as important.

1. Financial status.
2. Population density.
3. Quality of water.
4. Rate of consumption.
5. Sanitary survey of the area.
6. Source of water supply.
7. Topography of the area.
8. Trends of agglomeration.

Impurities in Water

It is not possible to find pure water in nature. The rain was as it drops down to the surface of earth, it absorbs dust from the atmosphere. This is further exposed to organic matter on the surface of land and by the time it reaches the source of water supply, it is bound to contain various other impurities also which require physical, chemical and biological tests.

Physical tests: Colour, taste and odour, temperature and turbidity.

Chemical tests: Chlorides, dissolved oxygen, hardness, pH, alkalinity, acidity, metals, nitrogen and its salts, total solids chemical elements, BOD and COD.

Bacteriological tests

The examination of water for the presence of bacteria is most important. The following are two standard bacteriological tests adopted to examine water.

a. **Standard Plate Count (SPC):** It is the enumeration of total bacterial count in the given sample and expressed as counts per ml.
b. **Most Probable Number (MPN):** It is nothing but determination of total coliforms present in the sample.

The bacteriological test is carried out in three phases.

1. Presumptive test.
2. Confirmative test.
3. Complete test.

The presumptive test is based on the ability of coliform group to ferment the lactose broth and producing gas. The confirmative test consists of growing cultures of coliform bacteria on a medium which does not supports the growth of other bacteria. The complete test is based on the ability of the culture growth in the confirmed test to again ferment the lactose broth.

Procedure

A definite amount of diluted sample of the waste was taken in multiples of ten in tubes such as 0.1 ml, 1.0 ml, 10.0 ml, 100 ml etc. The water is placed in standard tubes containing lactose broth (5ml). The lactose tube containing 5 ml of lactose broth with 0.1 ml of water; sample is incubated at 37°C for 48 hours. If the gas is formed in the tubes after 48 hours it indicates presence of coliform bacteria and the test is positive. If not the test indicates coliforms are absent and the test is negative. A negative result of the presumptive test indicates that the water is fit for drinking.

Confirmative Test

A small portion of the sample from the positive presumptive test is transferred to another fermentation tube containing brilliant green lactose bile. If gas is not seen in the tube after 48 hours the result is negative, if not the test is positive it requires complete test. Or sample from the positive presumptive test is streaked on EMB or Endos agar medium plates and incubated at 37°C for 48-72 hours. If the colonies with prominent metallic bright sheen and with dark spot on the tip of the colony appears then the test is positive.

Complete Test

The bacterial colonies from the confirmative test broth/plate are inoculated into the lactose broth or streaked on EMB agar plates and incubated for 48 hours. If the gas appeared in the tubes or bright metallic sheen colonies with apical dark spot appeared in the plates the result is positive and the water is considered not safe for drinking further high dose of chlorination is recommended. Recently use of membrane filters is practised. These membranes contain microscopic pores which retain bacteria. The water is filtered, through these membranes and incubated for 20 hours along with nutrients. The bacterial colony appeared if any can be enumerated with help of microscope. The flow chart for drinking water purification is shown as follows.

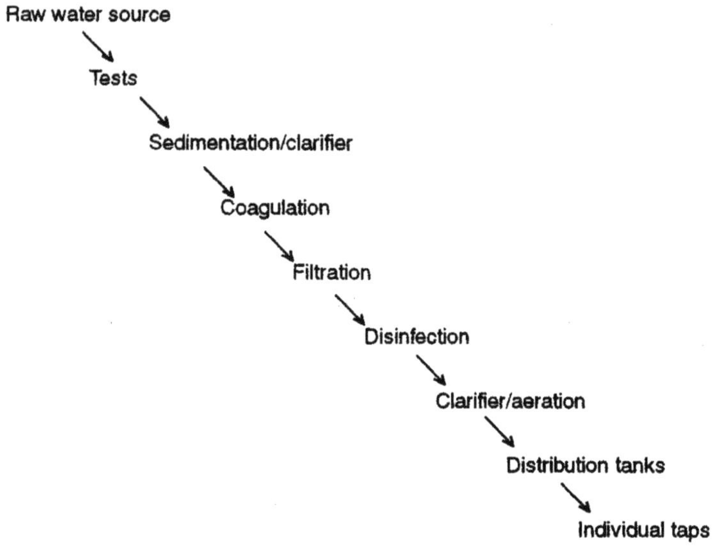

Source

The raw water is drawn from a river or a pond or a stream or from ground water source and is filled in the sedimentation tanks.

Sedimentation

To remove the suspended solids a provision is made in the treatment plant with sedimentation tanks and clarifier units before filtration.

Coagulation

The purpose of coagulation is to make the particles of bigger size to settle by adding coagulants to the water. The coagulants react with the water convert them into settleable size. The coagulants are added when the turbidity of the raw water is more than 40 units. This will be followed by sand filtration.

Coagulants used in the treatment:

1. Aluminum sulfate.
2. Chlorinated copper.

3. Ferrous sulfate and lime.
4. Magnesium carbonate.
5. Polyelectrolytes.
6. Sodium aluminate.

Filters

The process of filtration is most important in the purification of drinking waters. There are two types of filters commonly used for the purpose of filtration.

1. Slow sand filters.
2. Rapid sand filters.

Further rapid sand filters are divisible into two types.

1. Gravity type rapid sand filters.
2. Pressure type rapid sand filters.

Disinfection

When the water leaves the filters, it is still known to contain some impurities like dissolved inorganic salts, colour, odour, taste, iron and manganese etc.

Methods of disinfection:

1. Boiling methods: Suitable only for individual houses.
2. Excess lime treatment.
3. Iodine and bromine treatment.
4. Ozone treatment.
5. Potassium permanganate.
6. Silver treatment.
7. U.V. ray treatment.
8. Chlorination : commonly used.

Chlorination

Chlorine gas and its salts are employed in the disinfection of drinking waters. For large scale treatment chlorine is invariably used as disinfectant. This is because of its following important characteristics.

1. Chlorine is highly soluble in water up to 7000 mg/l.
2. It is readily available as gas, liquid and in powder forms.
3. It is toxic to microbial pathogens.
4. It leaves no harmful residues to man in solution.
5. It produces desired effects which lasts for long time.

6. The treatment by chlorine is cheap and reliable. However, chlorine is a poisonous gas requires careful handling and it may also gives rise to the problems of odour and taste to the distribution water.

Advantages

1. It accomplishes greater purification in few minutes which can be achieved by storage for long period of time. It, thus, avoids construction of costly storage tanks.
2. It is inexpensive and avoids the necessity of storage of raw water.
3. It provides extra security against waterborne diseases.
4. It serves as a convenient accessory to the process of filtration. Chlorine should be properly applied in respective methods and dosage. The contact time should be adequate and that the process should be well controlled.

Properties of Chlorine

Chlorine was first discovered in 1774 by Scheele and it was thought as a compound by the then chemists. But in 1810 Davy established it as an element and gave the present name. It is represented by the symbol Cl with an atomic weight of 35.46 and is 2.49 times heavier than air. It is highly soluble in water. Chlorine water is unstable and its decomposition is rapid on exposure to sunlight. Chlorine gas is greenish yellow in colour and pungent odour which causes irritation in the throat when inhaled.

Action

When chlorine is added to water the following reactions take place.

$$Cl_2 + H_2O \longrightarrow Hcl + HOCl \text{ (Hypochlorous acid)}$$

$$HOCl \longrightarrow H + OCl \text{ (hypochlorite ion)}$$

The hypochlorous acid dissociates by reversible equation into hydrogen ion and hypohlorite ions. It is the hypochlorous acid and hypochlorite ion which are responsible for the disinfection of water. Chlorine action will be more below the pH 7.0. The term free available chlorine is used to indicate the sum of hypochlorous acid and hypochlorite ions and molecular chlorine present in water.

The study on chlorine reveals that it reacts with the cell structure of the bacteria to form chloro-products which becomes poison to bacteria and kill them. The dosage of chlorine depend upon impurities present in the water. It usually varies between 0.2 to 1.0 ppm. If the chlorine is added in excess it may impart bad smell and taste to water. Temperature has also noticeable effect on the process of chlorination as it influences the solubility of gases and other products in water.

Application of Chlorine

1. As bleaching powder.
2. As chloramines.
3. As free chlorine.

Bleaching powder

It is the calcium hypochlorite $(Ca(OCl)_2$. The following reaction occurs when bleaching powder is mixed with water.

$$Ca(OCl)_2 \longrightarrow Ca + OCl$$

The hypochlotie ion so dissociated combine with hydrogen ion present in the water to form hypochlorous acid (HOCl). This process is known as hypochlorination. The bleaching powder is white in colour, it contains 35 per cent active chlorine. It is an unstable compound hence it requires careful handling. It continuously loose chlorine when exposed to atmosphere. In fact application of bleaching powder as a source of chlorine is a tedious job and is expensive. Normal water requires 0.5 to 2.5 kg bleaching powder per million litres of water. Now a days hypochlorination is carried out by commercial compounds such as high test hypochlorite (HTH), pittachlor, piucide, hoodchlor, perchloron etc., instead of bleaching powder. These commercial compounds have 65-70 per cent chlorine content.

Other Advantages of such compounds

1. These are available in powder form in small packets.
2. Their content does not decline after storage.
3. They can be applied to water in dried form or as solution.

Disadvantages

Since these compounds are costly hence these may be used

in small installations and in case of swimming pools but not for public water supplies.

Chloramines

It is understood that chlorine alone is not stable in water, but when mixed with ammonia it forms compounds known as chloramines. These chloramines are stable in water and are found to process disinfecting properties. They reduce odour from water. These reactions involved are shown below.

$$H_2O + Cl_2 \longrightarrow HCl + HOCl$$
$$NH_3 + HOCl \longrightarrow H_2O + NH_2Cl \text{ (Monochloramine)}$$
$$NH_2Cl + HOCl \longrightarrow H_2O + NHCl_2 \text{ (Dichloromine)}$$
$$NHCl_2 + HOCl \longrightarrow H_2O + NCl_3 \text{ (Trichloromine)}$$

These chloramines should be applied at pH range 4.5 to 7.0 Ammonia is generally added to water in the ratio 1 : 05 or 1 : 4. One precautionary measure is necessary when water is treated with the chloramines. That is the water after treatment should be supplied to the consumers only after a retention time of 30 to 60 minutes.

Advantages of Adding Ammonia to Water

1. It is more effective than chlorine alone.
2. The effects lasts for longer time.
3. The quantity of chlorine required is less.
4. There will be less irritation to nose and eyes. Hence it is more useful for treating water for small installations like swimming pool etc.

Free Chlorine Gas

The chlorine can be supplied in gaseous form. It is either dissolved in small quantity of water and then the solution is fed to the water tanks, or the chlorine gas is converted into liquid under pressure and applied. The liquid chlorine is highly corrosive hence care should be taken while handling the chlorine cylinders. The apparatus or device used to regulate the quantity of chlorine to be added to the water is known as chlorinator or chloronome.

Advantage for the use of free chlorine:

1. It can be stored for long time without any deterioration.
2. It occupies less space for storage.

3. Chlorine dose is precise and there is no changes or over dose.
4. Pure chlorine is a best disinfectant.
5. Initial cost of establishment of plant as well as maintenance is less.
6. The cost of transport is not more.
7. No sludge is formed.
8. The results obtained are uniform.
9. The working plant is simple and skilled supervision is not required.

Stages of Chlorination

Depending on the stage of application of chlorine and based on expected results seven types of chlorination are practised.

1. Plain chlorination.
2. Pre-chlorination.
3. Post-chlorination.
4. Double chlorination.
5. Break point chlorination.
6. Super chlorination.
7. Dechlorination.

Plain chlorination

In this process chlorine treatment is given to raw water to control the growth of algae and to remove bacteria. It also removes organic matter and colour from water. Thus, raw water supplied for consumption after giving plain chlorination only. Such process is followed in case of emergency and in the case of more clearer water. The dosage is 0.5 ppm or more.

Pre-chlorination

When chlorine is added to raw water before any treatment, it is referred as pre-chlorination. The chlorine is added in small quantity before the raw water enters into the sedimentation tanks. The dosage is adjusted to 0.1 to 0.5 ppm..

Advantages

1. It reduces taste and odour problems of the water.
2. It improves co-agulation and less quantity of co-agulants are required.
3. It protects filters from clogging.

4. It controls growth of algae in sedimentation tanks and filters.

Post chlorination

In this process chlorination is carried out at the end of treatment process. This is the standard treatment and chlorine is added to water after it leaves the rapid sand filter and before it enters into the distribution system. The dosage of chlorine should be such that, a residual chlorine of about 0.10 to 0.20 ppm appears in water at the end before entering into distribution systems. This small amount will disinfect the water when it flows through distribution system. Thus, it ensures final safety to water and hence it is recommended.

Double chlorination

When chlorine is added to water at more than one stage, it is known as double chlorination. When river water is highly contaminated and contains large amounts of bacterial biomass it becomes necessary to adopt pre-chlorination and post chlorination as well.

Break point chlorination

This term refers to the amount of chlorine required to treat the water. If water is pure and if it has no chlorine demand, any chlorine that added to such water will come out as residual chlorine. This is shown by line P. But in the case of water containing organic matter, chlorine has to perform two functions. That is to remove bacteria and it has to oxidize organic matter (Fig 9.8).

The chlorine added to water to perform the function of removing bacteria and oxidizing organic matter to certain extant. This is represented by point A on curve Q. At this stage if further chlorine is added in water, it is followed by a sudden decrease in residual chlorine content. This stage is some time accompanied by bad smell and taste. This naturally indicates that the extra quantity of chlorine added after point A on the curve Q has been utilised for further to oxidize the organic matter and remove smell and taste. If still chlorine is added up to point B where bad smell and taste disappears on the curve Q at point B and the residual chlorine tends to accumulate as represented line R. The point B on the curve B is known as break point chlorination. That is the

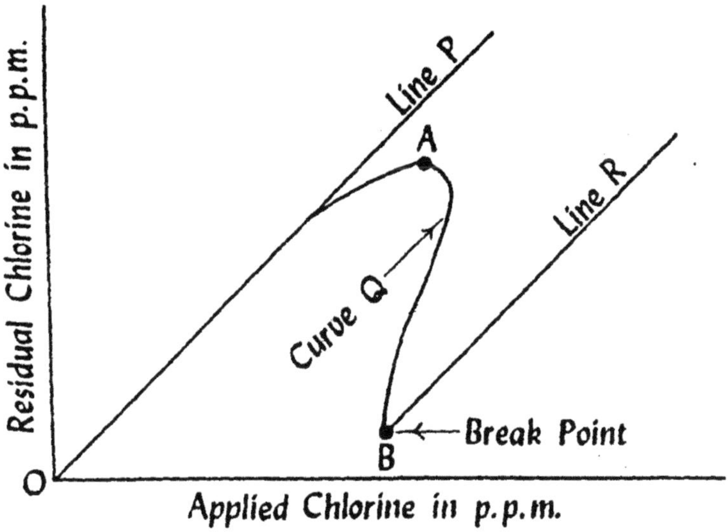

Fig. 9.8: Break point chlorination

chlorine added beyond point B breaks through the water and tends to accumulate and is represented line R. For normal drinking water break point is not necessary, however, swimming pool water requires this treatment.

Super chlorination

The application of chlorine beyond the stage of break point chlorination is called as super chlorination. The residual chlorine after break point may be 0.5 to 2.0 ppm. The excess chlorine may be added at any time or any stage of treatment. But usually it is added at the end of filtration. The super chlorination is adopted when there is epidemic in the locality and water is likely to contain high organic impurities. After super chlorination it is necessary to carry out dechlorination before the water is sent for distribution. In this case water is tested before and after super chlorination and dechlorination.

De-chlorination

It is a process to remove excess chlorine from water. The de-chlorination should be done in such a way that at the end chlorination process some chlorine remains in water. This residual chlorine will disinfect water while flowing through pipes. The compounds usually applied include the following.

1. Sodium thiosulphate.
2. Sodium bisulphate.
3. Sodium bisulphite.
4. Activated carbon.
5. Potassium permanganate.
6. Sulphur dioxide (gas or liquid).

De-chlorination may also be carried out by simple aeration or exposure to sunshine.

Tests for chlorine

The amount of free chlorine and combined chlorine available in water samples is determined by two types of tests.

1. Orthotolidin test.
2. Starch iodine test.

Among the two orthotolidin test is very common and the other, starch iodine test is costly.

Orthotolidin Test

In this test the water sample is collected 10 minutes after the chlorination and 1 ml of orthotolidin solution is added to 100 ml of sample in a glass tube. The colour developed is recorded. The formation of yellow colour indicates the presence of residual chlorine. The deeper is the colour, the greater is the amount of residual chlorine. The exact colour is determined by referring to standard graphs. For potable water lemon yellow colour is considered to be safe. Following points should be remembered while conducting Orthotolidin test.

1. **Blue Tinge Condition:** When water is highly alkaline a blue tinge appears by this test. In such case the amount of orthotolidin solution should be doubled.
2. **Chloramines:** If chloramines are present in water then the colour development will be delayed to 15 minutes instead of 5 minutes after chlorine addition.
3. **Free and Combined Residual Chlorine:** It is possible to distinguish between the free and combined residual

chlorine. That is the free chlorine develops colour within few seconds, where as combined chlorine develops colour after 5 minutes.

4. **Iron, Manganese and Nitrites:** Along with orthotolidin to the water containing above salts sodium arsenate is added.

   a. The sodium arsenate is added first which de-chlorinates the water. Then orthotolidin solution is added, the colour developed is recorded, which is due to iron, manganese and nitrites.

   b. Here orthotolidin is added first and after 5 seconds sodium arsenate is added to arrest the colour formed by the combined chlorine. The colour developed is due to free chlorine and due to iron, maganese and nitrites.

   c. Only orthotolidin is added and the colour developed after 5 minutes is recorded. This colour is due to free chlorine, combined chlorine, Iron, Manganese and nitrites.

The free residual chlorine = a–b
The combined chlorine = b–c

If water does not contain Fe, Mn and nitrites then test C is not required.

Standard Iodine test

The test is conducted as followed.

One litre of water sample is taken in a heat proof vessel. Add 10 ml KI solution to the above water sample. Mix properly, add 5 ml of starch, blue colour develops and the amount of chlorine is determined as follows.

Amount of chlorine = 0.354 x ml of thiosulphate solution (ppm)

C. Abatement of Wastewater Pollution

Basic purpose

The purpose of wastewater treatment is to remove contaminants, in such a way that the treated water can meet the acceptable water quality standards. The quality standards depend upon whether the water will be reused or discharged into a stream. The available wastewater treatment processes are grouped as physical, chemical and biological process.

The physical process are based on exploitation of physical properties of contaminants and is generally simplest form of treatment. The physical processes include, screening, grit removal, sedimentation, floatation and filtration. The chemical process utilize the chemical properties of pollutants in the treatment process. Commonly used chemical process are precipitation, agglutination and the disinfection. The combination of physical and chemical process include air stripping, carbon adsorption, oxidation-reduction, ion exchange reverse osmosis and electroidialysis. Biological treatment process utilize biochemical reactions and they include, biological filtration, activated sludge process, trickling filters, stabilization ponds, aerobic or anaerobic digesters lagooning, septic and imoff tanks.

The wastewater treatment process are generally grouped according to the water quality expected to produce. These process are carried out in three phases such as primary, secondary and tertiary treatment process. The primary treatment removes identifiable suspended solids and floating matter. The secondary treatment process is also known as biological treatment, where the organic matter, soluble or colloidal in nature are removed. The tertiary or advanced treatment also referred as integrated treatment process. These are costly and used to remove a variety of pollutants like, surfactants, inorganic materials, salts, residual soluble non-biodegradable organic compounds. A scheme of classification for wastewater treatment processes are shown in Table 9.10.

Table 9.10: Details About the Coagulants Used in the Treatment

Sl. No.	*Name*	*BOD removed %*	*Dose mg/l*	*pH*	*SS removed %*	*Remarks*
1.	Alum	60	40-90	6.0 to 8.5	80	Not widely used
2.	Chl-Cu	70-80	35-80	5.5 to 7.0	90	
3.	$FeCl_3$	90	30	5.5 to 7.0	90	Widely used
4.	$FeSO_4$	60	40	8.0 to 8.5	80	Used along with lime

Sewage Treatment

The most common form of water pollution control consists of a system of sewers and wastewater treatment plants. Sewage is collected from the sewerage network and pumped or delivered to the treatment plant where it is situated. The methods used to treat sewage before it is discharged into rivers, lakes or in the ocean are not expensive. The treatment is mostly restricted to the removal of bacteria, virus and the removal of heavy toxic metals is not thoroughly evaluated. The process of treating municipal sewage comprises three stages viz., primary, secondary and tertiary stages. Any of these process used are found to be dependent on the volume to be treated, location of the outfall, the dilution factor, the potential hazard to users and receiving streams and cost of the project.

Primary Treatment Process

Primary treatment is a mechanical process which simply removes solids. Metals screens remove large solids, sand and small stones which further settle in the grit chamber and the water passes into the sedimentation tanks. The small particles settle out as sludge in the sedimentation tanks. Scum at the top is removed. The sludge removed at this stage may be huge. Primary treatment is able to remove organic material about 25-30 per cent. The series of steps involved in the primary treatment are shown in Fig. 9.9. and they include, screening, grit removal and sedimentation.

Screens

The main purpose of the installation of the screens is to remove floating matter of comparatively large size. If such materials are not removed, they will choke up the pipes and channels and affect seriously the working of sewage pumps. These screens should be preferably be located just before the grit chambers. If the quality of the grit is not of much importance, as in the case of land filling, screens may be placed even after grit chambers. Screens are usually placed in an inclined position with an angle of about 30-60° with the direction of flow. Screens are sometimes accommodated in the body of grit chamber itself. These screens are of three types.

a. Coarse screens with spacing 40 mm and above.
b. Medium screens with 15 mm to 6 mm.
c. Fine screens with 1.5 mm to 6 mm.

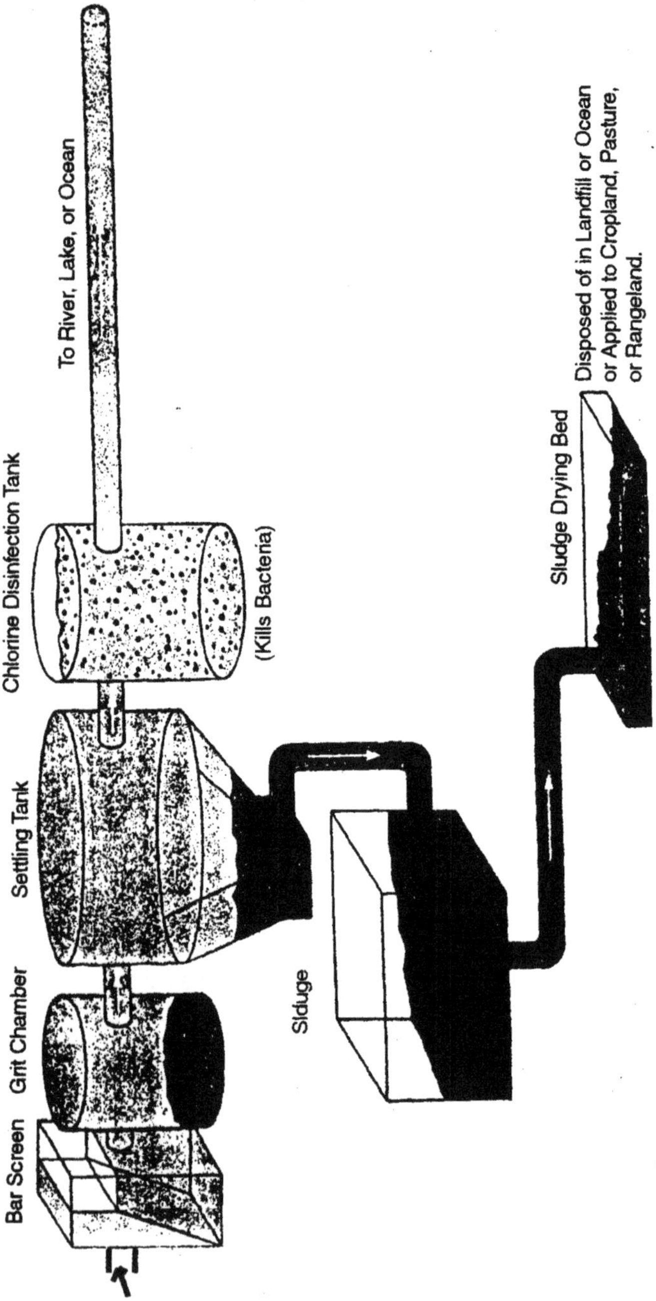

Fig. 9.9: Primary phase of sewage treatment and shaded areas show the sludge mass.

It is the spacing between bars which determines the type of screen. These may be fixed or movable type.

Screens require periodical cleaning which can either be done by hand-rakes or by mechanically operated rakes. A rake is a toothed bar. In hand operated screen laborers clean 2-3 times a day. The teeth of rakes are designed in such a way that they fit in between the spaces of screen bars. In case of mechanically operated screen cleaning is carried out at fixed regular intervals. The rakes are usually provided with time clocks or floating devices to control their working. If the screens are cleaned continuously and automatically clogging of screens will be reduced and the screens will work with more efficiency.

The design factors for screen include the following:

a. The velocity of the flow should be kept reasonably constant and it may be taken as 1m/sec at an average flow of sewage.
b. The slope of screens should be between 30-60°.
c. A perforated platform for temporary dewatering should be provided.
d. The maximum depth of screen should be 1m.
e. The retention period should be assumed zero.

Disposal of Screened Substances

The quantity of substance will depend upon the type of screen and size of the sewage system. Large sewage systems produce small quantity of per capita screenings, due to their long journey, they undergo degradation in the presence of high moisture content (90 per cent). Disposal of screened substances is carried out in four ways which include burial, digestion or disintegration, incineration and composting.

Burial

In this method screened substances are buried in the ground. This method is adopted for small sewage works and the results obtained by this method are not satisfactory. Screening are some time buried along with quicklime. These things should be buried deep into the ground to obtain complete decomposition and to avoid bad odours. Usually a cover of about 30-50 cm depth of porous earth should be provided on the top of the screenings. Digestion or disintegration or dilution of screenings is made with water up to 60-100 percent. Thus, the diluted material is again rejoined to the influent channel.

Incineration

The screened substances are dewatered by pressing or by centrifuging. Pressing is done in Imoff cone like vessel and piston is worked to press. In case of centrifuging method the screenings are placed in a cylinder and the cylinder is rotated. After dewatering the screenings are burnt in coal gas or oil furnace at high temperature.

Fertilizer

Screened materials are rich in plant nutrients and organic substances. Hence these may be used as fertilizer with some precautions. These should not be used as manure for the crops which are eaten raw.

Grit Chamber

Sewage contain both organic and inorganic substances. The purpose of providing grit chamber is to remove grit, sand and other inorganic matter from sewage. To achieve this velocity of flow in grit chamber is decreased to such an extent that the heavier inorganic materials settle down at the bottom of grit chamber and lighter materials are carried for further treatment. In general grit chambers are placed in the pumping stations before or after screens. But there is no fixed rule regarding the location of grit chambers.

Nature of Grit

Grit in sewage is obtained from domestic sources. Grit has a Sp. Gr. of about 2.0 to 2.5. The weight of dry grit is about 130 kg per m^{-3} and weight of wet grit is 1600 kg m^{-3}. Grit contains voids to the extent of 35 to 40 per cent. The grit is originated from domestic source, garages and service stations, storm water, utensil cleaning etc.

Volume of Grit

It will be affected by the following factors:

1. Areas of unpaved surfaces in the locality.
2. Characteristics of ground.
3. Design of grit chambers.

4. Intensity of cleaning the streets.
5. Location of grit chambers.
6. Cleaning methods (streets).
7. Occurrence of storms.
8. Provisions for catch basins.
9. Systems of sewage : combined or separate.

These are provided in duplicate so that any one can be stopped from working for the purpose of cleaning. A standby will be essential for the smooth and efficient working of grit removal by grit chamber. These may be horizontal flow type or vertical flow type. The horizontal type is more popular and is generally adopted in most of the sewage works. In horizontal type sewage enters and leaves the grit chamber in a horizontal direction. It may be with direct flow or with right angle flow as in vertical type.

In the horizontal type inlet and outlet are in one line and in latter case they are perpendicular to each other. In case of the vertical type sewage is brought into the grit chambers through a vertical pipe. It flows in upward directions and leaves at the periphery of chamber. This type is suitable for large installations.

Cleaning Devices

Grit accumulated at the bottom of the grit chamber is to be periodically cleaned. It is done by hand cleaning, hydraulic cleaning and by mechanical cleaning devices. The detention period (1 minute) with flow rate 20-30 cm/sec are designed.

Disposal: The grit collected is used to fill the low lying areas of sometime used as soil conditioner for gardens and crops.

Skimming Tanks

Sewage contain floating substances which is allowed to retain in sewage they may seriously affect the working efficiency of various treatment units. The object of installing a skimming tank is to remove floating substances like grease, soap, wood, pipes, fruits, cane, plastics etc. These floating materials are prominent in industrial effluents. The skimming tanks have a detention time 3 minutes and these are circular or elliptical in shape and 1 m deep.

Sedimentation Tank

These are also known as settling tanks. The particles which are heavier than water are likely to settle down due to the force of gravity in water. There are mainly two types of impurities found in sewage.

1. Inorganic suspended solids having Sp. Gr. 2.65.
2. Organic suspended solids having Sp. Gr. 7.04.

Particle having Sp. Gr. 1.2 will readily settle down at the bottom of the tank, but it is difficult to cause the settling of lighter particles. Stokes law is applicable for particles smaller than the diameter 0.10 mm.

The following objects are attributable for the installation of sedimentation tanks in sewage works.

1. The process of sedimentations reduces the BOD 30-35 per cent.
2. The quantity of settleable solids in sewage is reduced to about 80-90 per cent.

One to three minute detention time make the suspended solids to settle to the bottom. Materials collected at the bottom is called as sludge and treated sewage is called as effluent. Both partially treated sludge and effluents require additional treatment.

Clarifier Unit

These are the modified sedimentation tanks which help in clarifying liquids from settleable materials. Based on the location in the sewage treatment plant these are referred as primary clarifiers and secondary clarifiers.

In some cases coagulants are added to the sewage in sedimentation tanks to enhance the sedimentation of coagulants. The commonly used coagulants include, alum salt, chlorinated copper, line, ferric sulfate, ferric chloride, sodium silicate, and sulfur dioxide. These coagulants react with the colloidal matter in sewage and form the floc which can be removed.

Advantages

1. Chemical sedimentation is more effective than plain sedimentation.
2. It removes high levels of BOD, colour and turbidity.
3. It requires less space and it is a simple process.

Disadvantages

1. Lethal to microbes and responsible for self purification.
2. Cost of chemicals is expensive.

3. Produces large amount of sludge.
4. Requires skilled supervision.

The primary sludge is nuisance because it is bulky and must be removed. It contains more than 90 per cent water. In some cases sludge is dried in drying beds and some part of water is removed by filtering or it is stabilized in sludge drying tanks before drying. Sometime sludge digestion and sedimentation occur in a partitioned structure called Imoff tank. The digested sludge is dried and disposed.

The other products of primary treatment are gases and clarified wastewater. The gas is mostly methane burnt as fuel to provide heat for digesters in the temperate area or it may be wasted. The clarified wastewater is chlorinated and disposed or subjected to secondary treatment.

Secondary Treatment

During primary treatment those material that might be removed by some type of physical or mechanical action. Since much of the organic material in the wastewater may be colloidal or dissolved. The primary treatment process are largely ineffective in removing it. This organic material still represents a high demand for oxygen which must be reduced so that the effluent may be rendered suitable for discharge into water bodies. Secondary treatment (Fig 9.10) is essentially a biological process which is designed to remove most if not all remaining organic matter. Water leaving the secondary conventional treatment facilities normally is down to a BOD of 10 per cent or less of initial 95 per cent of the original. The bacteria are reduced by the tune of 90 per cent. Generally the treatment involves bacteria which are utilised for oxidation. The organic substances present in the effluents serve as food for the bacteria and, thus, within a short period of time they make the effluent free from organic substance by consuming them. The list of biological treatment systems are mentioned below:

1. Trickling filters (TF).
2. Activated sludge plants (ASP).
3. Aerobic and anaerobic digestion.
4. Aerated lagoons.
5. Waste stabilization ponds or oxidation ponds.

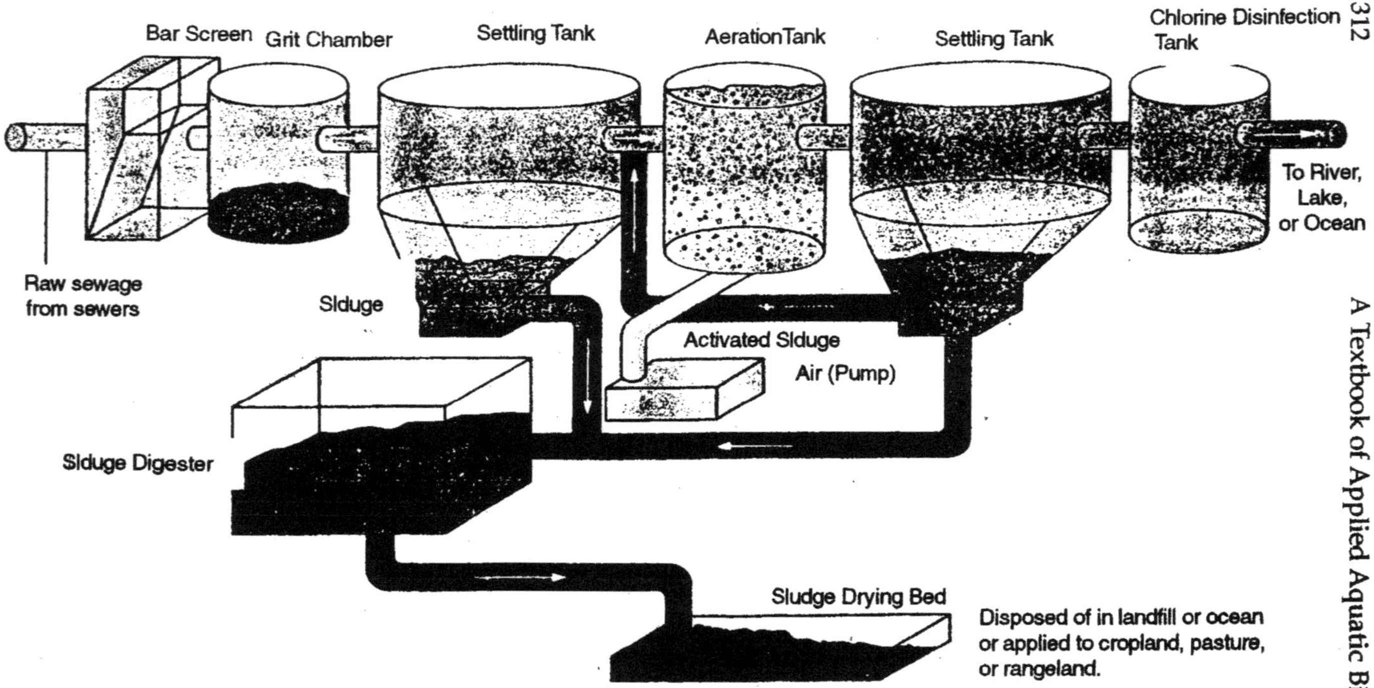

Fig. 9.10: Secondary phase of sewage treatment and shaded areas show the sludge accumulation.

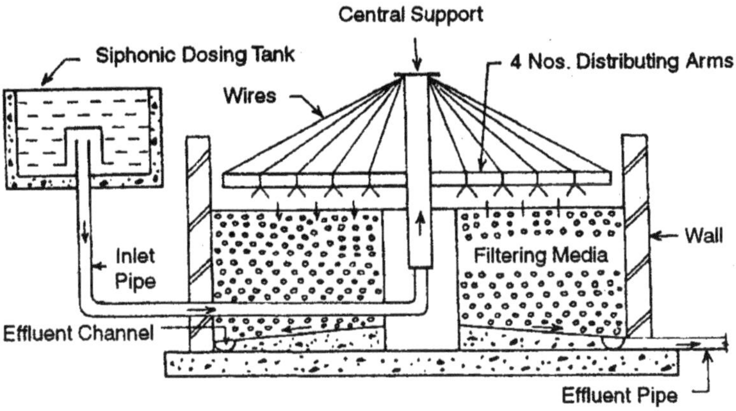

Fig. 9.11: Cross section of a Circular type Trickling filter.

The schematic figure for secondary phase of biological treatment is shown in Fig 9.10. The main purpose of secondary treatment is to remove BOD and solids beyond what is achievable by simple sedimentation. There are three commonly used approaches. In all those biological systems, the ability of microorganisms to convert complex organics into simpler products or low energy compounds is an important step. During the primary treatment process more than 45-50 per cent unstable organic matter is removed. Among the above mentioned devices the trickling filters and activated sludge plants are widely used in the secondary phase of sewage treatment in temperate areas of the world. Whereas in tropical countries waste stabilization ponds are widely employed in the secondary treatment process.

Trickling Filters

These are known as percolating filters or sprinkling filters (Fig 9.11). Sewage is allowed to sprinkle or trickle over a bed of coarse, rough, hard material and then it is collected through under drainage system. Oxidation of organic matter is carried out under aerobic conditions. A bacterial film known as bio-film or biological slime is formed around the particles of filtering media and for the existence of this film oxygen is supplied by intermittent working of the filter and by the ventilation facilities in the body of the filter. The colour of this film is blackish greenish or yellowish. It consists of bacteria, fungi, algae, lichens, protozoa, insect, larvae, rotifers etc.

The concept of tricking filter was made mainly to design a device which overcomes the limitations of a contact bed and the first trickling filter was put into operation in England in 1893. The first municipal installations of trickling filter in the United States established in 1908 since then, they have been used in the biological treatment of sewage elsewhere also.

Types

1. Standard rate trickling filter.
2. High rate trickling filters.

These may rectangular or circular in shape and the circular type or rotatory trickling filter are commonly used.

Filter Media

It may consists of crushed rock or clinker or specially manufactured material. The particles of filter media should be approximately cubical in shape. The filter should not contain any dirt or any other undesirable substances. The size of the particle will depend upon the type of trickling filter and it generally varies from 30 mm to 80 mm. The characteristics of filtering media are as in Table 9.11.

Table 9.11: Characteristics of the filtering media

Sl. No.	*Properties*	*Required level*
1.	Crushing strength	More than 100 N/mm^2 N/mm
2.	Hardness	More than 12 N/mm
3.	Per cent water	< 4
4.	Sp.Gr.	2.6 and above.

Floors

The floor of the trickling filter is generally made of R.C.C. and after trickling the liquid is made to flow towards the central drain or towards the periphery of filter as per design of underdrainage system.

Underdrainage System

Trickling filters are provided with suitable underdrainage to collect the sewage after it has passed through the filtering media. This liquid collected is sent for further treatment or disposal and the velocity of flow in the underdrainage should be adjusted to 60-90 cm/sec. Suitable ventilation should be provided for the successful working of the trickling filter. Ventilation is provided on the following lines

1. Forced ventilation is provided in case of deep filters with the help of fans.
2. Filters are provided with natural draft ventilation in various ways.
3. By constructing filters above the ground with open sides.
4. By keeping their ends of under drains open.
5. By providing vent pipes at the periphery of filters.

Distributor

Its function is to spray sewage evenly on the filter bed through distributors. These may be rotary type or fixed type. Rotary distributors are suitable for circular trickling filter.

Dosing Tanks

It is generally not applied on the surface of filter continuously. It is sprayed or applied 3-5 minutes and then it is stopped or discontinued for 5-10 min. For this automatic siphonic tanks are constructed based on the requirement of sewage treatment. It is done by primary sedimentation followed by filtration. In trickling filter the total suspended solids of the sewage are removed to some extent. The main function of it is to change the nature of suspended solids than to remove them. The design factors, surface area 50-65 m^2/m^3, size of the filter materials 50 mm to 100 mm, loading rate 15-20 kg BOD/day/100 m^3 of filter media, are usually incorporated.

Advantages

(1) The effluents are highly nitrified and stabilized and require little water for dilution.
(2) Mechanical wear and tear is less.
(3) Sludge contain 99 per cent moisture.
(4) Trickling filter operation require little power.
(5) These are flexible in operation. These can withstand the application of sewage with different strength and composition.
(6) These may be reduce BOD and colloidal matter 75-80 per cent respectively.

(7) Working is simple and it does not require skilled supervision.

Disadvantages

(1) Cost of construction is high.
(2) Installation of siphonic head is required.
(3) It may develop fly nuisance and bad order.
(4) The process require primary treatment of raw sewage before sprinkling on the trickling filter beds.
(5) These units require large land area.

Use

These are becoming popular especially in modern treatment plant of sewage.

Activated Sludge Process

This process was first developed in the UK in 1914, by Ardem and Lockett and it is so named because it involved the production of an activated mass of micro-organisms capable of aerobically stabilizing the waste. Various modifications of original process have been made. The term activated sludge is used to indicate the sludge which is obtained by settling sewage in presence of abundant oxygen. The activated sludge is biologically active and it contains a great number of aerobic bacteria and other micro-organisms which have an unusual property to oxidize the organic matter.

The activated sludge based on the degree of aeration looks light brown, golden brown and muddy brown in colour. The activated sludge is mixed with raw or partially treated sewage. Sewage is given the usual primary treatment and then passed into activated sludge process. When activated sludge is mixed properly with sewage containing sufficient quantity of oxygen, microbes present in activated sludge multiply rapidly. As a result organic solids present in sewage are rapidly oxidized and suspended and colloidal matter coagulate and they form a readily settleable precipitate. After such precipitate settles down the effluent obtained in clear. It contains low amount of organic matter and is generally not given further treatment except chlorination. A portion of settled sludge is set for recirculation and the remaining part is sent to sludge digestion tanks. The digested sludge is harmless and does not produce any nuisance.

In this system three basic process are involved:

(1) Mixing of activated sludge.
(2) Aeration.
(3) Settling in secondary clarifier.

The activated sludge is mixed properly with raw or partially treated settled sewage. The activated sludge is added to the influent of primary clarifier. The mixed liquor containing activated sludge and the effluents are agitated or aerated in the aeration tank. This is the most important operation of activated sludge process.

The method of aeration includes three methods:

(1) Diffused aeration.
(2) Mechanical aeration.
(3) Combined aeration.

In case diffused aeration compressed air is blown through the sewage. The diffused aerators are of two types. Those are porous plate diffusers and porous tube diffusers. The aeration period generally varied from 3 to 6 hours/day. The volume of returned activated sludge depends on the rate of BOD removal desired.

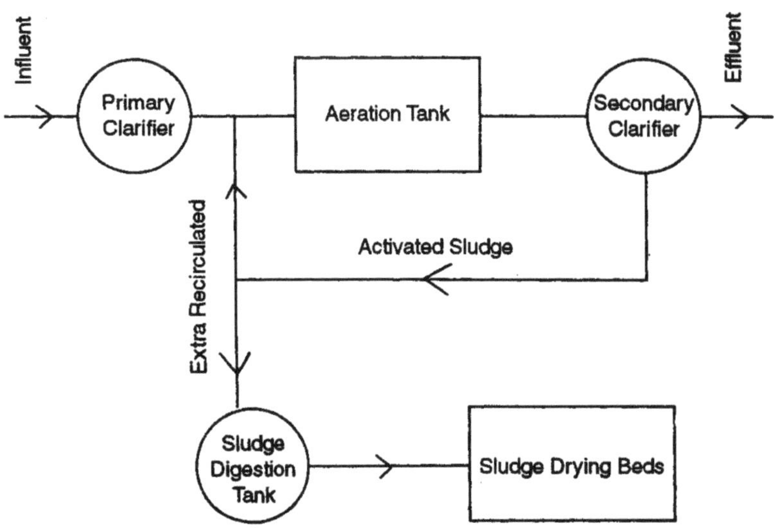

Fig. 9.12: Flow diagram showing activated sludge treatment process.

Table 9.12: Volume of recirculated activated sludge

Desired BOD removal (in ppm)	*Percentage of recirculated sludge*
150	25 per cent
250	30 per cent
300	35 per cent
400	40 per cent
500	48 per cent
600	53 per cent

On an average 6000 to 9000m^3 air will be required per million liters of sewage treated. There are two types aerators used in the ASP.

1. Ridge and furrow type aeration tanks.
2. Spiral flow type aeration tanks.

Ride and Furrow Type Aeration Tank

The ridge and furrow type air diffusers plates are placed in the furrows or depressions and the air is supplied to the diffusers by main pipe through air distributing pipes. The air diffusers plates are placed at right angle to the directions of flow and the compressed air released from the plates forms air bubbles which brings the required aeration. This type of tank is generally 10 m to 12 m long and 3-5 m wide and 3-5 m deep.

Spiral Flow Type Aeration Tank

This type aeration tank requires small quantities of compressed air at low pressure. Spiral motion is setup by the air released through the tube diffusers. The dimension of tube tanks are similar to the above type.

Mechanical Aeration

This is also known as surface aeration and in this method the surface of sewage in the aeration tank is agitated with the help of mechanical aerator to enhance the aeration. This method of aeration continue 6-8 hours/day. In the mechanical aeration system various processes have been developed to achieve the required agitation. Most of such processes are patented and they are shown below.

(1) Hertley process.
(2) Haworth process.
(3) Kessener process.
(4) Link belt process.
(5) Simplex process.

Sludge Bulking

An activated sludge when overloaded, becomes sick and it results in the swelling of sludge in volume. This phenomena is known as sludge bulking. This bulking indicates malfunctioning of the system. Sludge bulking occur due to the following reasons.

1. Accumulation of sludge at the bottom of aeration tanks.
2. Aeration below required degree.
3. Presence of harmful industrial waste.
4. Sudden change in the character of sewage.
5. Unusual detention of sludge in the calrifier.

Sludge bulking leads to degradation in the quality of effluents and hence extra care should be taken to control the development of sludge bulking. It can be controlled by following methods:

1. Chlorination of sewage.
2. Prolonged aeration.
3. Raising of pH of sewage above 8 by adding lime.
4. Re-aeration of returned sewage.
5. Reduction of sewage flow to aeration tank for a short time.
6. Reduction of suspended solids in the sewage.

Advantages

1. Cost of installation is low.
2. Good quality effluents are produced.
3. Process requires small area of land.
4. No–nuisance of fly and midges.

Disadvantages

1. Shock loading and change in volume of sewage will affect the treatment.
2. Cost of operation is high.
3. Bulk sludge obtained requires disposal.
4. It is sensitive to industrial wastes.
5. It requires skilled supervision and more power supply.

Advanced or Tertiary Treatment

The effluents from conventional biological treatment plants still contain some pollutants and they are as follows:

1. Suspended and collided solids.
2. Organic substances escaped from secondary treatment or by products of bacterial metabolism.
3. Nutrients namely phosphorous and nitrogen.
4. Dissolved solids such as chlorides and other mineral salts which are usually present in original water supply and increased during usage.
5. Bacteria and viruses.

The further treatment or polishing of effluents obtained from secondary treatment to meet the higher standards by removing suspended matter and other pollutants is achieved through various integrated processes (Fig 9.13). Further treatment by removing ammonia, nitrogen and phosphorous there by enabling the effluents to become useful directly for domestic purpose or industrial use or to meet special requirements before discharging into a body of water is possible through tertiary phase. Nitrification is considered to be associated with secondary treatments similarly denitrification is also but if additional stage is provided to achieve this, denitrification is considered to be a part of advanced treatment. A schematic diagram for the advanced treatment is shown in Fig 9.13.

Tertiary and advanced treatment techniques are generally understood as the techniques which are applied after conventional secondary treatment in order to attain higher rate of removal of pollutants than is possible with the conventional secondary treatment methods. The suspended solids from a secondary treatment plant comprise of colloidal and non–settleable solids which escape from the clarifier. An approximate 20 mg suspended solids represent about 15 mg BOD. The solid-liquid separation commonly attained through microstraining, mixed media filtration and chemical coagulation.

Microstraining

The microstrainers are mechanical filters that consist of a horizontally mounted drum, the cylindrical surface which is made up of special metallic filter fabric and that rotate slowly in a tank

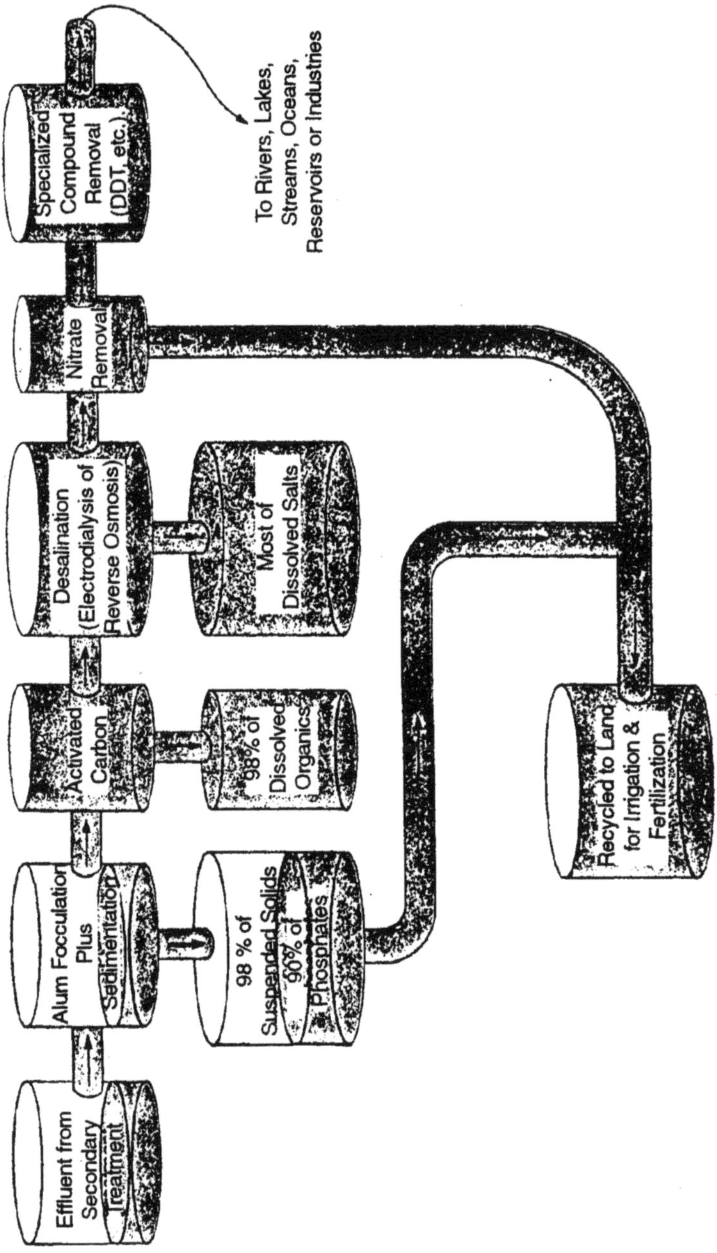

Fig. 9.13: Tertiary or Advance Phase of Sewage Treatment Processes.

with two compartments, so that water enters the drum from one end and flows through the filtering fabric of the drum. The drum is usually submerged 2/3 of its depth. The solids are retained on inside the screen. It has very fine opening of 23-60 microns and are washed from the fabric through a row of jets fitted on the top of the machine. The wastewater containing solids is flushed from the screen is collected in hopper or inside the drum for return to the secondary plant.

Microstrainers are used in plain filtration application can reduce secondary effluents solids from 20-25 mg to 6-10 mg. Microscreens have advantage over granular filters in that they operate continuously without the need of a separate back washing cycle. They have the disadvantage of being more sensitive to reactions in the incoming suspended solids and are not used for removal of chemical floc.

Filtration

In this case the wastewater is passed through a filtering medium such as fine sand or coal to remove suspended or colloidal matter. The effluents from coagulation and sedimentation may contain 3-5 mg suspended solids and 0.5-1.0 mg phosphorous. Efficient filtration of chemical effluents can reduce suspended solids to zero levels and phosphorous to 0.01 kg/l. Plain filtration of activated sludge plant effluents will reduce the solids from 20-25 mg/l to 5-10 mg/l. Filtration of wastewater is achieved by passing the through a granular bed of 75-90 cm deep, composed of 1-5 mm size particles.

Conventional filters such as those used widely in water treatment are made up of only one gradation of sand. Modern multimedia filters are usually made up of 2-3 different materials like coal, sand and garnet etc. Varying size and specific gravities coarse particles and which become fine with the depths.

Chemical Coagulation-Flocculation

Both microstraining and filtration process are capable to remove effectively all the solids. The most effective method of removing residual colloidal solids is through chemical coagulation-flocculation. Solids in wastewater effluents are negatively charged dispersed particles. They can be made to agglomerate into larger particles which may be removed by liquid-solid separation techniques. Coagulation refers to agglomeration of particles by charge reactions and flocculation refers to agglomeration by interparticle bridging.

Dispersed colloidal solids may be induced to coagulate by adding counter ions to the solution. The particle in a coagulation appear as tight pockets which settles rapidly, while a flocculated suspension with fluffy floc settles slowly. Lime, alum and aluminum salts and synthetic organic polymers are used to promote coagulation-flocculation.

Removal of Algae from Oxidation Pond Effluents

The effluents from Waste Stabilization Ponds or Oxidation Ponds is coagulated by the addition of alum and pH is controlled by addition of acid. Air is added through small bubbling. The bubbles attach to coagulated flocks of suspended algae, bacteria and colloidal matter. After further flocculation in the floatation chamber these flocs rise to the surface where they can be scraped off. After thickening the solid concentration of the flock is 7-9 per cent. It is dewatered by centrifuging then the solid concentration is increased to 15 per cent. The float is then dried to 90 per cent solids before it is used for the preparation of some useful by-products like livestock food.

The dissolved air floatation methods harvest 85-90 per cent of the algal biomass. Normally a good success is obtained at pH 6.0-6.5 and to use 100-150 mg/l alum depending on the concentration of suspended solids, alkalinity and phosphorous in the liquids. Some polyelectrolytes and chitosan a natural flocculate taken from the shells of a marine crustacean offers low cost benefit ratio.

Autoflocculation

A term used for the settling of algae with the help of natural coagulation or bioflocculation without addition of chemicals. Algae appear to autoflocculate under stress conditions. Intensive sunlight, high pH and low nitrogen levels have been associated with autoflocculation. The mechanism most probably involves formation of magnesium hydroxide flocs and excretion of polysaccharides by algae and bacteria.

Dissolved Solids Removal

The dissolved organic solids are originated from three

sources. They are, the influent organics which escape the treatment, compounds which cannot be degraded biologically and metabolic by-products of sludge. The treatment techniques which will also reduce soluble fractions of phosphorus and nitrogen thereby reduce the potential of eutrophication. The process commonly used for the removal of dissolved solids from wastewater are activated carbon and ion exchange water sorption phenomena, while electrodialysis and reverse osmosis are membrane phenomena.

Carbon Absorption

Even after secondary treatment coagulation-sedimentation and filtration and soluble organic materials that are resistant to biological breakdown will persist in the effluents. Activated carbon will remove organic contaminants and the colour from water by adsorption. It is the attraction and accumulation of one substance on the surface of another. The amount of carbon surfaces available is most important because adsorption is surface phenomenon. The activated carbon in its manufacture process produce many pores within the particles and there is a vast area of walls on these pores that account for most of total surface area of the carbon that makes it so effective in removing organic substances.

After the capacity of carbon for adsorption is exhausted it can be regenerated by thermal oxidation or chemical treatment. The activated carbon used for wastewater treatment may be in a granular form (0.8 mm) or in a powder form. By combining biological treatment with co-agulation sedimentation-carbon adsorption, a colourless and bacteria free effluent with COD less than 3.0 ppm and BOD 1.0 ppm can be produced.

Ion Exchange

It is a process in which ions are exchanged in between a liquid and a solid surface phase. The ion exchange driving force is defined as the relative affinities of the ions to solid surface. This process is widely used for softening and deionization of potable waters has been found of little use in the secondary treatment, it is because of organic fouling.

Using new microporous anion exchange resins, it has been shown that secondary sewage effluent after passing through microstrainers could be demineraliased satisfactorily. The inorganic salt content of the effluents may be reduced to very low levels and the organic content was reduced by 70-90 per cent. Suspended solids contained in water will clog ion exchange contractors and cause operating problems. A major problem involved in the ion-exchange process is the disposal of concentrated regenerative brines. In effluents the ion exchange process concentrates the undesirable ions into a small waste stream and it is necessary to dissolve off this brine waste.

Electrodialysis

This method of removing part of salt contents of water has been developed extensively for the treatment of brackish waters but has not been found significantly useful in effluent treatment. An electric current is passed through a unit cell consisting of an anode and an ionic permeable membranes. Anions and cations from the water migrate to their respective electrodes resulting in partial desalination of water. In practice many unit cells are created by arranging several alternative anion and cation permeable membranes to form a series of cells between a pair of electrodes.

Since the electrical efficiency of the system requires a low resistance, the salt content of the water can not be reduced to very low level. The method removes most ions to about the some extent and is mot suitable for the removal of complete ionic content. Electrodialysis could be applied to wastewater effluents only if the organic matter were first removed by passing through activated carbon.

Ultrafilteration and Reverse Osmosis

The electrical potential is the driving force in electrodialysis reactions and pressure is utilised in the reverse osmosis process to force a solvent through a semipermeable membrane. The passage of solute is retarded by the membrane so that a purified solvents obtained.

A membrane can be regarded though inaccurately as sieve passing only material having size less than the diameter of its pores. It is now possible to obtain ultrafilteration membrane which will arrest not only suspended particles but also colloidal matter and larger organic solute. Still finer membrane can be made which arrest even univalent ions. By mounting the membrane in a suitable structure and applying pressure to the solution the flow of water may be reversed on the basis of reverse osmosis.

The pressure required must exceed the osmotic pressure of the solution and in practice several times this pressure is needed to achieve a satisfactory rate of flow. At higher pressure the membranes is irreversibly damaged and flow diminishes. In some cases where removal of suspended matter and large organic molecules is inadequate, an ultrafilteration process may be employed. Both process essentially involve separation and yield a product of reclaimed water and a concentrate stream amounting of 10-15 per cent of the total flow which must be disposed off.

The concentrated brines from reverse osmosis, electrodialysis and ion exchange are commonly deposed in evaporation ponds. Thus, ultimate disposal of dissolved matter is on land. The material removed in the activated carbon process is oxidized in the regeneration furnace to carbon dioxide and water and which may be disposed to atmosphere. The activated carbon technique has been accepted as a viable wastewater treatment. However, reverse osmosis, electrodialysis and ion exchange have not been widely applied to wastewater treatment inspire of their high treatment efficiency.

Nitrogen Removal

The inorganic form of nitrogen in wastewater effluent contribute to eutrophication of receiving water bodies. Ammonia can be utilized by some bacteria as an energy source thereby depleting the DO. In the chlorination process of wastewater ammonia produces chloramines which are less effective in disinfection. The micro-organisms metabolize the organic nitrogen and exert an oxygen demand on the receiving water. The two techniques commonly employed for nitrogen removal from secondary effluents are, ammonia stripping and biological nitrification and denitrification.

Ammonia Stripping

The pH of wastewater is elevated to 9.5 or more, at which the ammonia nitrogen exists in solution as NH_3. Although ammonia is very soluble in water, it can be removed by exposure of wastewater to sufficiently large volume of air in a stripping tower. The ammonia then escapes to the atmosphere.

$$NH_3 + H_2O \xrightarrow{\text{hydrolysis}} NH_4 + OH \text{ (K = 2 x } 10^9 \text{ at } 20^{\circ}\text{C)}$$

Biological Nitrification

Aerobic chemo-autotrophic bacteria, *Nitrosomonas*, *Nitrococcus* and *Nitrobacter* in domestic wastewater catabolize ammonia and nitrites to produce nitrates.

$$2\,NH_4^+ + 3O_2 \longrightarrow 2\,NO_2^- + 2H_2O$$

$$2\,NO_2^- + O_2 \longrightarrow 2NO_3^-$$

This type of biological conversion is impossible up to 90 per cent at pH 7.8 to 8.9. Where as the facultative chemo-autotrophic bacteria accomplish biological denitrification and convert nitrate to nitrogen gas.

Phosphorous Removal

Domestic sewage contains both organic and inorganic phosphates which serve as nutrients to algae. The algal blooms utilize only 20-50 percent of the total available phosphorous and the remaining amount is discharged with the secondary effluents. The phosphorous in secondary effluents mainly occurs as orthophosphate which can be readily removed by the treatment with lime salts of ammonium or ferric ion. The phosphorous content in the final effluents may be reduced to 90-95 percent. The cost may be more than conventional treatment methods.

Disinfection

The use of ozone as disinfectant of water is well practised. The ozone, apart from destroying bacteria, it also kills viruses in the wastewater and clarify the colour of the effluents. Chlorination of sewage is not recommended. It is because the residual chlorine in a river may inhibit natural process of self purification. On the other hand the disappearance of residual chlorine, there is evidence that bacterial growth would proceed rapidly leading to extensive

Table 9.13: Chlorine dosage during disinfection

	Effluent Source	*Chlorine dose mg/l*
1.	Primary sedimentation	5-20
2.	Chemical precipitation	2-6
3.	Secondary treatment plant	3-15
4.	Tertiary treatment	1-5

oxygen sag. Another effect of residual chlorine in water that it forms chlorinated phenols, which leads to objectionable taste and odour in water supplies.

Land Treatment

This approach is nothing but use effluents for land irrigation. It is considered as a source of nutrients and water, rather than considering it as disposal problem.

Advantages

Land treatment provides a very advanced degree of treatment without generating any chemical sludge. It recycles nutrients and water contained in the wastewater for productive use and even it may help in reclamation of unproductive land while reducing pressure on irrigation water. Operating costs are less than other tertiary processes.

Disadvantages

1. Large areas is required.
2. Spray irrigation consumes more power.

Sludge Disposal

There are different types of sludges produced in different treatment units such as chemically precipitated sludge. Sludge from primary settling tanks, from activated sludge plants and oxidation ponds. The disposal of sludge do not create any special problem, but it has to be carefully planned. Sludge has objectionable odour. It is bulky and contains large amounts of water. If the moisture content of the sludge is reduced from 96 per cent to 70-80 per cent, it becomes viscous, at 10 per cent of moisture content it becomes dry assumes powder form. Sludge is present in bulk in any particular sewage treatment unit. The quantity of sludge produced depends on total suspended solids incoming from the sewage, likely settlement of solids and moisture content of the sludge.

Methods of Sludge Disposal

Various methods are adopted for the sludge disposal. They include, disposal on land, distribution by pipeline, sludge drying beds, dumping in the ocean, heat drying, incineration, lagooning, vacuum filtering and digestion followed by drying.

Disposal on Land

It can be made either by ploughing or by trenching method. The sludge is mixed with lime and spread on land and ploughed. Then crops can be raised or trenches of 90 cm wide, 60 cm deep are dug in parallel rows at a distance of 150 cm. The trenches are filled with sludge and thin layer of earth is covered on it. The process is repeated by digging new trenches in between old ones. It is a useful method. Under normal conditions one hectare of good land is required to disperse the sludge from about 10000-12000 people.

Distribution by Pipe Line

In this method the sludge is conveyed through the pipe line on nearby farm and is utilised as fertilizer. The success of this method depends on sufficient land available and cooperation from private land owners.

Sludge Drying Beds

In this case the sludge is dried by spreading over the land. Sludge is applied to surface of drying beds to a depth of 20-30 cm. Depth of the sludge depends upon quality and local weather conditions. After about a week or so the surface of the sludge cracks and at this stage sludge is ready for removal. Sludge at this stage contains 70 per cent moisture and is taken out from the beds and dumped into pits for further drying. The so formed sludge cakes are rich in NPK. It can, therefore, utilised as fertilizer and or buried or incinerated or used for filing low land areas.

Dumping into Sea

This is possible where sea is nearby. Sludge should be taken deep into the sea from the shore line, so as to avoid nuisance of sludge being washed back to the shore. It is more economical but lethal to sea fauna.

Heat Drying

In the USA sludge is heated to produce fertilizer which is produced by the Activated sludge plants. It is costly and not practicable in the developing countries.

Incineration

In this method sludge is burnt in incinerators and 5-10 per

cent ash is obtained. The gases produced are used as burning fuel. Incineration proved to be the best to dispose the sludge for areas having limited space for the disposal.

Lagooning

Wet sludge is brought to the lagoon and dumped for the drying by the natural process for 2-6 months through evaporation and percolation. Dried sludge is removed and used as manure. However, by lagooning or ponding develops offensive odour, fly nuisance and possibilities of underground water contamination. The lagoon may be covered with lime or fine soil to reduce the odour and fly nuisance.

Vacuum Filters

This is the dewatering method and cakes may be used as manure.

Reuse of Sewage Effluents

Indirect use of wastewater occurs, when the effluents are discharged into a river or stream and withdrawn at the downstream and used for irrigation or for public water supply. The practice of discharge of sewage water into streams and withdrawal of them for reuse provides dilution and separation in time and space, allowing the process of natural purification to take place. This is true in most of the Indian rivers. In contrast to this, planned reuse of reclaimed waters is practised for several applications with dilution in natural waters.

The sewage effluent reuse include the following:

1. Agricultural irrigation.
2. Urban irrigation.
3. Industrial reuse.
4. Ground water re-charge.
5. Aquaculture.

Water Quality and Regulations

The US EPA regulates treatment plant discharges but does not directly governs water reuse or applications. Regulations for wide reuse applications have been adopted in individual states in USA. But no state has regulations that include all the potential applications. In India wastewater irrigation is not regulated but a similar situation exists. The regulations specify the water quality requirements and treatment process to protect the public health and maintain water quality of the receiving water body.

Other requirements include, storage, monitoring and loading rates for irrigation systems. For example a properly designed agricultural irrigation system is restricted for non-food crops (fodder, fiber, seed and pasture) and also restricted from public access, can reuse secondary effluents without health or environmental risk. In contrast, irrigation of food crops, parks and residences are unrestricted from public access and require advanced treatment for pathogen removal. Types of reuse may be grouped as restricted agricultural irrigation, unrestricted urban irrigation, industrial use and use for recharge of aquifers.

Criteria to be Monitored

The treatment and water quality limits are based on the review of requirements for individual states. In some allow stabilization pond or some other secondary treatment and chlorination before used for irrigation. In general the degree of treatment, disinfection and pathogen removal are related to possible exposure to public. Small treatment systems utilize stabilization ponds for effective removal of helminth eggs and protozoa. Fecal bacteria are also reduced efficiently in tropical areas in ponds with a detention time more than 15 days.

This type of water is suitable for restricted agricultural use and for some industrial applications.

Agricultural Irrigation

Agricultural reuse of wastewater is advantageous because the wastewater contains plant nutrients and solids. Agricultural areas may be adjacent to treatment plants and income is gained by growing cash crops like sugarcane and tree plantations. In arid regions wastewater may be the only source of water available. The water which is not lost in evaporation and percolates to the ground water and drains to surface impoundments with no discharge from irrigation site.

The major contamination are percolation of nitrates to ground water. Retention of heavy metal in the solid and pathogenic hazards to farm workers is objectionable. Storage is required to balance the supply and demand for irrigation. While supply of wastewater is continuous, demand for irrigation depends on growing seasons, crops, application rates and climatic conditions. In areas where treatment is designed for irrigation and not for the increased water quality, storage must ensure an irrigation use without overflow during peak flows. Detention of storage water removes protozoa cysts, viruses and pathogenic bacteria and the retention time is fixed 16-20 days based on the climatic conditions.

Restricted Agricultural Use

Reuse of water irrigation for fodder, fiber and seed crops presents the least chance of human contact. In this case treatment requirements and water quality varies in different states of USA.

Unrestricted Agricultural Use

Unrestricted agricultural use includes water both for processed food crops and food eaten raw. Spray irrigation requires higher degree of water reclamation. Such application in Colorado and Hawaii for processed food growing, consumed raw and vineyards is observed. The secondary treatment with disinfection do not remove helminth eggs, but it requires storage and further treatment like coagulation and filtration.

In urban areas the water may be used for irrigating golf, playgrounds, land scapes and parks. But in such place there should be warning boards stating that irrigation with reclaimed water not for drinking. Even in such situations it is difficult to avoid the risk as still we have poor folk who are uneducated (in India), spend most of their free hours in public areas.

Chapter 10
Waste Stabilization Ponds

History

Since the setting up of the River Pollution Commissions in 1865 and 1868, as a result of foul conditions of many rivers in England, the studies on waste water treatment have been initiated and many treatment methods like dilution method, chemical treatment, land irrigation, septic tanks, biofilters, activated sludge process and stabilization ponds or oxidation ponds have been developed. The oxidation ponds have developed more interest among the sanitary engineers because of their simplicity in operation, maintenance and effluent quality. Oxidation ponds have spread all over the world and became more popular specially in tropical developing countries. The first oxidation pond "The San Antonio Pond" or "Mitchell Lake" aroused due to practice of land disposal of sewage and industrial effluents for irrigation in large area of deficient rainfall in USA.

Importance

These are important biological treatment systems mostly utilized as secondary treatment units to improve the effluent quality. Their position in the treatment process is showed in a Flow Chart 10.1. The domestic sewage alone or domestic sewage mixed with industrial wastes are routinely treated in effluents treatment plants (ETP) or sewage treatment plants (STP) in three phases, viz. preliminary and primary, secondary and tertiary or integrated treatment phase. In the primary phase, the physical properties of the liquid effluents are exploited. That is mainly the removal of colour, removal of solids through screening, and sedimentation is carried out. Later the partially treated wastes are disposed directly into near by aquatic streams. In the secondary phase the effluents are treated in the first phase is further upgraded in the biological treatment systems, like, biological filters, activated sludge plants, waste stabilisation ponds and aerated lagoons. The tertiary phase includes costly measures, by a group of physico-chemical and biochemical techniques. Those include aeration, flocculation, alum treatment, reverse osmosis, ozonation, ion exchange and disinfection processes. The finally treated effluents are reused for irrigation or in the brick formation.

Chart 10.1: Position of oxidation ponds in the treatment process

Sewage →	Preliminary Treatment →	Primary Treatment →	Secondary Treatment →	Tertiary Treatment →	Effluents
	By screening using Grit Chambers, Skimming Tanks	By using Sedimentation Tanks.	By using Biological filters, Activated sludge process & oxidation ponds	Disinfection Chlorination Iodination Bromination	Aeration and use for Irrigation Aquaculture etc.
	(To remove plastic and pieces of wood, grit grease, oil etc).	(To remove 80-90% of settleable solids)	(To remove 45-50% unstable organic matter 80-90% BOD5)	Ozone Treatment and UV Ray treatment	

Principles

Basically the stabilization pond is a complex system comprises of actions and interactions between different groups of microbes viz., bacteria, algae, fungi, rotifera, protozoa and crustacean larvae. These organisms coexist and compete with each other for food and survival. Bacteria being saprophytes decompose the organic matter into simple forms like phosphates, ammonia, carbon dioxide and water. The autotrophic algae utilize much of these simple compounds as nutrients and synthesize fresh biomass through the process of photosynthesis. During photosynthesis sufficient amount of molecular oxygen is released by algae, which is utilized by bacteria for the oxidation of organic matter and,

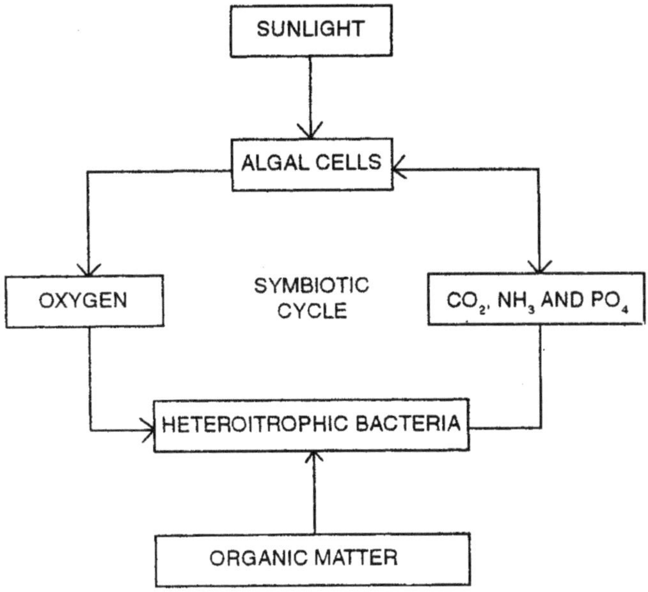

Fig. 10.1: Algal-bacterial symbiosis

thus, complete algal bacterial cycle (Fig 10.1). Weidemann and Bold (1965), also included fungi in the symbiotic cycle and explained that the nutritional aspects of bacteria, algae and fungi are interrelated.

Types of Ponds

Waste stabilization ponds are classified on the basis of the availability of oxygen for the degradation process into 3 major groups. They are aerobic ponds, anaerobic ponds and facultative ponds. According to Hosetti and Frost (1998) the stabilization ponds are classified into 10 types.

Aerobic Ponds
Anaerobic Ponds
Polishing Ponds
Facultative Ponds
Night Soil Ponds
Macrophyte Ponds
High rate algae ponds

Maturation Ponds
Fish Ponds
Storage Ponds

Aerobic Ponds

Aerobic ponds are shallow structures of about 0.3 m depth or less, so designed as to maximize light penetration and growth of algae. Photosynthetic action, aerobic conditions are maintained throughout the depth of the pond at all times.

Anaerobic Stabilization Ponds

These do not require dissolved oxygen for microbial activity as the anaerobic and facultative organisms use oxygen from compounds like nitrates and sulfates as their hydrogen acceptors and give end products like methane and CO_2. Such ponds can, therefore, accept higher organic loadings and operate without algal photosynthesis.

Facultative Stabilization Ponds

These ponds are partly aerobic and anaerobic, they are often about 1-2 m in depth, favouring the growth of facultative algae, fungi and bacteria. Such ponds are predominantly aerobic during the sunshine hours as well as for some hours of the night. In the few remaining hours, the pond bottom waters may turn anaerobic. Most of the existing waste stabilization ponds in the world are of facultative type with varying degree of aerobicity and anaerobicity conditions. A schematic figure explaining the functioning of waste stabilization pond is printed in Fig. 10.2.

Biological Activities

The stabilization pond contains various bacteria, algae, fungi, viruses, protozoa, rotifera, nematodes, insects and crustaceans. These groups of organisms commonly found in the pond effluents are shown in Plate 10.1 to 10.3. These ponds often acquire aquatic weeds. The biology of a stabilization pond is normally explained with reference to facultative types of ponds with diverse groups of organisms. Stabilization ponds are anaerobic when algae and green plants are absent.

Initial research on stabilization ponds describe pond activity in terms of the mutualisitic behaviour of algae and bacteria via

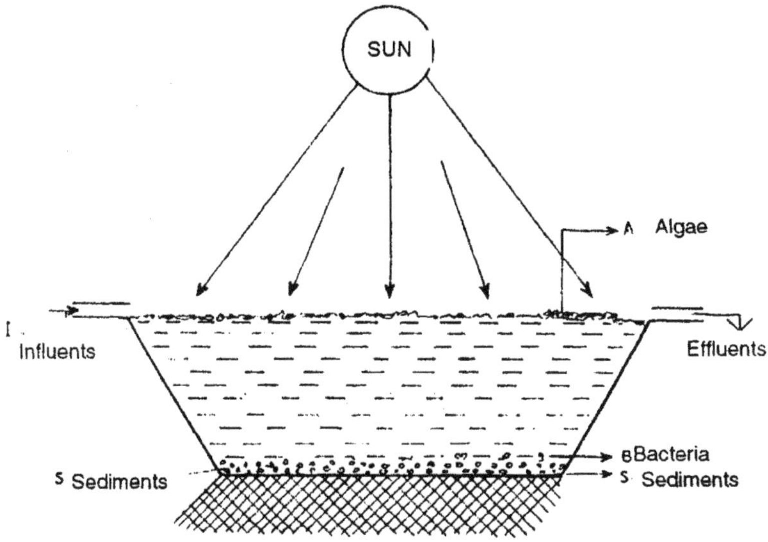

Fig. 10.2: Schematic figure of an oxidation pond

photosynthesis. It is only after 1970, that the other members of the community, namely fungi and protozoa are considered important for the stabilization process. Like bacteria, fungi also degrade organic matter to release nutrients. Protozoa and invertebrates graze on suspended particles and bacterial cells which subsequently help to upgrade the water quality. Each ecological group interacts during the treatment of wastewater in stabilization ponds. Excess algae and bacteria produced in the ponds leave with the effluents or settle to the bottom.

Bacteria

The role of bacteria in wastewater purification was first reported by Johnson (1914). Bacteria necessary for the stabilization of organic load are abundant in soil, washing and in sewage. According to the conditions in the pond aerobic, facultative and anaerobic bacteria grow and stabilize the organic substances present in the wastes. The ability of bacterial flocculation is one of the important phenomena on which all biological wastewater treatment systems depend for separation of colloidal organic substances from liquid phase through bacterial action. The following genera are reported to be commonly present in the oxidation ponds.

Achromobactor sp.
Proteus alkeligens
Chromomonas zoogloea
Pseudomonas sp.
Thiospirillus sp.
Thiopedia sp.
Rhodothecae sp.
Nitromonas sp.
Nitrobactor sp.

Algae

These are oxygenators and important protein resources. Study of algal types and density provides a useful indication of pond status and effluent quality. These algae include the following genera. *Chlorella* sp., *Scenedesmus* sp., *Micractinium* sp., *Chlorococcum* sp., *Chlorogonium* sp., *Coelastrum* sp., *Gonium* sp., *Ankistrodesmes* sp., *Eudorina* sp., *Actinastrum* sp., *Pandorina* sp., *Pyrobotrys* etc. Blue green algae common to stabilization ponds include. *Oscillatoria* sp., *Spirulina* sp., *Arthospira* sp., *Phormidium* sp., *Merismopedia* sp., *Anabaena* sp., *Anacystis* sp. and *Microcystis* sp. Some flagellates commonly found are, *Euglena* sp., *Phacus* sp., *Trachelomonas* sp. and *Chlamydomonas* sp. The most common diatoms are *Navicula* sp., *Cyclotella* sp., *Asterionella* sp., *Tabellaria* sp., *Melosira* sp., *Fragillaria* sp., *Synedra* sp.

Algal biomass fluctuates with the environmental changes associated with the diurnal and seasonal variations in the climate. The independent factors, like zooplankton grazing and transient chemical toxicity and pathogenicity will also affect algal density. Algae concentrate and tend to microstaratify throughout the volume of the pond (Hosetti *et al.*, 1984).

Protozoa

In wastewater treatment process protozoans have been reported to be present in more numbers particularly during prestabilization period. These include colourless microflagellates and ciliates. Among ciliates, holotrichales are the most prominent forms, mainly feed on bacteria and suspended solids. These all finally assist in the improvement of the effluent quality.

Patil (1979) carried out experiments on protozoa in the removal of califorms and pathogenic bacteria and suggested that before algal bloom begins and increases, the pH is beyond detrimental level, it is protozoa that cause the reduction in bacterial population. Also the presence of *Vorticella, Microstoma, Colpodium, Paramecium* and *Podohyra* in the scum formed at the surface of oxidation ponds, during prestabilization and over loading periods was reported. Various other genera such as *Stylonchia, Burasaria* and *Perispira* help in the purification of sewage in pond.

Fungi

The most common fungi found in stabilization pond are, *Mucor fregilis, Saprolegnia* sp., *Monilia* sp., *Candida albicans, Trichoderma koningi, Aspergillus niger, Achlya* sp. and *Candida tropicans.*

Factors affecting the Pond Performance

Temperature and light intensity are the major environmental factors that govern the growth and activities of microbes in the ponds. In addition to the above, detention time, loading factor, depth and succession of organisms also influence the pond performance.

Temperature

Temperature greatly influence the type and number of saprophytes, oxygen production and various biochemical reactions that taking place in stabilization ponds. Temperature below 15°C is found to lower the pond performance and temperature between 15-30°C enhance the pond functioning. Different groups of micro-organisms prefer different range of temperatures. For example, bacteria can grow in a wide range of temperature. According to the temperature variation bacteria are classified as psychrophils, mesophiis and thermophils.

The saprophytic bacteria are particularly found to be more active in oxidation pond. Where as psychrophilic bacteria are reported to grow at temperatures range 0.0 to 4.0°C in ponds of cooler areas. *E. coli* grows well at temperature range between 10°C to 45°C and optimum is 37°C. The minimum and maximum limits of temperature for the algal growth are reported to be 14°C and 37°C respectively. Different groups of algae have different temperature optima. The blue green algae grow well at 35°C; while 33°C to 35°C have been shown to be favourable for the green algal growth and 27-37°C for euglenoids.

Light Intensity

Algae are the main source of oxygen production required for the oxidation of organic matter in the ponds. The photosynthesis by algae is governed by light intensity and day length. Hence it forms one of the major factors that influences the overall performance of the pond. The optimum requirements of light intensity differs from species to species. A light intensity of 200 to 400 fc was shown to be essential for the growth of most common freshwater algae. Light intensity above 1000 to 2000 fc may even inhibit the photosynthetic activity. The photoperiod also influences the algal growth rate.

Detention Time (DT)

The principal factor in the design of aerobic ponds is DT. Meron *et al.* (1965) based on the studies on ponds proposed that the overall purification process can be divided into 3 phases as a function of detention time. This study indicates that 7 to 10 days detention time is found to be optimum for maximum BOD reduction and chemical changes resulting from the intense biological activity and algal production in the tropical areas. The DT in a pond is determined primarily by the degree of bacterial purification required. Larger DT facilitates maximum removal of BOD, bacteria and nutrients.

Depth

Among the various factors depth seems to play a vital role in the pond performance. A range of 4-6 feet depth has been universally accepted as standard depth for facultative stabilization ponds. The findings of earlier workers show that shallow ponds do not provide the best stabilization of organic substances and are conductive to emergent vegetation and mosquito breeding. In addition the sludge layer in shallow ponds may turn into septic and produce bad odour.

Chemical Factors

Effluents from biological treatment units are usually monitored in terms of DO, BOD, bacteria, pH, phosphates, sulfates, ammonia-nitrogen and suspended solids. Recently enzymes are also included to define the quality of effluents.

Hydrogen Ion Concentration

The pH of the algae ponds, in general remains on the alkaline range. Variation in pH from 7.0 to 11.0 has been shown to be associated with the algal growth and activity in sewage ponds. Seasonal and diurnal variations in the pH of oxidation pond have also been reported. This increase in pH is due to the extraction of CO_2 and bicarbonates by autotrophs from the medium. The pH range of 6.5-8.0 is shown to be optimum for most of the bacterial growth. Algae have shown to grow well at a wide range of pH. Protozoans were affected to a little extent by the variations in pH. Vorticellids were much sensitive to pH variations. These are found to grow well between the pH range 6.4-8.3

Biological Oxygen Demand (BOD_5)

The biological or biochemical oxygen demand (BOD) is an important parameter used to measure the pollution load or organic strength of wastewaters. BOD is measured by incubating the BOD bottles containing the known aliquot of waste sample at 20°C for 5 days. Hence it is referred to as BOD_5. The BOD of domestic sewage ranges between 200-400 mg/l. Effluents of oxidation ponds are measured for BOD. Oxidation ponds are able to remove the BOD to the extent of 80-90 per cent in tropical countries and 50-60 per cent in temperate countries. A study conducted by Hosetti *et al.* (1988) on the BOD removal efficiency of pond community individually and in combination revealed that *Chlorella vulgaris (83.3 per cent), Scenedesmus quadricauds (80.8 per cent), Tolypotrix tenvis (74.1 per cent), Lhygbya allorgei (78.2 per cent), Nostoc punctiformae (82.5 per cent), Escherichia coli (55 per cent), Shigella sonnei (51.6 per cent), Staphylococcus aureus (47.5 per cent), Bacillus subtillus (50.6 per cent), Paramecium caudatum (68.1 per cent)* and combination of all the members (79.4 per cent) removed varied levels of BOD. The comparative study revealed that the role of pond community in the removal of BOD is found to be beneficial than individual organisms (Table 10.1).

Table 10.1: BOD removal capability of certain micro-organisms grown in animal wastes

Micro–organism	*Days*								*Per cent*
	0	*3*	*6*	*9*	*12*	*15*	*18*	*21*	*removal on day 21*
Complex pond community	320	355	280	220	184	96	78	66	79.4
				Algae					
Chlorella vulgaris	320	290	280	276	244	90	67	53	83.3
Scenedesmus quadricauda	320	300	284	192	146	96	72	61	80.8
Tolypothrix tenvis	320	232	212	188	155	145	121	83	74.1
Lyngbya allorgei	320	232	192	144	118	94	82	70	78.2
Nostoc punciformae	320	242	186	104	97	67	63	56	82.5
				Bacteria					
Escherichia coli	320	264	482	290	240	216	180	144	55.0
Shigella sonnei	320	278	436	232	220	245	186	155	51.6
Staphylococcus aureus	320	336	354	322	280	185	154	152	47.5
Bacillus subtilis	320	348	402	266	205	234	180	158	50.6
				Protozoa					
Paramecium caudatum	320	353	252	190	172	112	106	102	68.1

Source: Hosetti *et al.* (1988).

Nitrogen

Unicellular algae are effective in removing nitrogen from waste waters. *Spirulina maxima* absorbed 100 per cent nitrogen from domestic sewage under laboratory conditions. The nitrogen removal in stabilization ponds may reach 85 percent or more according to environmental conditions.

Phosphorus

Algae of different species contain variable amounts of phosphorus, collectively, planktonic algae contain the largest amount of organic phosphorus fraction in the water column. Some algae obtain phosphorus from detergent wasters and many species store phosphorus granules within their cells. The most effective removal of phosphate occurs when stabilization ponds are operated in series.

Carbon

Carbon in the gaseous form and dissolved CO_2 or as bicarbonates in solution may be utilised by algae in photosynthesis. Carbon removal and recycling in pond system is linked with transformation of nitrogen, phosphorus and sulfur.

The flora of the bottom layers of the facultative ponds and anaerobic ponds include a group of mesophilic bacteria such as *Escherichia coli, Flavobacterium sp., Alkaligenes sp.* and *Pseudomonas sp.* They decompose proteins, fats and carbohydrates into the respective aminoacids, glycerides and monosaccharides. These are further reduced to methane and CO_2 by *Methanobacterium* and *Methenococcus*. The compounds are converted into elemental nitrogen, water and CO_2 by bacteria and fungi.

The facultative bacteria have an acid vacuolar degradation systems which is similar to a lysosomal system and proteolytic enzymes. Proteins are degraded in all living cells to respective amino acids and further into water and CO_2 or converted into organic cellular structures. During anaerobic conditions methenogenesis is the principal mechanism for stabilization of organic substances in anaerobic pond. Methenogenic organisms requires specific pH (6.0-7.0), and warm temperatures which may be achieve artificially in temperate zones. The methenogenesis is prevented at temperatures below 15°C.

Pilot model Oxidation ponds operated at 4, 6 and 8 ft. depth had a benthic pH 7.0. Maximum reductions in BOD and production of DO were recorded in the shallow pond (4 ft.). The performance of deeper ponds (6-8 ft) was significantly less. Variations in pH associated with BOD, DO and catalase activity at the eutrophic layers were significant (Hosetti and Patil, 1992) to flora as they assimilate and oxidize soluble organic matter. Oxygen is also utilised by algae during the hours of darkness for assimilation of CO_2. Many algal genera of stabilization ponds function both autotrophically and heterotrophically.

Ammonia Nitrogen

Fitzgerald and Rohlich (1958) observed 45.95 per cent ammonia-nitrogen removal in stabilization ponds. Higher levels of ammonia are toxic to algae within certain pH conditions. Ammonia toxicity may completely inhibit algal photosynthesis. This may depend on the species and previous exposure to ammonia. Inhibition of algal photosynthesis by high ammonia levels can give rise to anoxic conditions in ponds even with low BOD.

Ray (1978) succeeded in removing 95 per cent NH_4-N by passing 5 stage pond series system containing a filter, water hyacinth and duck weed in I, Zooplankton in II, glass shrimp in III and fish in IV pond respectively.

Sulfide

Sulfide is toxic to algae and bacteria and penetrates into the algal cell membrane in undissociated form (H_2S). In contrast to ammonia-nitrogen the hydrogen sulfide level increases as the pH is reduced. Sulfide levels above 8.0 mg/l inhibit algal photosynthesis. Prolonged higher levels of hydrogen sulfide will destroy green algae, and will be replaced by purple sulfur bacteria. They utilize sulfur as an electron donor, while photosynthesizing and assimilating organic substances. These bacteria photosynthesize, but there is no oxygen produced and the pond remains anoxic. Sulfide oxidation occurs more rapidly than it is generated in the ponds. When all the oxygen is used for respiration the photosynthetic bacteria are reduced and the algae populations are re-established. Hydrogen sulfide also inhibits the metabolic activity of the facultative and anaerobic bacteria.

Sulfur

Proteins and amino acids are degraded to ammonia nitrogen by proteolytic bacteria. In doing so hydrogen sulfide is released from sulfur containing amino acids (methaonine and cystine). The amount of H_2S released from sewage in this way varies between 5-8.0 mg/l and more when wastes are rich in proteins and animal slurry. Sulfide production in ponds is related to the sulfate contents of the wastes. Sulfide is converted into elemental sulfur by anaerobic sulfur bacteria and by sulfide oxidizing organisms namely *Beggiota* and *Chloroflexus species*. Purple and green sulfur bacteria perform similar functions in aerobic conditions.

Removal of Bacteria

Scientifically designed and well maintained ponds have been shown to be efficient in removing coliforms and indicator organism. Bacterial reduction to an extent of 99 per cent have been recorded in sewage ponds. It is experimentally proved that the increase in pH due to algal photosynthesis (above 9.0) has shown to be responsible for bacterial kill in oxidation pond.

Enzymes Activities in Oxidation Ponds

During the past five decades the routine parameters like pH, DO, BOD, NH_4-N, permanganate values, total suspended solids and bacterial assay have been used to monitor the quality of wastewaters with other factor like COD and inorganic elements. The bacterial count and BOD indicate the organic strength of the waste and analysis of these parameters requires maximum time for results. Hence to find out a simple analytical procedures using wastewater hydrolytic enzymes was suggested.

Biological wastes containing complex organic substances are degraded by the action of microbes. These microbes release several enzymes in the medium which degrade organic substances. A review on enzymatic evaluation of oxidation ponds performance includes the decomposition of H_2O_2 a toxic chemical produced in the cells by oxidative metabolism removed by catalase. This may be taken as an indirect measure of organic strength and microbial status of the waste.

Catalase, Protease, Phosphates, Amylase, Urease, Dehydrogenates, Keratinase and Lipase, found in sewage stabilization ponds are associated with the bacterial density. The per cent loss in catalase, Protease, Phosphates and Amylase, decreased with the increase in depths and organic load of the ponds. It also indicated that with the increase in aerobicity of the liquid, bacteria become more active and start removing hydrogen peroxide at a faster rate. As a result more catalase is produced in the cells and BOD is reduced. During anaerobic conditions more H_2O_2 was accumulated in the cells and hence high catalase was recorded. As the green algae grow and produce large amounts of oxygen which is used by the bacteria and increases their activity. It is responsible for the reduction in enzymes in effluents. This concept is evidenced by the reductions in BOD and nutrients in the effluents.

Measurements of phosphates activity may be used to determine changes in the composition of waste in the system. The measurement of protease in waste samples will help to monitor the levels of proteinaceous materials. High levels of amylase activity in water represent the quantity of glucose, maltose, dextrose, and starch in the wastes and reduction in amylase activity in the effluents has been reported. The enzymatic technique may not work properly in the ponds fed with strong natured chemical wastes. Under such conditions microbes don't sustain and release enzymes.

Pathogen Removal

Stabilization ponds are designed to kill pathogens and remove nutrients and BOD of water through an extended detention time. Water stabilization ponds are efficient in pathogen removal. It is also documented that instead of using single unit ponds, anaerobic, facultative and maturation ponds in series yield good results. A retention of 20 days will remove helminth eggs and protozoan cysts from the ponds. The per cent bacterial removal from sewage ponds varied between 95.8-99.0 in a tropical stabilization ponds in India. The study on virus removal in stabilization ponds was started in 1967. The reduction in enterovirus and *Vibrio cholera* from the maturation pond effluents was reported.

Heavy Metal Removal

Heavy metal levels which are toxic in fresh waters are detoxified in wastewater ponds. The wastewater stabilization ponds posses chelating and buffering activities. Heavy metals accumulate in the sludge when their concentration increases with the increase in sludge volume. This will affect the pond performance. In terms of detoxification the ponds function similar to other biological treatment systems. The long retention time, slow rate of sludge accumulation, neutral or alkaline pH and aerobic condition promote the complexation and precipitation of metals. Stabilization ponds are popular means of producing effluents with low levels of metals. The wastes rich in heavy metals are pretreated in anaerobic ponds whose performance is not dependent on algal activity. There are many species of algae tolerant to heavy metals.

Economic Uses of Effluents and Sludges

Oxidation pond is an ideal system for treating waste waters rich in organic substances. Stabilization or oxidation pond presents a habitat with a great ecological diversity, high pH and DO near the surface and low pH (around neutral) and anaerobic conditions at the bottom. Hence, the stabilization ponds offer versatile environmental conditions for developing and maintaining a stable biodegradation process under fluctuating conditions imposed by varied influent wastes. Possible uses of effluents and sludges are shown in Chart 10.2

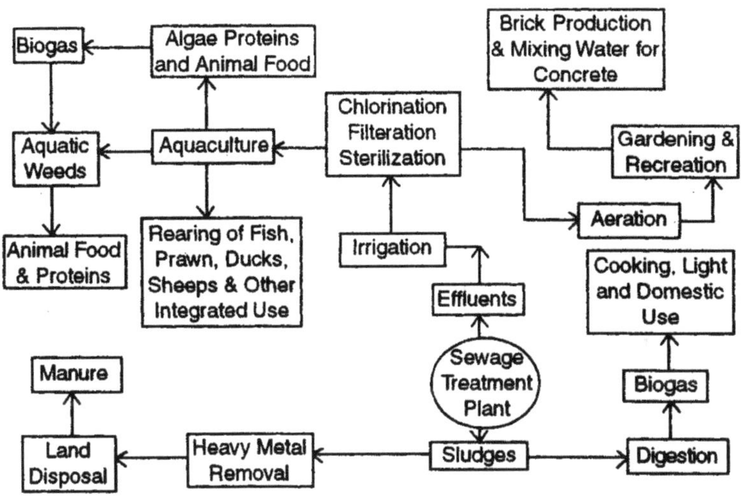

Chart 10.2: Flow chart showing use of utilization of effulents and sludges

Sewage Sludge

The normal purpose of a sewage treatment system is to receive the collected sewage and sewer transmitted wastewater, purify it and return the processed water to waterways already burdened by untreated discharges and other wastewater disposal facilities. The solids are retained as sludge treated to reduce its mass and volume.

Sludge from stabilization ponds in dried, stabilized in oxidation pits or mixed with animal manure and city garbage and applied on the land. The sludge used on cropland is in a liquid, sticky or crumbly state. The manuring effect of sewage sludge is governed by an unbalanced and variable ratio of nutrients. There is usually a deficiency in potassium and a greater proportion of nitrogen. Application of sludge enriches the soil and its fertility. To avoid damage to cropland and the contamination of ground water, sludge should be applied in small amounts at short intervals to balance sludge water infiltration and evaporation. The phosphorus content increases when sewage rich in organic material is added to soil. There is also a possible increase of microbial activity and reduction in soil erosion.

Effluent Irrigation

Sewage farming or transportation of sewage to rural areas or irrigation and disposal was practiced in Europe as early as 1559. It is common practice in Maharasthtra and Karnataka states of India, for more than 40 sugar mills to use effluents after treatment in ponds to irrigate sugarcane and coconut plantations. Stabilization pond effluent is the most immediately available water resource for irrigation in semiarid areas. The evaluation of stabilization ponds has been directed by irrigation needs which led to enlarged storage capacity and improvements in treatment efficiency.

Aquaculture

Aquaculture means growing animals and aquatic plants such as rice and sugarcane in irrigated tropical areas. Plankton, fish, prawn, swan, duck, goat and other integrated programmes for live stock production help resource utilization, reduce financial risk by diversifying crops and live stock whilst providing additional food and income. Philippines have a system of rearing goats and swans in fertilized ponds with sewage for rearing.

Production of fishes and prawns in sewage fertilized ponds is practiced in China, India, Malaysia, Hongkong, Taiwan, Singapore, Israel, Yugoslavia, Hungary and Germany. In India, fertilized pond polyculture rear native carps, marigala (*Cirrhina mrigala*), the rohu (*labio rohita*), the catla (*Catla catla*) and hybrid Chinese carp species.

Wastewater Fisheries

One of the main purposes of passing sewage through a pond is to oxidise the organic matter and to reduce the BOD of the pond effluents. The operational capacity of the pond is reduced during periods of O_2 depletion. Fish rearing is expected to overcome three cycles of high and low DO conditions without the expense of mechanical aeration of the water. Plankton density is reduced by this indirect means of harvesting. The fish occupy a position at the end of the nutrients chain that started with the introduction of organic wastes. Fish species are selected in relation to their ability to survive in a sewage fed pond environment and to feed on the various plankton growths. *Tilapia sp* were selected for growing in sewage fed waters because they tolerate low DO.

The other advantages are waste disposal problems are reduced, waste stabilization is achieved and phytoplankton feeding fish can be raised. Fish effectively use the plankton and bottom fauna of oxidation ponds. Ponds stocked with herbivores fish have a more biologically balanced system. Fish reduced plankton, BOD, benthic populations, average DO, whilst increasing pH levels. Sewage fed fish ponds recycle organic matter into algae and fish through the food web. The density of phyto and zooplanktonic organisms and water chemistry reflect the variety of the fishing.

Biogas

Anaerobic digestion is the most common process of producing energy from sewage sludge. The use of the biomass for biogas production has long been practised and India learnt the art from China. Biogas production covers the simultaneous production of multiple use of energy for cooking, lighting, waste pumping as well as an excellent compost to improve soil fertility. Biogas plants in rural areas run on animal wastes on in combination with night soil sludge obtained from oxidation ponds. Gas production is related to the size of the digester pit.

Algae Harvesting

Micro-algae harvested from high rate algal ponds are used for the extraction of proteins. Harvested algae have been used as proteins, animal or fish or poultry food and mechanical paddle are used to stir the ponds to improve algal growth. Separation of algae from final effluents is important both to improve the quality of effluents and to harvest the algae. The main problem in waste–water ponds is the high cost of separating algal cells from the effluents. Depending upon the climate, the high rate algal ponds can operate at the loading rate of 300-1000 mg/l and give over 1000 tonne dry biomass annually. This contains over 45 per cent protein and 40-80 per cent algae, the balance is bacteria and protozoa.

Animal Feed and Algal Proteins

The drum dried algal biomass concentrate is used for raising *Cyprinus carpino* and *Tilapia gallilea* and wastewater algae are also used for feeding freshwater shrimp, *Macrobrachium hainanence.* High rate algae ponds produce high quality animal feed as well as proteins. Possibilities of utilizing the algae obtained from high rate oxidation ponds is shown in Chart 10.3.

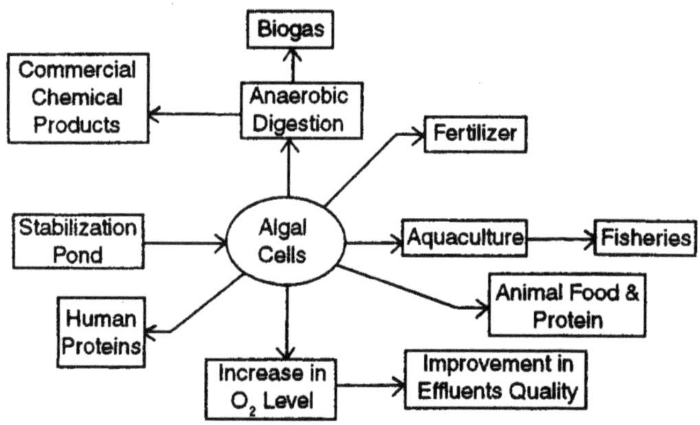

Chart 10.3: Algal utilization

Algae Removal

A number of methods of removal of algae have been investigated and it also include chemical treatment methods, sedimentation, floatation, straining, filtration and centrifuging. Among these chemical treatment was considered as the most promising.

With the efficient separation and recovery, algal biomass could be used for energy production as well as for producing animal supplements and fertilizers. Every method of algae removal from pond effluents has specific advantage and disadvantage, selection must be governed by the particular treatment situation.

Oxidation ponds are primarily used as tertiary treatment facilities especially to polish the effluents from conventional treatment plants. However, they are also used to treat the raw sewage or settled sewage and for treatment of industrial effluents.

Oxidation ponds have attracted the interest from all environmental scientists and sanitary engineers, because of their simplicity in operation, easy maintenance and high quality effluents. These ponds occur in more than 20 countries worldwide and are particularly popular in India, Israel and Canada.

The treatment occurs by natural, physical, chemical and biological processes in a shallow stabilization pond without any external interferences. Stabilization ponds are low cost waste recycling dynamic ecosystems comprising of bacteria, algae, virus, protozoa, rotifers, insects, crustacea and fungi. As these organisms can co-exist and compete with each other for survival, the organic content of the waste stabilized to produce high quality effluents. The economic value of the stabilization ponds is determined by the quality of effluents for irrigation, fish rearing, ducks, prawn production and algal protein extraction.

The disadvantage of these systems is the requirement of large area for construction of ponds. There is also a risk of potential odour problems, mosquito and midge nuisance and little control over the speed of effectiveness of the treatment process occurring in the ponds.

Plate 10.1: Bacteria and Protozoa found in waste oxidation pond effluents and sediments

Bacteria

1. *Zooglea ramigera*
2. *Spareotilus natans*
3 & 4 *Beggiota sp.*
5. *Thiotrix sp.*

Protozoa:

6. *Nitzscchia plaea*
7. *Fragillaria intermedia*
8. *Alternaria triticicola*
9. *Drecellaria ellici*
10. *Zoophagus insidians*
11. *Naegleria gruberi*
12. *Acanthamoeba hyalina*
13. *Rugipes bilzi*
14. *Amoeba gorgonia*
15. *Amoeba proteus*
16. *Chrysamoeba radians*
17. *Amoeba radiosa*
18. *Amoeba vespertilio*
19. *Pelomyxa palustris*
20. *Arcella vulgaris*
21. *Amoeba polypolid*
22. *Centropyxis aculeata*
23. *Leptochlamys ampullacea*
24. *Chlamydophrys stercocrea*
25. *Mastiamoeba longifilum*
26. *Thylacomonas compressa*
27. *Cryptomonas oveta*
28. *Ancyromonas contorta*
29. *Oikomonas termo*
30. *Oikomonas socialis*
31. *Egulena gracilis*
32. *Euglena acus*
33. *Phacus accuminatus*
34. *Bodo caudatus*
35. *Cercoboda angustus*
36. *Parenema trichophorum*
37. *Coleps hirtus*
38. *Spathidium spathula*
39. *Chilodonella uncinata*
40. *Chilodonella caudata*
41. *Didinium balbianii*
42. *Dileptus anser*
43. *Colpoda steinii*
44. *Balantidium coli*
45. *Ophryoglena collini*
46. *Paramecium aurelia*
47. *P. multimicronucleatum*
48. *Ctredoctema acathocrypta*
49. *Hypocomella phoronopsidis*
50. *Metopus fucus*
51. *Bryometopus sphagnii*
52a. & 52b. *Coenomorphamedusula*
53. *Lacrymaria coronata*
54. *Stylonchia mytilus*
55. *Onchodromopsisflexilis*
56. *Litonotus fasciola*
57. *Euplotes patella*
58. *Aspidisca lynchus*
60a. *Vorticella companula (adult)*
60b. *Vorticella companula (telotroch larva).*

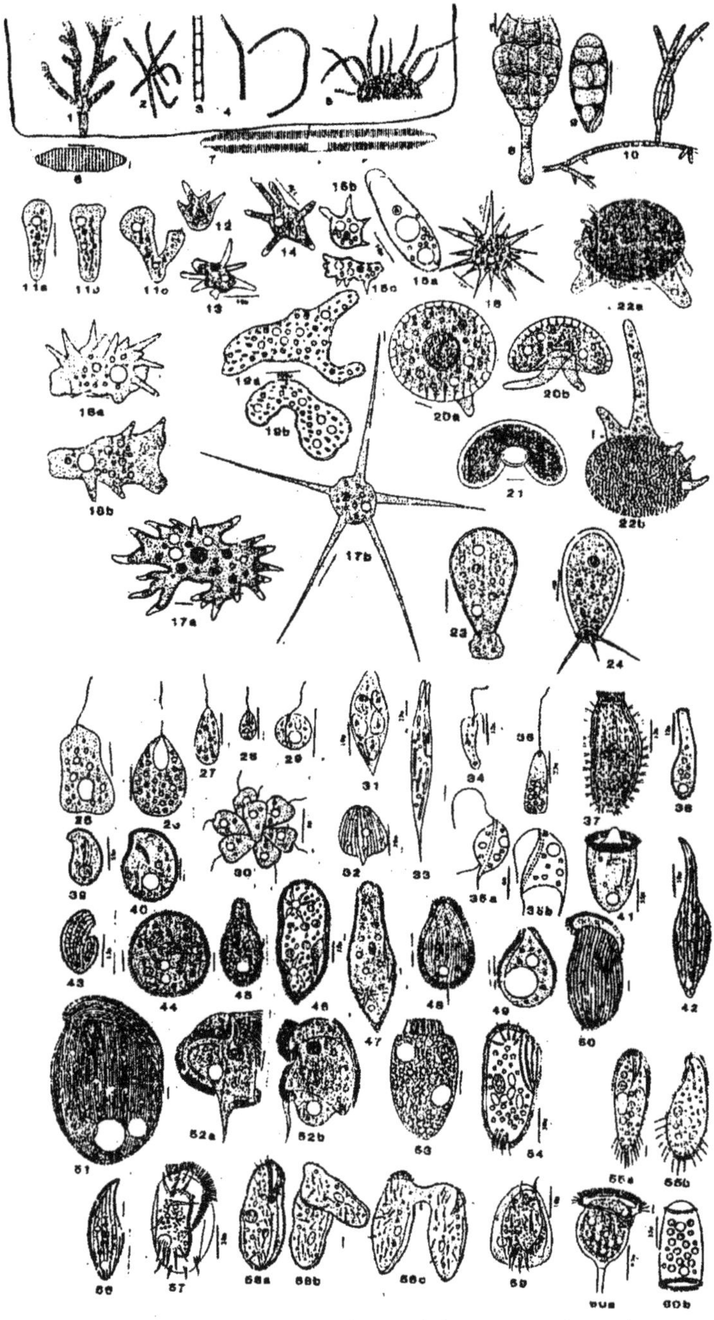

Plate 10.1: (Bacteria 1 to 5 and remaining are protozoa)

Plate 10.2: Wastewater algae commonly found in oxidation pond samples.

1. *Gleocapsa*
2. *Chlorella*
3. *Oscillatoria*
4. *Pyrobotrys*
5. *Phormidium*
6. *Phacus*
7. *Tetraedron*
8. *Chalmydomonas*
9. *Stigeoclonium*
10. *Lyngbya*
11. *Gomphonema*
12. *Chlorococcum*
13. *Arthrospria*
14. *Chlorogonium*
15. *Spirogyra*
16. *Nitzschia*
17. *Lepocinclis*
18. *Merismopedia*
19. *Carteria*
20. *Anabaena*
21. *Euglena*
22. *Schizothriz*
23. *Golenkinia*
24. *Chlamydomonas*
25. *Schroederia*
26. *Closterium*
27. *Scenedesmus*
28. *Closteriopsis*
29. *Massartia*
30. *Pteromonas*
31. *Cosmarium*
32. *Cryptomonas*
33. *Dictyosphaerium*
34. *Chodatella*
35. *Chromomonas*
36. *Ankistrodesmus*
37. *Closteridium*
38. *Ocrococcus*
39. *Chromulina*
40. *Vacuolaria*
41. *Spirulina*
42. *Planktosphaeria*
43. *Elakatothrix*
44. *Diacanthos*
45. *Polyedriopsis*

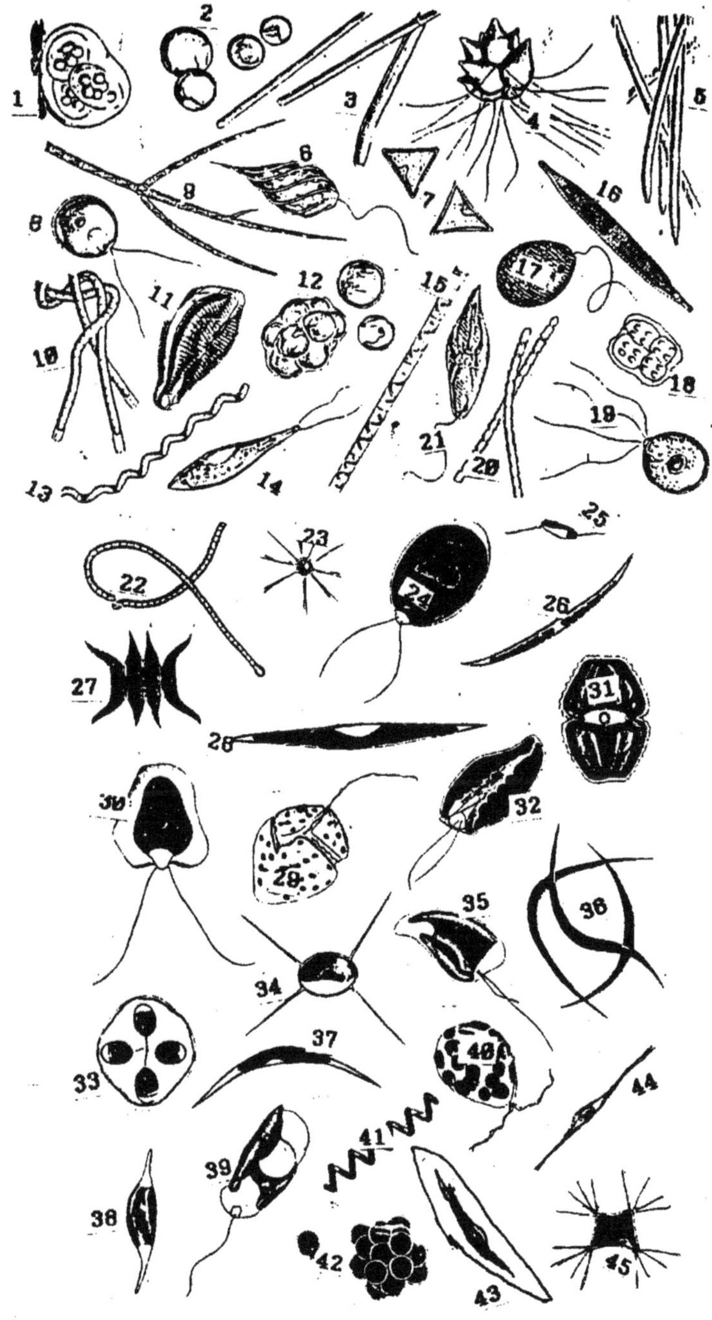

Plate 10.2: Wastewater algae

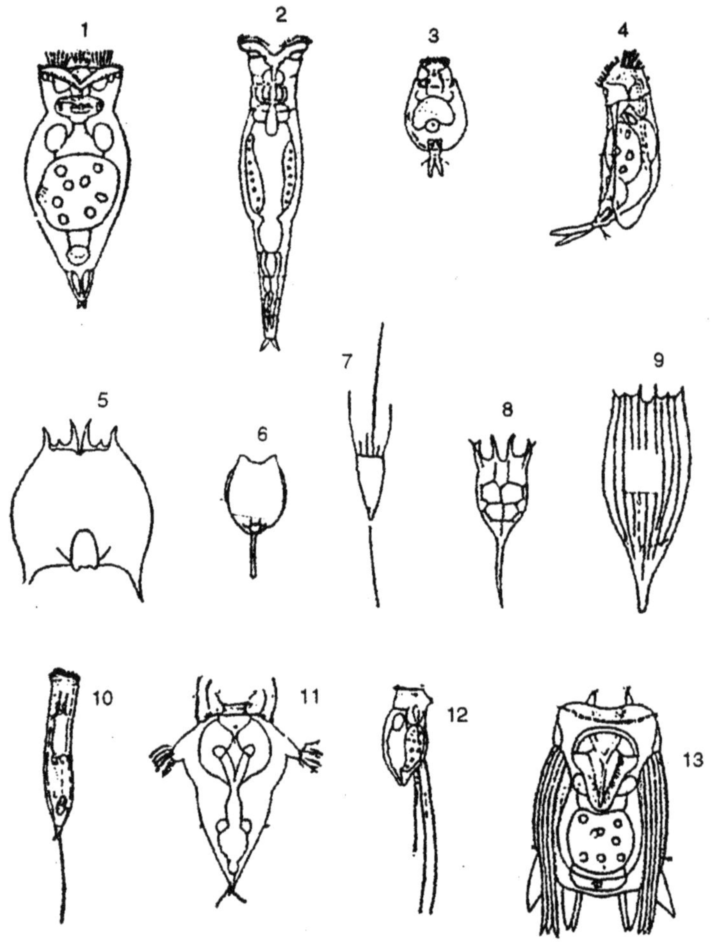

Plate 10.3: Wastewater rotifers

1. *Epiphanes (600 µm)*	2. *Philodina (400 µm)*
3. *Euclanis (250 µm)*	4. *Proales (450 µm)*
5. *Brachionus (200 µm)*	6. *Monostyla (150 µm)*
7. *Keratella (200 µm)*	8. *Nothoica (200 µm)*
9. *Trichocera (600 µm)*	10. *Synchaeta (260 µm)*
11. *Filina (150 µm)*	12. *Polyarthra (175 µm)*

Chapter 11
Marine and Freshwater Biotoxins

Introduction

This Chapter deals with some of the diseases caused by eating aquatic food sources. It has been known that certain fish and shellfish are poisonous and can cause diseases and death when they are eaten. The earliest report of this kind is from the Chinese Pharmacopoeia dates back to 2800 BC, recorded in injections used against eating pufferfish (Kao, 1966). The chemical and biological basis for these aquatic food borne toxicities have been studied over the last 50 years beginning with the pioneering work of Mayer and his school on the etiology of paralytic shellfish poisoning in California (Mayer *et al.* 1928). It is now clear that certain microscopic algae produce very potent toxins (Phytotoxins or algal toxins) which are chemical compounds of relatively lower molecular weight. The concentration of these toxins will be more in the marine or fresh water during algal bloom formations or red tides. The algal bloom or red tide is a phenomenon characterized by sudden, rapid proliferation of algae due to some environmental factors.

The toxin producing algal cells are taken up by the consumers directly as in the case of bivalve mollusks, or through several trophic levels as in fish. These food items are then eaten by man. Algal blooms and their toxicity were commonly observed throughout the world during the last 2-3 decades. The reasons are not yet deciphered completely, but it is believed that some climatic and hydrographic factors and manmade pollution of the water are responsible for the growth of toxin forming algae. The surveillance, detection and reporting systems of these have been improved in recent years. Since fish and shellfish constitute important component of the food supply and also the main source of protein for many inland and coastal communities, the apparently increasing contamination constitutes a specific chemical hazard which deserve special concern.

The toxins produced by cynophyceae formed altogether different problem which do not involve any vectors, in between the food chain. In this the microscopic toxic cells are brought into direct contact with the human skin during swimming in the sea or in the fresh water or the toxic cells may be transferred to human body through drinking water. Thus, the growth of blue green algae in drinking water sources pose added risk to human health.

The chemistry of these toxins is still only partly understood. The composition of Paralytic Shellfish poisoning (PSP) complex has been studied in detail and some individual components and their structure of PSP, Diarrhoeic Shellfish Poisoning (DSP) and Neurotoxic Shellfish Poisoning (NSP) complex are characterised. In the absence of sufficient knowledge of chemistry in the past, most measurements have been made by a bioassay using mice, a procedure that is non-specific in nature but is still the only method in practical use for the quality control of sea food. This method is non-sensitive enough for the analysis of clinical samples and specific and sensitive methods for detection should be developed based on new approaches.

The clinical features of intoxication of biotoxins are variable and the neurological and gastrointestinal symptoms are involved. In some cases the symptomolgy of poisons due to different biotoxins could be similar and specific analytic methods to help diagnosis are desired. The scope of this topic does not cover the aspects like, clapeotoxism, hallucinatory fish poisoning, scrombotoxin and shellfish allergen reactions, however, it explains only the chemistry and analysis of marine biotoxins (dinoflagellate toxins) and freshwater biotoxins (Cyanophycease toxins).

1. Paralytic Shellfish Poisoning (PSP)

Acute intoxication after shellfish consumption in a syndrome that has been known to man for several centuries. The etiology of the syndrome was studied in California in the 1920s following several episodes of the shellfish intoxications.

Chemical Properties

The chemistry of PSP has been reviewed by Shimizu (1978) and Schantz (1980). The PSP are group of toxins that are produced by dinoflagellates of the genus *Gonyaulax*. The first agent to be characterised was saxitoxin, which though initially discovered in shellfish in California, has also been found in the Alaskan butterclam, *Saxodomus giganteus* from which the chemical was derived. Saxitoxin has shown to be a dervivative of tetrahydropurine (Fig 11.1).

Saxitoxin is a white hygroscopic solid, rapidly soluble in water, sparingly soluble in ethanol and methanol but practically insoluble in non-polar organic solvents. It is a very basic substance with two titratable groups pK_a 8.2 and 11.5 and a relative molecular mass of 299 (Schantz *et al.*, 1961). Subsequently several other groups of PSP complex have been characterised chemically. Those include, 1-hydroxy saxitoxin (neosaxitoxin), 11-hydroxy saxitoxin sulfate and the 11-b-epimer of this compound. The last four compounds named as gonyautoxins II, III, I and IV respectively. In general PSP saxitoxins are heat stable at acidic pH, but very unstable and easily oxidised under alkaline conditions. Recently a novel group of PSP compounds with a sulfocarbomoyl group has been isolated from both dinoflagellates and shellfish (Kobayashi and Shimizu,

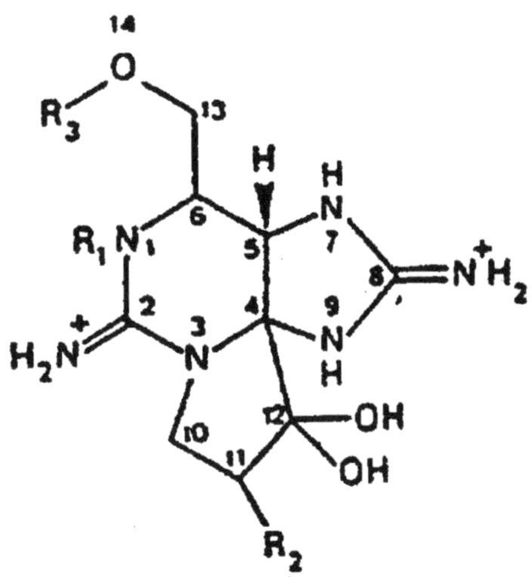

Fig 11.1: Structure paralytic shellfish poisons (PSP)

1981). These toxins have low toxicity until hydrolyzed to a more potent forms like gonyautoxin V and Sulfocarbomoyl gonyautoxin IV. Now totally about 13 PSP toxins have been known to us.

		R1	*R2*	*R3*
(1)	saxitoxin	– H	– H	$-C(=O)-NH_2$
(2)	neosaxitoxin	– OH	– H	$-C(=O)-NH_2$
(3)	gonyautoxin-I	– OH	– αOSO_3^-	$-C(=O)-NH_2$
(4)	gonyautoxin-II	– H	– αOSO_3^-	$-C(=O)-NH_2$
(5)	gonyautoxin-III	– H	– βOSO_3^-	$-C(=O)-NH_2$
(6)	gonyautoxin-IV	– H	– βOSO_3^-	$-C(=O)-NH_2$
(7)	gonyautoxin-V	– H	– H	$-C(=O)-N(H)-SO_3^-$
(8)	gonyautoxin-VI	– OH	– H	$-C(=O)-N(H)-SO_3^-$
(9)	gonyautoxin-VII	– H	– βOSO_3^-	$-C(=O)-N(H)-SO_3^-$
(10)	gonyautoxin-VIII epimer	– H	– αOSO_3^-	$-C(=O)-N(H)-SO_3^-$
(11)	sulfocarbamoyl gonyautoxin-I	– OH	– βOSO_3^-	$-C(=O)-N(H)-SO_3^-$
(12)	sulfocarbamoyl gonyautoxin-IV	– OH	– αOSO_3^-	$-C(=O)-N(H)-SO_3^-$
(13)	decarbamoyl-saxitoxin	– H	– H	– H

Methods of Analysis of PSP Compounds

The most common procedure for determination of PSP compound is a bioassay using mice, but this assay is not completely satisfactory, because of lack of sensitivity and pronounced variations. Several other alternative chemical methods are being developed.

Mouse Bioassay

During the investigation Sommer and Mayer (1937) have developed a bioassay method for PSP toxins. It consisted of intraperitoneal injection to mice with an acidified extract of shellfish tissue, and the determination of rapidity of death following the injection. By following the conditions for bioassay (Mouse weight, pH of the extract and salt concentration) and introducing a purified saxitoxin standard a fairly reliable routine procedure was established. Now so for it is the only method for assaying PSP that is in routine use by regulatory agencies all over the world.

Because different strains of mice differ in their sensitivity and susceptibility to the PSP toxins, the sensitivity of the mouse colony used in the bioassay must be determined by calculating a correction factor (CF, values) after intraperitoneal injection of the saxitoxin standard. The acidified extract of the shellfish are screen tested in a few mice in order to determine the dilution of the extract that will kill mice of 19-21 g body weight within 5-7 min., the conditions under which bioassay is most sensitive. In the main test, time to death is converted into mouse units (MU) from the concentration of toxin can be calculated using the CF values, assuming that PSP toxins are saxitoxins or its derivative. Saxitoxin levels as low as 400 μg/kg can be detected by the procedure, and the sensitivity is reduced with increased salt concentration (NaCl) in the extract. Other toxins can also kill the mouse within 6-10 minutes after intraperitoneal injection. Thus, the mouse bioassay can not distinguish PSP from tetrodotoxin but confusion between PSP and other toxins is unlikely in the cases where the origin of sample is known.

Immunological Assay

Johnson and Mulberry (1966) have developed an assay in which purified PSP (Saxitoxin) was conjugated with proteins by formaldehyde condensation, and the antitoxin to the conjugate was produced in rabbits. The antisera was used in hemoaggultination and benotonite flocculation tests, with PSP extracted from spiked samples from butterclams, causing variable inhibition in the tests. The hemoagglutination– inhibition test is slightly more sensitive than mouse assay, where the detectionlimit of the bentonite flocculation-inhibition test was comparable to that of mouse assay. Puffer fish did not react in the tests. The method suffers from some saturation phenomenon and is not useful for a quantitative determination of PSP in shellfish because, increasing amounts of PSP in extracts caused almost identical reactions.

Chemical Methods

Several spectrophotometric procedures involving colour reactions of PSP have been developed. The earliest procedure of McFarren *et al.*, 1958 was based on the Jaffes reaction, which involved a calorimetric reaction with the guanidine moiety. This procedure was inadequate, with a limit of detection of 1000-15000 μg/kg. A fluorimetric method has been developed for the detection of saxitoxin (Bates *et al.*, 1978). The fluorescent purine derivative of saxitoxin thus obtained was measured spectrophotometrically. Recently a high pressure liquid charomatrographic (HPLC) procedure is also uses the separation of toxins on a bonded phase cyano column and detection by fluorescence following alkaline oxidation with periodate. Six PSP components were identified and quantified with this method.

Sources and Occurrence

The PSP toxins occur in, and are produced by, certain unicellular marine algae known as dinoflagellates, members of the phylum Dinophyta. Most of the PSP producing dinoflagellates are found in the genus Gonyaulax. This includes: *G. tamerensis, G. catenella, G. acatenella, G. monilata, G. polyedra, G. excavata* (var of *G. tameransis*) and *Pyrodinium bahamense* (var *compressa*).

Dinoflagellates are single celled organisms 40-50 μ in diameter and are propelled by two flagellates. Some are bioluminescent. In addition to motile form *G. excavata* produces resting cysts (zygotes) as a result of sexual reproduction (Dale, 1977). Lacking flagellates these cysts sink and accumulate at the sediment water interface, they over winter there. Motile cells reproduce asexually by binary fission.

The toxin producing varieties of Gonyaulax vary in their toxic potential (Table 11.1).

Table 11.1: Relative toxicity of dinoflagellates of the genus Gonyaulax

Species	*Minimum number of cells required to produce one mouse unit of PSP (About 0.18 μg of PSP)*
G. polyedra	1.7×10^5
G. catenella	7×10^4
G. acatenella	6×10^3
G. tamarensis	4.5×10^3

Source: Prakash, 1967.

Form the biochemical point of view, these observations might indicate that PSP toxins are secondary metabolites similar to toxins produced by microscopic fungi but much is not known about the pathway how these are produced. PSP toxins are found not only in the motile cells *Gonyaulax* species, but also in the resting cells. Saxitoxin and neosaxitoxin have been isolated from strains of a cyanophyte, *Aphanizomenon flos-aquae*. In addition PSP components gonyautoxin I, II, III have been detected in a microalga *Jaina sp.* belonging to red algae (Rhodophyta). These red algae are eaten by crabs and snails, and PSP has been detected in these mollusks.

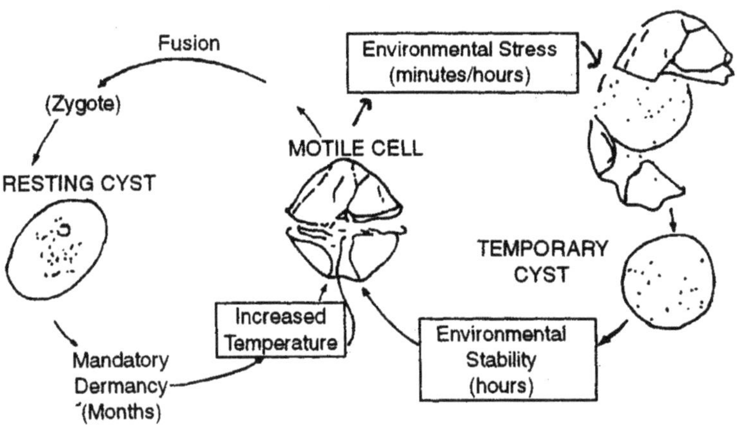

Fig. 11.2: Cyst Development in *Gonvaulax exceave*

Oceans and Blooms

Contamination of shellfish with PSP toxins has traditionally been associated with the appearance of algal blooms, the so called red tide. Dinoflagellates are able to reproduce asexually at high rates, under the influence of particular environmental conditions. When a population of dinoflagellates developed quickly forming a dense mass of cells to about 10^1 to 10^6 cells/litre, it is termed as a bloom. At levels of 10^6 cells/litre, the water can become coloured depending on the algal species, hence the name red tide, but it is important to realize that not all algal blooms are red coloured.

It is also to be noted that not all red tides are associated with toxic blooms and contamination of shellfish. As red tides are essentially a coastal phenomenon, it has been suggested that land drainage might play a role in their initiation. Concentration of chelators and trace metals may be involved. Thus, Anderson and Morrel (1978) reported that *G. tamarensis* was more sensitive to copper (II) ions than other members of the phytoplankton. Blooms may result directly from the rapid growth of algal populations, physical factors may transport the blooms to specific areas, where biological behaviour such as positive phototoxis can result in the formation of dense blooms. These phenomena may be triggered by meteorological factors such as rainfall and wind.

Occurrence in Sea Food

Mollusks

The PSP components are transferred to shellfish (mussels, clams and scallops) during filter feeding, a characteristic feature of bivalves. During this process food organism in the seawater such as *Gonyaulx* cells are transported from gills in the mantle cavity to the oesophagus and stomach and its associated diverticula. The highest concentration of PSP have been found in the digestive organs, also found in other organs like soft tissue of bivalves. Shellfish are generally not harmed by the presence of PSP toxins though some depression was observed in the filtration rate of bivalves exposed to dense mass of *Gonyaulax* cells. Black mussels (*Choromytilus meridionialis*) and white mussels (*Donax sera*) in South Africa have been found paralyzed during red tides associated with *G. catenella*.

Gastropods

Amounts of PSP have been detected in the digestive glands

of carnivores gastropods, such as rough Whelk, *Buccinum undulatum* under natural conditions as well as during experimental conditions. The species of turban cells, *Turbo argyrostoma, T. marmorata* and two species of top shells *Toctus nilotica maxima* and *T. pyramis* inhabiting coral reefs have been found to contain PSP toxins in the visceral regions. Saxitoxin, neosaxitoxin were predominant in the toxin profile. The highest toxicity recorded was 4000 µg/kg. As the PSP containing alga, *Jaina* sp. was present in the stomach of gastropods, presumed to be the source.

Crustacea

PSP was not found in the lobsters, *Homarus americanus* during red ride episodes, however, when lobsters were fed with PSP containing clams PSP was noted in the gut but not in the tissues. PSP may accumulate in crabs Xanthid crabs inhabiting coral reefs have been found to cause intoxication with high fatality rate in Fiji, Japan, Palau and in the Philippines. The species most frequently implicated in the poisoning in the crab, *Zosimus aenus* which accumulates saxitoxin and neosaxitoxin on eating the *Jaina sp.* algae.

Transmission from Zooplankton to Fish

Death episodes have been observed in fish (herring and sandlance) associated with blooms of *G. exacavata* in North sea and in the Bay of Fundy, Canada. As these fish feed on zooplankton but not on dinoflagellates, it has been suggested that the zooplankton feeding on dinoflagellates may act as vectors of PSP. Zooplankton collected during a bloom of toxic *G. excavata* contained PSP, even three weeks after the bloom peak.

Accumulation in Fish

PSP as measured by the mouse bioassay, has been detected in sand launce (970 µg/kg, involved in mass death of sea birds) (Nisbet, 1983). Pufferfish collected from areas with occasional episodes have been found to contain saxitoxin in the liver and amounting to 0.2 per cent of the total toxicity the component being tetrodotoxin (Yasumoto, 1980).

Exposure

All reported cases of PSP intoxication in free living animals and in human beings have been associated with alimentary canal to contaminated food. The shellfish most often to contain PSP are clam mussels and include members of the families, Mactridae (*Spisula solidissima*), Mycidae (*Mya arenaria*), Mytilidae (*Mytilus californianus, Mytilus edulis, Modiolus modiolun*) and Veneridae (*Protatheca staminea, Saxidomus giganteus, Saxidomus nuttalli*) (Halsted, 1978). Occasionally Scallops and oysters may be involved. In the pacific area some toxic crabs have been reported to contain PSP. Several countries have developed surveillance programmes for PSP contamination of shellfish. In the USA the shellfish areas are closed if the PSP level is more than 800 μg PSP/kg in the body of shellfish as measured by mouse assay. This 800 μg is ten times more than lowest level PSP that caused intoxication in the earlier outbreaks.

Effects on Animals

Fish

Fish kills by PSP produced during G. exacavata blooms, have been reported from Europe (North sea) and from Northeast coast of America (White, 1977). The Sand eel (*Ammodytes sp.*) and Herrigns (*Clupea harengus*) do not feed on dinoflagellates, and zooplankton appears to have acted as vector of PSP. No reports on the pathology and histochemistry are made, but Experimental oral administration to herrings was found to be rapidly fatal with oral LD_{50} for herrings, Salmon and cod in the range of 400-755 μg PSP/kg (White, 1977).

Sea Birds

Two items in the history death of sea birds due to an algal bloom has been recorded in the Northeast coast of England. That is in May 1968 a bloom of *G. excavata* associated with the mass death of shags, *Phalcrocorax aristotelis* relatively less number of terns, *Sterna spp.* and Cormorants, *Phalcrocorax carbo* (Coulson *et al.*, 1968) is reported. The pathological lesions observed in dead birds include extensive inflammation of alimentary canal and hemorrhages at the base of the brain and elsewhere.

As sea birds do not eat mussels, fish (sand eel) appear to have been the vector for PSP. A second episode in 1975 was recorded by monitoring collar ringed birds (Armstrong *et al.*, 1978). Thus, 62 to 64 per cent mortality rate of shags was associated with G. excavata bloom in the spring of 1975. Compared with the annual average mortality rate of 16 per cent, increased mortality rate were also observed in the herring gull *larus argetatus*, the cormorant and the fulmar, *Fulmarus glacialis*. A monitoring programme of PSP in the north east coast of England in mussels over a period of 9 years is given in the Table 11.2

Table 11.2: Maximum values of PSP in the mussels (*M. edulis*) from the northeast coast of England (PSP level µg/kg).

Year	*Number of mussels*
1968	1,000,000
1969	12,250
1970	8,000
1971	1,000
1972	400
1973	400
1974	5,500
1975	12,300
1976	1,750

Mass death in common terns *Sterna hirundo* was observed off Massachusetts coast in June 1978 (Nisbet, 1983). PSP was detected in the principal food of the terns in sand launch and the same time high level of PSP were also reported in the shell fish in the area.

Experimental Studies

The principles of the experiment was mouse assay developed by Sommer and Mayer 1937 is the measurement of time to death. The experiment was conducted to determine the LD_{50} for various animals. It is found that the PSP was less toxic when administered by oral route than parenterally. Increasing pH also reduced the toxicity.

Table 11.3: LD_{50} single dose of PSP of administration 4 hours test on mouse

Intravenous	*Male (3.4)*	*Female*
Intraperitioneal	10.0	8.0
Oral	263	

Table 11.4: Comparison of LD_{50} in different animals

Animal	LD_{50}
mouse	420
rat	212
monkey	400-800
cat	280
rabbit	200
dog	200
guinea pig	128
pigeon	100

The experiment were conducted by injecting the extract of PSP from Alaskan butter clam and the toxin was suspected to be saxitoxin.

Mode of Action

Both saxitoxin and tetrodotoxin have been used in the experiment. It include a wide spread blockage of impulse generation in peripheral nerves and skeletal muscles. Direct cardiac effects are minimal. In mammals these effects lead to Paralysis, respiratory depression and circulatory failure.

Effects on Man

The sign and symptoms of PSP in man may range from a slight tingling and numbness around the lips to complete paralysis and death from respiratory failure. Typically the tingling sensation around the lips gums and tongue develop within 5-10 min. of consumption. In moderate and severe cases this is followed by a feeling of numbness in the finger tips toes, and within 4-6 hours the same may progress to the arms, legs, and neck so that voluntary movements can made with great difficulty. In fatal cases death is usually caused by respiratory paralysis within 2-12 hours of consumption of the PSP containing food.

Mild

Tingling sensation or numbness around lips, gradually to spreading to face and neck, prickly sensation in fingertips and toes, headache, dizziness and nausea.

Moderate

In coherent speech, progression of prickly sensation to arms and legs, stiffness and incoordination of limbs general weakness and feeling of lightness slight respiratory difficulty and rapid pulse rate are recorded.

Severe

Muscular paralysis, pronounced respiratory difficulty choking sensation, high probability of death in the absence of ventilatory support. Sensitivity to PSP is variable that estimates of the human dose resulting in death range from 500 to 1000 µg/ kg body weight.

Ciguatera Toxins

A variety of fish inhabiting tropical and subtropical seas may become toxic and the ingestion of them cause an intoxication in human beings named as ciguatera. It is characterized by neurological and gastrointestinal symptoms. The term ciguatera is of Spanish origin derived from cigua, which is a Caribbean trivial name for a marine snail, *Tubro pica* that when eaten said to cause indigestion. The principal toxin is ciguatoxin, but some other toxic compounds were also identified. This disease has been known to man since 17th century and appears to be the most commonly associated with sea food toxins.

The chemical structure of the ciguatoxin are still not completely known. The toxin has been extracted from the moray eel liver and after elaborate purification, it has been obtained in pure crystalline form as a white solid (Scheuer, 1982). Ciguatoxin is a highly oxygenated liquid soluble in polar organic solvents but insoluble in water. The relative molecular mass is estimated to be 1111.7 and possible molecular formula is $C_{53}H_{77}NO_{24}$ or $C_{54}H_{78}O_{24}$ Other toxic components extracted from ciguatoxin fish are maitotoxin and scaritoxin. The chemical structure of the two compounds is unknown. However, the scaritoxin resembles ciguatoxin in chemical and some chromatographic properties.

Methods of Analysis for Foodstuffs

Biological Methods

For ciguatoxin a mouse injection test has been modified and the method consists of injecting serially diluted semipurified toxin extracts into mice and observing the mortality rate for 24 hr. The results obtained as mouse units and one mouse unit is defined as the amount of toxin that kills a mouse (20 g body weight) in 24 hours. The method does not distinguish between ciguatoxin and scaritoxin. A bioassay for ciguatoxin in fish has been developed. It is on the basis of feeding cats or mongoose a ration containing 100 g of fish to be tested per kg ration. The cat is less satisfactory because it may regurgitate part of the test meal. Test animals were observed for 48 hour with the response rated from 0 (No response) to 5 deaths in 48 hours. Recently a bioassay using mosquitoes (*Aedes aegypti*) has been developed. The procedure involve intrathorasic injection in mosquitoes of serially diluted extract from fish and the toxicity of the fish is expressed as mosquito LD_{50}.

Chemical Methods

A radioimmuno assay for ciguatoxin has been developed using antibodies produced against a conjugate of human serum albumin and ciguatoxin isolated from toxic morey eel. The results of the assay were correlated with those of assays on mouse, mongoose, guinea pig and all the three assay procedures showed good correlation, when ciguatoxin was present in fish tissues in high concentration.

Source Occurrence and Exposure

Algae

A dinoflagellate, *Gambierdiscus toxicus* has been identified as the source of ciguatoxin and maitotoxin. It is an armored dinoflagellate with two flagella living around coral reefs, closely attached to microalge, such as *Turbinaria ornata, Amphiroa sp.*, and *Jaina sp.* Ciguatoxin and amophitoxin have been isolated from the biodetritive layer on coral reefs from *G. toxicus* using the mouse assay and some biochemical characteristics as identification procedures. Fishes such as parrot fish (*Scarus gibbus*) and sturgeon fish (*Ctenchateus striatus*) are likely to contain these toxins which feed on layers of micro-organisms and detritus colonizing coral beds and thereby accumulate toxins. Subsequently strains of *G. toxicus* able to produce ciguatoxin and maitotoxin have repeatedly been isolated from microalgae such as *Halimeda sp. Penicillus sp. Acetabularia sp.* and *Gracillaria sp.* and from coral reefs of off the coast of Florida, USA.

Fish

More than 400 species of bony fishes have been reported to have caused ciguatera (Halstead, 1978). In general ciguatoxic species are limited to fish that feed on algae and the detritus of coral reefs, particularly the sturgeon fish (*Ctenochaetus striatus*) parrot fish (*Scarus gibbus*) and the larger reef carnivore that prey on these herbivores. Thus, larger carnivores such as morey eels, snappers, groupers, carrangs, spanish mackerels, certain inshore tunas are most toxic. Ciguatoxin has been detected in the contents of gut, liver and flesh of sturgeon fish. The highest content of ciguatoxin was found in the liver and viscera. It is also found in the turban shell a marine snail a food item that has occasionally caused ciguatera like intoxication in man. The natural disturbances of coral reefs include hurricanes and storms, manmade anchors or some constructions provide conditions for the growth of macroalgae to which *G. toxicus* cells are attached, resulting in increased dinoflagellate production. In addition there may be some other environmental factors which influence the growth of dinoflagellates.

Human Exposure

The only known pathway of human exposure is through the consumption of contaminated fish and marine snails. The intoxication may spread to other parts of the world through inter regional transport of fish. The data on absòrption, retention, distribution and metabolism of ciguatoxin in human beings or animal are not available.

Experimental Studies

Ciguatoxin extracted from fish is purified to some extant by solvent extraction and injected intravenously, or intra-peritoneally into mice produced acute effects characterised by diarrohea, retching, inactivity and death after convulsive spasms. Similar effects were observed after 48 hours when cats and mongoose were fed a ration containing ciguatoxin. Thus, ciguatoxin has a LD_{50} in mice of 0.45 µg/kg body weight. The pharmacological action of ciguatoxin is related to its direct effects on excitable membranes rather than to its anticholinestrase properties.

Effects on Man

The clinical picture is variable. Typically symptoms occur within 1-6 hours of ingestion of toxic fish. Initial symptoms include nausea, malaise and numbness and tingling of lips, tongue and throat. Patients may later develop vomiting, abdominal cramps, diarrohea, paraesthesia of the extremities, itching, and arthralgia. In more severe cases, ataxia, weakness, blurred vision, insomnia and hypertension may develop. The duration of illness is variable. Most of the patients recover within three days, but the symptoms persist for weeks or even years in severe cases. It is said that as little as 0.1 μg ciguatoxin can cause illness in an adult. In most severe cases death results from circulatory collapse or respiratory failure.

Epidemiology

Cases of ciguatera are commonly encountered throughout the Caribbean area and much of the Pacific area. The annual incidence of ciguatera intoxication on the virgin island are 3.6 per cent and household survey reveals 7.3 percent per thousand. Ciguatera outbreaks also occurred in Maryland, USA due to fish meal brought from Florida. In France through a fish meal imported from China, in Canada it was associated with the consumption of barracuda brought by tourists from Jamaica.

3. Tetrodotoxin

In comparison to other two biotoxins studied Tetrodotoxin is not produced by algae, but it may be by certain fishes and other animals. On the basis of number of human victims involved yearly, it is said that Tetrodotoxin poisoning is not an important public health problem. However, in contrast with most other algal intoxication studied below the illness caused by Tetrodotoxin is more severe and mortality rate is high. In recent years considerable interest in the tetrodotoxin has developed among the natural product chemists and neurobiologists. It is used as most important tool for selective blocking the sodium channel. The compound has been obtained from an extract of pufferfish viscera in the form of colourless crystal prisms that are slightly soluble in water. It is an amino perhydro quinazoline with a relative molecular weight of 319.

The toxin is unstable at pH levels above 8.5 and below 3.0

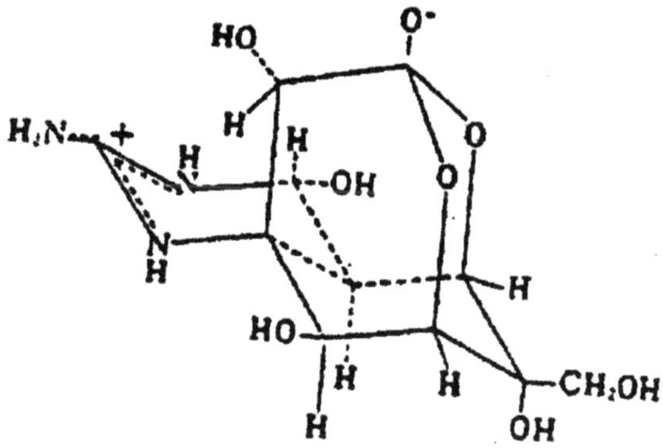

Fig. 11.3: Structure in Tetrodotoxin

Methods of Analysis in Food

Though having entirely different structure the tetrodotoxin induces toxicity similar to saxitoxin, and the mouse bioassay developed for PSP has also been utilised for Tetrodotoxin. In Japan a modification of the mouse bioassay is now in operation as an official method and a chemical assay has been developed on the basis of production of fluorescent compounds of Tetrodotoxin by alkali treatment.

Occurrence and Human Exposure

It is mainly found in the ovaries, liver and intestines of various species of puffer fish, lesser amount being found in the skin, the body muscle is usually free from toxin with the exception of *Lagocephalus lunaris* with muscle containing very high amounts of toxin. The most toxic members of the puffer fish are belonging to family tetrodontiadae, but not all the fishes in this family contain toxin. The most toxic ones are caught along the coasts of China and Japan and the meat of these fishes is considered as a delicacy. The amount of toxin present is related to the reproductive cycle and is greatest before spawning in the early summer. Tetrodoxin has also been found in the skin of a group of newts of the genus *Taricha* native to the Northern California and Southern Oregon in the USA. It has been detected in the skin of frog in the Central America of the genus *Atelopus*. It is also identified as a poison in the blue ringed Octopuses, *Hapalochlaena maculosa* and *H. lunata* of the Southern Australia involved in human fatalities through bites. It is also found in Japanese ivory shell *Babylonia japonica* and in the trumpet shell. There has not been any linking of tetrodotoxin in these animals with algae or microbes. But it is noteworthy that puffer fish raised in the ponds do not contain tetrodotoxin. Human exposure is generally limited to consumption of certain fish species. Identification is more difficult with frozen fish flesh. Its mode of action is similar to that of saxitoxin.

Effects on Animals

The effects of Tetrodotoxin either in contaminated fish or in pure form have been tested experimentally on a large variety of species. In animals with few exceptions the signs of intoxications are generally the same and comparable with those caused by PSP compounds. The effects involve peripheral neuromuscular system, which is paralyzed to different extent. There are three manifestations of tetrodoxin intoxication that appear to be some what different from PSP compounds. Tetrodotoxin is a highly potent agent, so that vomiting is frequently observed in both cats and dogs and also in man. It followed by neuromuscular paralysis.

Table 11.5: Comparative lethality of tetrodotoxin in differentanimals.

Animals	*Minimum lethal dose* *µg Tetrodotoxin /kg body weight*
Carp	2.0
Pigeon	2.7
Rat	2.7
Sparrow	4.0
Guinea pig	4.5
Frog	5.0
Hen	6.0
Mouse	8.0
Dog	8.0
Cat	9.0
Turtle	46

Source: Kao, 1966.

Effects on Man

In man onset of symptoms occurs from 10 to 45 minutes after ingestion, but may be delayed for 3 hour or more. Parasthesia may appear in the face and extremities and may be followed by sensations of lightness, floating or numbness nausea, vomiting, Diarrohea and epigastric pain may be present. Later respiratory symptoms become predominant with dyspnoea, rapid respiration and use of auxiliary muscles. Cyanosis and hypotension follow and convulsions and cardiac arrhythmia may occur. In most cases the victim remain consciousness until before death, which usually takes within 6 hours. In Japan during 1974-79 average was 60 cases with 20 deaths. In USA during 1970-74 two cases, in Italy 3 deaths are reported. Samples of puffer fish from China contained 0.5 to 30 mg tetrodotoxin per kg of tissue.

Neurotoxic Shellfish Poisoning (NSP)

A disease in human being associated with red rides involving the dinoflagellates *Gymnodinum breve* has been encountered around the coast of Florida, USA named as neurotoxic shell fish poisoning (NSP). According to symptoms and mode of exposure two syndromes are identified. The first one is NSP associated with the consumption of shell fish containing the cells or metabolites of *G. breve*. The symptoms are predominantly neruotoxic and resemble PSP except that paralysis does not occur. Second syndrome is NSP associated with aerosols of *G. breve* characterised by respiratory symptoms.

There is, however, much less data is available on this disease, compared with other diseases caused by dinoflagellates toxins. The *G. breve* toxins have never been chemically identified in food and in episodes involving human beings and only a limited number of toxicity studies on animals have been conducted.

Analytical Methods

More recently four chemical substances have been isolated from cultured cells of *G. Breve* and the structure was determined for three of them named brevitoxin B, Brevitoxin C, and GB-3. These chemicals have the same carbon chain structure containing 11 continuous ether rings.

The substances are soluble in organic solvents and are unstable in chloroform. The toxins are not fluorescent and these

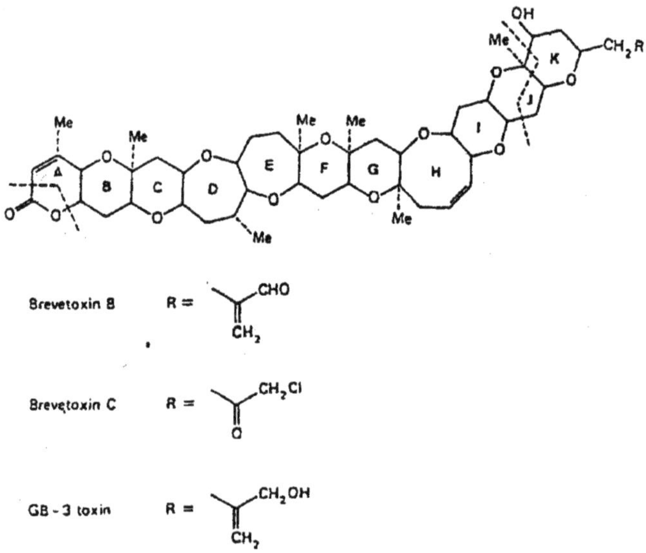

Fig. 11.4: Structure of *Gymnodinium breve* toxin

lack any chemical methods to analyse them. During purification process a fish bioassay has been employed.

Sources and Occurrence

The NSP toxic compounds have been isolated exclusively from *G. breve* a non–thecate dinoflagellates encountered around the coast of Florida, USA particularly during red tides which are initiated in the off shore waters in the late summer and autumn months. The iron content of water may be used as predictive guide as maximum iron has been observed immediately preceding to the red tides (Kim and Martin, 1974). Recently the alga is renamed as *Ptychodiscus brevis.*

Effects on Animals

A disease has been observed in fish and birds that is thought to be caused by *G. breve* toxins, because the episodes have occurred in close association with blooms of *G. breve* cells and similar symptoms were observed after feeding with *G. breve* cells to birds. Every 3-4 years blooms of *G. breve* occur on the west coast of Florida, causing massive fish kills. The fish mainly involved are tomtate fish, *Haemulon aurilineatum* and striped mullet, *Mugil cephalus*. The fragile necked cells of the *G. breve* rupture on passage through the bill processes of the fish releasing the toxins. Fish that swim into a red tide area will continue actively for a while, then will suddenly loose balance gasping at the surface before becoming passive on the bottom. Death occurs without pathologic lesions. Mass death of sea birds and fishes have been observed associated with red tides of off the west coast of Florida. The birds involved were double crusted cormorants, *Phalacrocorax auritus* and redbreasted mergansers, *Mergus merganser* and lesser scaup, *Aythya affinis*. The signs were weakness, reluctance to fly, clear nasal discharge, viscous oral discharge, oil gland dysfunction, diarrohoea, tachycardia and hypotension.

Experiments on Animals

In a study using white pekin ducklings force fed with tissues of clams contaminated with *G. breve* cells toxins. The birds showed ataxia and spastic movements within 3 days and died within 5 days (Forrester, *et al.*, 1977). Similar symptoms and leading to death were observed within 6-22 hour in male white leghorn chicks fed with tissues of Oysters, *Crassostrea virginica* that has been filter fed in the laboratory on toxic *G. breve* cells (Ray and Aldrich, 1965). A bioassay of mosquito fish *Gambusia affinis* was developed. The acute effects of crystalline preparation of brevitoxin B (named as T_{34}) have been observed during in vivo and in vitro studies.

Effects on Man

In human beings consuming shellfish contaminated with *G. breve* cells symptoms like parasthesia, alternating sensation of hot and cold, nausea, vomiting, diarrohoea and ataxia occur within 3 hours. Paralysis has not been observed and the NSP appears to be milder than PSP. In the USA during the period 1970-74 two outbreaks a total of 5 cases of NSP were recorded. Both associated with the consumption of clams and no death was occurred. The shellfish containing any detectable level of NSP per 100 g is determined by mouse assay and is considered as unsafe for human consumption.

Diarrhoeic Shellfish Poisoning

Introduction

An intoxication characterized by gastrointestinal disturbances, often occurring as outbreaks associated with the consumption of shellfish is named as diarrohoeic shellfish poison (DSP). DSP have been reported from several places of the earth, including the far East, Europe and South America. *Dinophysis fortii* is an armored dinoflagellates, it has been identified as a causative agent of DSP in Japan where *D. acuminata* is suspected of being the toxin producer in recent outbreaks of Netherlands. DSP has not been detected in the cells of *D. acuminata* because attempts were made to cultivate the organism isolated from dutch waters have failed. Cases of DSP in Chile were associated with the *D. acuta* though the detailed information is not available, occurrence of DSP toxins, okadaic acid has been confirmed in a benthic dinoflagellate, *Prorocentrum lima* (Murakami, *et al.*, 1982). It has been noted that in the presence of *D. fortii* at a low cell density of 200 cells/litre, mussels and scallops become toxic enough to affect man. The infestation period in Japan ranges from April to September.

Chemical Properties

It is suspected that there may be about 9 chemical components act as toxins of DSP in shellfish tissue. About five of them have been studied in detail. These toxins are classified into two groups namely, okadaic acid and its derivatives named as dinophysistoxins and the other group, polyether lactones named as pectonotoxins.

The chemical structure of dinophysistoxin-2 is not known because of its limited availability, while pectonotoxin-3, -4, -5 closely related to pectonotoxin-1 in chemical structure.

Analysis

A mouse bioassay using intraperitoneal injections of the toxin extracts and a 24 hour observation periods is being used as a regulatory measure to monitor shellfish toxicity in Japan and shellfish with DSP toxin more than 50 μg/kg are banned for harvesting or sale.

Effects on Animals

The mice injected with toxic extracts of DSP shellfish intraperitoneally show inactivation and general weakness, and die within 30 minutes to 48 hours depending on the dose given. Chick are less sensitive. Vomiting was observed in cats fed with

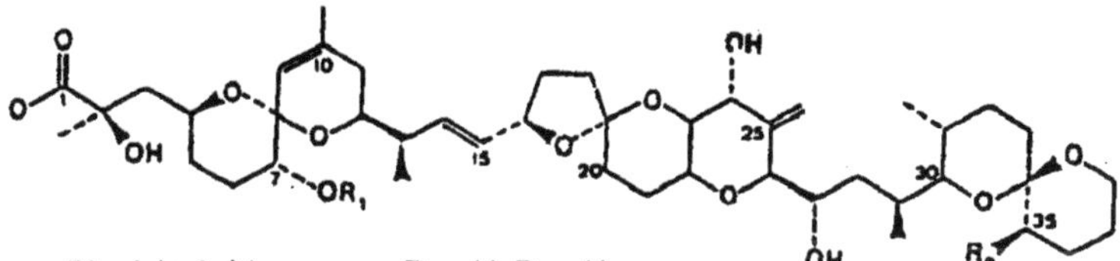

Okadaic Acid : $R_1 = H, R_2 = H$
Dinophysistoxin – 1 : $R_1 = H, R_2 = CH3$
Dinophysistoxin – 3 : $R_1 = acyl, R_2 = CH_3$

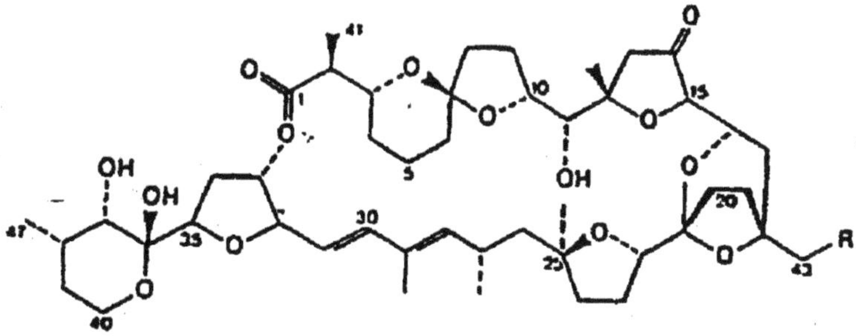

Pectenotoxin – 1: R = OH
Pectenotoxin – 2: R = H

Fig. 11.5: Structure of Diarrohoeic Sehllfish Poison (DSP)

toxic mussels and scallops. In rats (*Rattus norvagicus*) DSP toxic shellfish induced diarrohoea and reduced appetite.

Effects on Man

During the period 1976-82 more than 1300 people were diagnosed as DSP cases in Japan. The sign and symptoms were diarrohoea (92 per cent), nausea (80 per cent), vomiting (79 per cent), abdominal pain (53 per cent) and chill (10 per cent). The time from consumption of shellfish to onset of the symptoms is 30 minutes to several hours. Cases of gastrointestinal disorder were encountered in Chile in 1970 and 1971 associated with the blooms of *Dinophysis sp.* (Averia, 1979).

Cynophyte Toxins

Sources and Properties

Contact with the filamentous cyanophyte *Lyngbya majuscula* when swimming in the sea can result in a type of dermatitis called

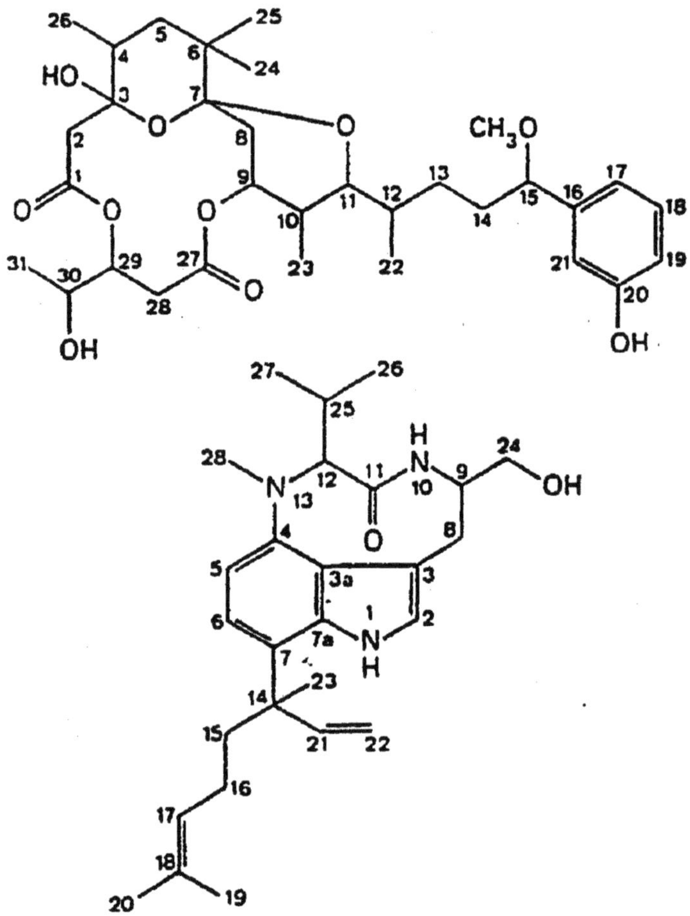

Fig. 11.6: Structure of Cynophyte Toxins

Debromoaplysiatoxin (Top) and
Lyngbyatoxin (Below)

swimmers itch or seaweed dermatitis. This was reported from Hawaii and Japan. Two skin toxic components have been isolated from *L. majuscula*, those are debromoaplysiatoxin and lyngbyatoxin. The crystalline form of debromoaplysiatoxin is a colourless niddle with a MP 105-107°C and molecular mass of 592. Lyngbyatoxin is a tan coloured gummy solid in crystalline form with a relative molecular mass of 437 (Fig. 11.6). Debromotoxin is also isolated from two other species from Oscillatoriaceae i.e. *O. nigroviridis* and *Schizothrix calcicola*.

Effects on Animals

The toxicity of debromoaplysiatoxin was studied on the skin of mice and rabbits. In the mice there were petechial hemorrhages with 1 hour of application of 0.5per cent debromoaplysiatoxin; by 24 hour the area was pale and oedematous, and by 5 days there was a firm crust which took 2-3 weeks to heal. Histologically the application of 5 percent toxin include changes over 24 hour was almost destruction of epidermis, edema, coagulation of collagen and infiltration through dermis. The dermal and subdermal muscles showed mural fibrinoid necrosis. In the rabbit also there a dose effect relationship and the lowest dose induced a reaction that was mild erythema. Histologically the changes are similar to those seen in mice. No details about the skin testing of lyngbyatoxin are published.

Effects on Man

Cases of acute dermatitis after contact with *L. majuscula* have been reported from Hawaii and Japan. In Hawaii more than 125 cases received for treatment and hundreds of mild cases were suspected in the year 1958. The clinical picture included gradual onset on itching and burning within a few minutes to few hours after swimming in the sea where algal presence is suspected. Visible dermatitis and redness develops after 3-8 hours followed by blisters and deep desquamation. The eruption affects the region of the body not covered.

Histologically, the lesions were described as acute, vesicular dermatitis with superficial desquamation, oedema of the epidermis and occasionally the eruptions contain polymorphonuclear leucocytes and erythrocytes. These lesions have been reproduced by applying solutions of debromoaplysiatoxin on the skin. It is alcohol soluble compound the lowest concentration developed dermatitis was 0.05 per cent solution.

Freshwater Cyanophyte Toxins

Although the diseases induced by fresh water cyanophytes in drinking water are known. The first known chemical substance was anatoxin in 1977 (Devlin *et al.*, 1977). There are indications of intoxications induced in animals as field cases in farm animals and as experimentally induced disease in a variety of animal species are reported.

Most of the information on toxic freshwater cyanophytes is concerned with the three of the following species. *Microcystis aeruginosa, Anabaena flos aquae* and *Aphanizomenon flos aquae. M. aerruginosa* is a coccoid blue green algae and the onset of microcystis bloom in lakes is correlated with the temperature (19-20°) is also supported by experiments. The other two species are filamentos blue green algae. All the three species are ubiquitous and toxic bloom of these have been reported from many countries.

Sources Properties and Exposure

Microcystis aeruginosa toxin: Several toxic preparations have been isolated from the alga which contain peptides, carbohydrates and other compounds. Two compounds of the low relative molecular masses have recently been isolated, one is a peptide with a hepatotoxic properties and the other is causing respiratory arrest in mice.

Anabaena flos-aquae: Anatoxin-a extracted from the alga. When this was administered intraperitoneally caused acute effects including signs of paralysis and death are reported. Chemically it is alkaloid with a relative molecular mass 165. It is analysed by HPLC. Further three more toxic preparations viz., anatoxin b, c and d were also isolated. Anatoxin -a acts as potent nicotinic against paralyzing peripheral muscles.

Aphanizomenon Flos-aquae

Recently two toxins similar to saxitoxin and neosaxitoxin on the basis of proper chromatography were isolated. The possibility of human exposure to this toxin is through recreational and municipal water supplies, however, no reports are available.

Effects on Animals

Field cases of *Microcystis* intoxication in farm animals particularly in cattle have been reported. These cases were characterised by hemorrhages, photosensatization and liver damage including necrosis of hepatocytes and moderate proliferation of bile duct epithelium. Similar symptoms were observed on experimental animals like monkey. Gonadotoxic and embryotoxic effects as well as mutagenic effect were observed in the rat bone marrow.

Anabaena Flos-aquae

Field cases of sudden death of cattles associated with the drinking of water from lakes containing blooms have been reported (Hammar 1968). Death occurred with few hours of ingestion of the algal mass and the characteristics indicated respiratory failure. These cells administered orally to calves, rats, ducks and gold fishes caused death as a result of respiratory failure.

Aphanizomenon Flos-aquae

The peritoneal injection of a toxic extract of the cells to sheephead minnows (*Cyprinodon varigatus* 0.5 µg/kg), in mice (8 µg/kg) caused death, however, it has not affected the copepods, cladocerans and ostracodes.

Effects on Man

During the blooms of cyanophytes (*Microcystis sp., Anabaena sp.*, and *Apahanizomenon sps)* in Canadian lakes in 1959 many cases of acute death in domestic animals like dogs, cats, cattles, horses associated with drinking of lake water have been reported. In addition 12 people became ill after swimming in the lakes, with headache, nausea, and gastrointestinal upsets. In August 1975 a water borne outbreak of gastrointestinal disease occurred in Sewickley, Pennsylvania, USA affecting 62 per cent of the population (size 8000). This water contained more than 100000 cells per ml, dominated by *Schizithrix calcicola* and *Lyngbya spp.*

Bibliography

Aalgarswami, K. 1987b. Cultured Pearls-prodcution and quality. *Bull. Cent. Mar. Fish. Res. Inst.*, Vol. 39. pp. 107-115.

Adoni, M. 1985. *Work Book on Limonology*. Pratibha Publications, Sagar, MP. India.

Alagarswami, K. 1974. Development of Cultured Pearls in India, *Current Science, vol. 43*. pp. 205-207.

Alagarswami, K. 1987a. Tehcnology of Cultured Pearl Producton. *Bull. Cent. Mar. Fish. Res. Inst. Vol. 39*. pp. 98-105.

Alagarswami, K. 1991. *Production of Cultured Pearls*. Indian Council of Agricultural Research, New Delhi. pp. 1-112.

Alagarswami, K., Dharmaraj, S., Chellam, A. and Velvayudhan, T.S. 1989. Larval and Juvenile rearing of black lip pearl Oysters *Pinctata margaritifera* (Linnaeus). *Aquaculture, vol. 76*. pp. 43-56.

Alagarswmi, K., Dharmraj, S., Velayudhan, T.S., Chellam, A., and Victor, A.C.C. 1983a. On controlled spawning of Indian pearl oyster Pinctada fucata (Gould). *Proceedings of the symposium on Coastal Aquaculture, vol. 2*. pp. 590-597. Marine Biological Assoc

Allen, E. 1934. Neglected factors in the development of thermal springs. U.S. Nat. Acad. Sci Proc., 20: 345-349.

Allen, E. and A.L. Day 1935. *Hot springs of Yellowstones National Park*. Carnegi Instit. Washington Publ. No. 466-525.

Anderson, D.M. and Morrel, P.M.M. 1978. Copper sensitivity of *Gonyaulax tamarensis. Limnol Oceanogr., 23:* 283-295.

Ansari, Z.A. and Parulekar, A.H. 1988. Community structure of meiobenthos from a tropical estuary. *Ind. J. Mar. Sci. 27:* 362-366.

Armstrong, I.H., Coluson, J.C., Hawkey, P. and Hudson, H.M. 1978. Further mass seabird deaths from paralytic shellfish poisoning. *Br. Birds, 72:* 58-68.

Avaria, S.P. 1979. Red tides off the coast of Chile. In *Toxic dinoflagellate blooms,* Ed. by Taylor, D.L. and Seliger, H.H. Amsterdam Elseiver, 161-164.

Bates, H.A. and Rapoport, H. 1975. A chemical assay for saxitoxin the paralytic shellfish poison. *J. Agri. Food Chem. 23:* 237-238.

Bates, H.A., Kostraiken R., and Rapoport, H. 1978. A chemical assay for saxitoxin : Improvements and modifications. *J. Agri. Food Chem. 26:* 252-254.

Bhimachar, B.S. and George P.C. 1950. *Proc. Indi. Acad. Sci. 31:* 339p.

Bhosale, L.J. 1978. Distribution of trace elements in the leaves of mangroves, *Indian J. Mar. Sci., 8:* 58-59.

Birasal, N.R. 1996. Primary productivity of the Supa Reservoir during the first filling stage, Dandeli. *J. Envi. Biol., 17:* 127-132.

Boyle, P.R. 1969. The survival of osmotic stress by *Sypharochiton pelliserpentis. Biol Bull 136:* 154-185.

Brooks, J.L. 1950. Speciation in ancient lakes. *Quart. review Biol., 25:* 30-60.

Brooks, T.L. 1950. Specification in lakes. *Quart. Rev. Biol. 25:* 30-60 and 131-176.

Champan, V.J. 1976. Mangrove biography. In *Wet coastal systems,* Elsevier Scientific publishing Co., Amsterdam. 1-29. Eds. Chapman, V.J.

Chellam, A., Dharamraj, S., and Velayudhan, T.S. 1987c. Biofouling, Boring and Predation of Pearl Oyster. *Bull. Cent. Mar. Fish Res. Inst. vol. 39.* pp. 92-97.

Chellam, A., Dharmaraj, S., and Victor, A.C.C. 1987b. Expeimental Sea-Ranching of Pearl Oyster in the Gulf of Mannar. *Bull. Cent. Mar. Fish. Res. Inst., vol. 39.*

Chellam, A., Velayudhan, T.S., and Victor, A.C.C. 1987a. Pearl Oyster Farming. *Bull. Cent. Mar. Fish. Res. Inst., vol. 39.* pp. 72-76.

Chellam, A., Victor, A.C.C., and Gandhi, A.D. 1983b. Larval rearing and production of spat of pearl oyster *Pinctata fucata* (Gould) *Aquaculture. Vol. 34:* 287-309.

Clesceri, L.S., Greenberg, A.E. and A.D. Eaten, 1998. *Standard methods for examination of water and wastewater* 20th Ed. APHA, AWWA and WEF Washington D.C. USA. (Edited))

Coulson, J.D., Potts, G.R., Deans, I.R. and Fraser, S.M. 1968. Mortality of Shags and other sea birds caused by paralytic shellfish poison. *Nature (London), 220:* 23-24.

Cull, B.C. 1977. Ecology of Marine benthos, *University of South Carolina Press, Columbia* 467p.

Dagaonkar, A. and D.N. Saxena 1992. Physico-chemical and biological characterization of a temple tank, Kaila Sagar, Gwalior, India. *J. Hydrobiol 8:* 11-19.

Dale, B. 1977. Cyst of toxic red tide dinoflagellate, *Gonyaulax excavata* (Braarud) *Balech from Oslof Jorden, Norway, Sarsia, 63:* 29-34.

Daniel, S.D., and Durairaj, S. 1993. On commercial aspects of cultured pearl production from pearl oyster *Pinctata fucata* (Gould) in India, *Sea Food Export Journal. vol. 25*(8). pp. 25-38.

Datta Munshi, J. and Datta J.S. Munshi 1995. *Fundamentals of Freshwater Biology.* Narendra Publishing House, Delhi, India.

Datta Munshi, J.S., Biligrami, K.S., Datta Munshi, J. and R.N. Yadav, 1984. Gases from thermal springs of Bihar. *Ind. Nat. Acad. Sci. Letters 7:* 291-297.

Davis, P.S. 1969. Physiological ecology of Patella III. Desication effects. *Mar. Biol. Assoc., UK 49:* 291-304.

Dayton, P.K. 1971. Competence, disturbance and community organization: The provision of and subsequent utilization of space in a rocky intertidal community. *Ecol. Monogr., 41:* 351-389.

Delsman, H.C. 1939. Preliminary Plankton Investigations in the Java Sea. *Treubia 2:* 139-181.

Devlin, J. P. , Edwards, O.E., Gorham, P. R., Hunter, N.R., Pike, R.K. and Stavric B. 1977. Anatoxin-a, a toxic alkaloid from *Anabaena flos aquae* NRC 44th *Can. J. Chem. 55:* 1367-1371.

Dharmaraj, S., Velayudhan, T.S., Chellam, A., Victor, A.C.C., Gopinath, G. 1991. Hatchery Production of Pearl Oyster Spat: *Pinctata fucata. CMFRI Special Publication. No. 49.* pp.1-31.

Doubilet, D. 1997. Black Pearls of French Polynesia. *National Geographic, vol. 191*(6). pp. 30-37.

Doubilet, E. 1991. Australia's Magnificent Pearls. *National Geographic. vol. 176*(12). pp. 109-123.

Edmondson, W.T. 1974. Book review. *Limnol. Oceanong. 19:* 369-375.

Esser, W. 1978. Uber die sessiler organisman auf die selb streinigungsgeschwindigkeit in Fliessgewassern, *Gas-u Wasserfach, 119:* 582-586.

Fenchel, T. 1978. *Ecology of micro and meiobenthos.* Ann. Review. Ecol. and Systematics, 9: 199-121.

Fenchel, T. 1968. *The ecology of marine microbenthos. II. The food of marine benthic ciliates.* Ophelia, 5: 73-121.

Fitzerald, G.P. and C.A. Rohlich, 1958. An evaluation of stabilization pond literature, *Sew. Ind. Wastes, 30:* 1213-1224.

Forrester, D.J., Gaskin, J.M., White, F.H., Thompson, N.P., Quick, J.A., Henderson, G.E. and Woodard, J.E. 1977. An epizootic of waterfowl associated with a red tide episode in Florida. *J. Wild Dis., 13:* 160-167.

Ganapathi, S.V. 1974. Biochemical studies on algal-bacterial symbiosis in high rate algae ponds with varying detention time and algae, Final Tech. Report Amer. PL-480 Project, Baroda.

Ganapthi, S.V. 1972. Organic produciton in seven types of aquatic ecosystems in India. In *Tropical ecology with an emphasis on organic production.* Eds. Golley, P.M. and Golley F.B. Athens. Grece.

Geldrich, E.E. 1972 : Waterborne pathogens, In *Water Pollution microbiology,* Ed. Mitchell, R. 207-241. Wiley Interscience, N.Y.

Gerlach, S.A. 1971. *On the importance of marine meiofauna for benthos communities,* Oecologia, 6: 176-190.

Goldman, C.R., Mason, D.T. and J.E. Hobbie 1967. Two antartic desert lakes. *Limnol. Oceanogr. 12:* 295-311.

Gunter, G., Williams, R.H., Davis, C.C. and F.G.W. Smith 1948. Catastrophic mass mortality of animals and coincident phytoplankton bloom on the west coast of Florida. *Ecological Monographs.* 18: 310-324.

Halstead, B.W. 1978. *Poisonous and venomous marine animals of the world.* Princeton, Darwin Press, U.K.

Hammer, U.T. 1968. Toxic bluegreen alage in Saskatchewan. *Can. Vet. J., 9:* 221-229.

Holm, N.A. and A.D. McIntyre (Eds) 1984. Methods for the study of marine benthos. *II Ed., Blackwell Scientific Publications, Oxford.*

Hora, H.L. and T.V.R. Pillay 1962. *Handbook of fish culture on the Indo-pacific region,* FAO Fish. Biol. Techn., No. 14.

Hosett, B.B. and H.S. Patil 1992. Enzyme activity : An indicator of effluent quality of sewage stabilization ponds, *Bioresopurce Technology, 39:* 215-220.

Hosetti B.B. and S. Frost, 1998. A review of the control of the biological waste treatment in stabilization ponds, *Critical reviews in Envrionmental Science and Technology, 28:* (2) 193-218.

Hosetti, B.B. 1987. *Studies on the use of some important groups of microbes in the treatment of wastewaters*. Doctoral thesis, Karnatak University, Dharwad.

Hosetti, B.B. and H.S. Patil 1988. Use of catalase activity in the evaluation of physico-chemical parameters in a polluted river. Cited in *Threatened Tebitats*. Jagmandir Book Publishers, New Delhi 393-404, Eds. Agarwal, U.P. and L.D. Chaturvedi.

Hosetti, B.B. and H.S. Patil, 1987. Performance of sewage stabilization ponds at different depths. *Water, Air and Soil Pollution, 34:* 191-198.

Hosetti, B.B. and Patil, H.S. 1988. Enzymatic evaluation of oxidation pond performance. *Int Revue.* ges Hydrobiol., 73; 641-650.

Hosetti, B.B., Patil, H.S. and S.S. Rodgi, 1984. Vertical distribution of microbes in oxidation ponds in relation to physico-chemical parameters, *Environment and Ecology, 2:* 159-161.

Hosetti, B.B., Yalamani, A.H. and S.M. Gaddad 1988. BOD removal capabilities of some microorganisms. *Environment and Ecology, 6:* 503-507.

Hulings, N.C. and J.S. Grey (Eds) 1971. Proceeding of the first international conference on meiofauna. *Smithsonian contributions of Zoology, No.76* 205p.

Hutchenson, G.E. 1957. *A treatise on limnology Vol I. Geogrpahy, physics and chemistry,* John wiley and sons Inc., N.Y., USA.

Hutchinson, G. E. 1967. *A treatise in Limonology. II introduction to lake biology and limnoplankton.* John Wiley, N.Y. USA.

Hynes, H.B.N. 1960. *The biology of polluted waters,* Liverpool University Press, Liverpool, U.K.

Ingole, B.S. and A.H. Parulekar 1988. Role of salinity in structuring the meiofauna of a tropical estuarine beach: Field evidence. *Ind. J. Mar. Sci. 27:* 356-361.

James, P.S.B.R., and Narasimham, K.A. 1994. Saltwater Pearl Fisheries and Pearl culture in India: An update. *Aquaculture Magazine. vol. 20*(4). pp. 41-49.

Johanson, M.W. 1939. The correlation of water movements and dispersal of pelagic larval stages of certain littoral animal sand crab *Emerita. J. Mar.* Res. *2:* 236-245.

Johnson H.M. and Mulberry, G. 1966. Paralytic shellfish poison : serological assay by passive haemoagglutination and bentonite flocculations. *Nature (Lond), 211:* 747-748.

Joseph, M.M. 1990. *Aquaculture in Asia.* Published by the Asian Fisheries Society, Indian Branch, Mangalore. pp. 77-78.

Joshi, G.V. 1997. To Coax a Friendly Osyter. *Deccan Herald.* 22nd April, Tuseday. pp. 19.

Kant, S. and A.K. Raina 1990. Limonological studies on two ponds in Jammu II. Physioc-chemical parameters. *J. Environ. Biol., 11:* 137-144.

Kanwisher, J. 1957. Freezing and dying in intertidal algae. *Biol. Bull., 113:* 275-285.

Kao, C.Y. 1966. Tetrodotoxin, saxitoxin and their significance in the study of excitation phenomena. *Pharamcol. Rev., 18:* 997-1049.

Khatri, T.C. 1984a. Diurnal variations in physico-chemical parameters during summer season in Lakhotia lake of Pali Rajasthan. *Environ and Ecol., 2:* 95-97.

Khatri, T.C. 1984b. Seasonal variation in ecosystem of the Lakhotia lake in Rajasthan. *Ind. J. Fish., 1:* 122-129.

Khatri, T.C. 1985. Physico-chemical features of Idukki reservoir, Kerala during premonsoon period. *Environ and Ecol., 3:* 134-137.

Khatri, T.C. 1985. Seasonal distribution of phytoplankton in a tropical lake of western Rajasthan. *J. Econ. Taxon. Bot., 7:* 697-707.

Khatri, T.C. 1987. Seasonal distribution of phytoplankton in Idukki reservoir of Kerala, India. *Environment and Ecol., 5:* 71-73.

Kim, Y.S. and Martin, D.F. 1974. Interrelationship of Peace River parameters as a basis of the iron index. A predective guide to the Florida red tide. *Water Res., 8:* 607-616.

Kiran, B.R. Research fellow: Kuvempu University, Personal communication.

Krishnamoorthi, S. R. and S.G. Bharathi 1994. Distribution of iron in the surface waters of the polluted river, Kali around Dandeli area, Karnataka. *Envrion and Ecol., 12:* 275-278.

Kusuma, M.S., Neelakantan, B. and R.G. Konnur 1988. Plankton distribution in the Kali estuary, Karwar, Central Westcost of India, *Environment and Ecology 6:* 115-119.

Lubchenco, J. 1978. Plant species diversity in marine intertidal community: Importance of herbivore and algal competition abilities. *Amer. Not. 112:* 23-29.

Lubchenco, J. and Menge B. 1978. Community development and persistance in a low rocky intertidal zone. *Ecol. Monogr., 59:* 67-94.

Marimuttu, R. and V. Krishnamurthy 1985. Diurnal studies of phytoplankton ecology in two tropical ponds. *Proc. Nat. Symp. Pure and Appl. Limnol., Ed. Adoni, Bull Bot. Soc. Sagar, 32:* 240-243.

Matsui, Y. 1958. *Aspects of the environment of pearl culture ground and the problem of hybridization in the genus Pinctada. In: A.A. Buzzati (Ed.)* Perspective Press, Barkeley. pp. 519-531.

Mayer, K.E., Sommers K.F. and Schoenholz P. 1928. Mussel poisoning. *J. Prev Med., 2:* 365-394.

Mcfarren, E.F., Schantz., E.J., Campbell, J.E. and Lewis, K.H. 1958. Chemical determination of paralytic shellfish poison. *J. Assoc. Off Anal. Chem., 41:* 168-177.

McIntyre, A.D. 1969. *Ecology of marine meiobenthos.* Biol. Review. 44: 245-290.

Meron A., Rebhum, M. and B. Sless 1965. Quality changes as function of detention time in waste stabilization ponds, *J. Wat. Pollut. Cont. Fed., 37:* 1617-1670.

Metcalf, T.G. 1978. Indicators for viruses natural waters. In *Water pollution microbiology* Ed. Mitchell, R. Vol 2. 301-324. Wiley N.Y.

Mookarjee, D. 1971. The Kosi responds to treatment. *Ind. J. power and river valley Deptt.* 317-318.

Murakami, T. Oshima, Y. and Yasumoto, T. 1982. Identification of okadaic acid as a toxic component of a marine dinoflagellate, *Prorocentrum lima. Bull. Jpn. Soc. Sci. Fish., 48:* 69-72.

Muus, B.J. 1967. The fauna of Dannish estuaries and lagoons. *Meddr. Danm Fish-og Hauvenders 5:* 1-316.

Nag, N.C. 1931. Investigations on the radioactivity of the hot springs of Rajgir. *Trans. Bose Instit., 7:* 319p

Navalkar, B.S. 1951. Succession of mangrove vegetation of Bombay and salsete islands, *J. Bombay Nat History Soc., 50:* 157-160.

Neelakantan, B., Naik, U.G., Prasad, P.N. and M.S. Kusuma 1988. Hydrobiology of kali estuary and the adjoing brakish water systems of Karwar. *Envrionment and Ecology 6:* 139-147.

Nirmalkumari, K.R. and N. Balakrishnan 1984. Relative abundance of predatory aquatic insect in the Checkai canal, Trivendru, India. *Comp. Physiol. and Ecol. 9:* 105-113.

Nisbet, I.C.T. 1983. Paralytic shellfish poisoning effects on breeding terns. *Condor, 85:* 338-345.

Nygaard, G. 1944. Hydrobiology of some Danish ponds and lakes. K. *Daniske Vidensk Susk*1: 1-293.

Odum, E.P. 1969. Strategy for ecosystem development. *Science 164:* 262-270.

Padmavathi, P. and M.K. Durgaprasad 1997. Impact of unwanted fish on the water quality, plankton and fish production in a carp culcture ponds in the environs of Kolleru lake. *J. Environ. Biol., 18:* 79-93.

Palmer, C.M. 1969. *Keys to water quality indicator organisms.* Department of Interior Federal Matter. Central Admn. Ohio USA.

Patil, H.S. 1979. *Comparative ecology of various species of algae with a view to select the best for increasing the efficiency of sewage oxidation ponds.* Ph.D. Thesis, Karnataka University, Dharwad, India.

Pillai, V.K. 1995. *Manual in Pearl Oyster Seed Production Farming and Pearl Culture.* CMFRI Special Publication, No. 63. pp. 1-41.

Prain, D. 1903. Flora of the Sunderbans, *Rec. Bot. Surv. India. 2:* 231-240.

Prakash, A. 1967. Growth and toxicity of a marine dinoflagellate, *Gonyaulax tamarensis. J. Fish Res. Board Can., 24:* 121-128.

Prasad, R.R. 1958. Notes on the occurrence and feeding of *Noctiluca* and their effects on plankton community and fisheries. *Proc. Ind.* Acad. Sci. 47: 331-337.

Raman, R.P. 1991. Enchanting Pearls, *Science Reporter, Vol. 28* (3). pp. 19-23.

Ramesh, A.M., Katti, R.J., Hariharan, V., Bhatt, C. and T.R.C. Gupta 1992: Phytoplankton of the coastal waters of Mangalore. *Environment and Ecology 10.* 310-316.

Ray, D. 1978. Upgrading stabilization pond effluents by river hyacinth culture. *J. Wat. Pollut. Cont. Fed. 50:* 833-845.

Ray, S.M. and Aldrich, D.V. 1965. *Gymnodinium breve* : induction of shellfish poisoning in chicks. *Science, 148:* 1748-1749.

Refaele, P. 1997. In Search of the Perfect Pearl. *Reader's Digest. vol. 151*(906). pp. 80-90.

Reid, G.K. 1961. *Ecology of inland water and estuaries.* Von Nostrand Reinhold Publishing Co., N.Y.

Rodhe, A. 1969. Crystalization of eutrophication concept in Northern europe. In *Eutrophication, causes consequences and corectives* 50-64. National Academy of Science Washington, D.C.

Ruttner, F. 1931. Hydrographiche and hydrochemische beobactungen auf Java, Sumatra and Bali. *Arch Hydrobiol. Suppl., 8: 197-454.*

Ruttner, F. 1953. *Fundamental of limnology.* University of Tornoto Press, USA.

Saha, S.K. 1980. Limnological survey of thermal springs of Bhimbandh, Monghyr. Ph. D., thesis Bhagalpur University, Bhagalpur, Bihar, India.

Saha, S.K.and J. Datta Munshi 1980. Biota in relation to physico-chemical environment along thermal gradient of hot spring of Bhimbandh, Bihar. Pol rch. Hydrobiol., 27: 337-347.

Salvanayagam, M., Peterraj, T. and A. Jebaneasan 1988. Interactions of hydrological parameters with the distribution of aquatic hemipterans in river Coovum, Madras. *J. Envrion. Biol.*, 9: 361-370.

Schantz, E.J. 1980. Phytotoxins from dinoflagellates. *Pure Appl. Chem., 52:* 183-188.

Schantz, E.J., Mold, J.D., Howard, W.L., Bowden, J.P., Strnager, D.W., Lynch, J.M., Wintersteiner, O.P., Ducher, J.D., Ealters, D.R. and Riegel, B. 1961. Paralytic shellfish poison. VIII. Some chemical physical properties of purified clam and mussel poison *Can. J. Chem., 39:* 2117-2123

Scheuer, P.J. 1982. Marine ecology : Some chemical aspects. *Naturwissenschaften, 69:* 528-533.

Sharma, K. 1994. Glittering Peral. *Science Reporter. vol. 31*(7). pp. 55.

Shimizy, Y. 1978. Dinoflagellate toxins. In *Marine natural products,* Edited by Scheuer, P.J.N.Y. Academic Press, 1-42.

Shivappa, D., Ravi, G.S. and B.B. Hosetti 2000: Ecological cahracteristics of water from Savalanga tank, Shimoga. Ind. J. Envrion and Ecoplan., 3: *(1) 137-141.*

Sommer, H. and Mayer, K.F. 1937. Paralytic shellfish poison. *Arch. Pathol., 24:* 560-598.

Subramanyan, R. 1946. A systematic account of marine plankton of Madras coast Proc. Ind. Acad. Sci. 24: 85-197.

Theinemann, A. 1925. Die Binnengewassre Mitteleuropas: Eine Limnologische Einfuhrung Die Binnengewasser. 1: 1-225.

Thompson, K. 1975. Productivity of Cyperus papyrus in photosynthesis and productivity in different environments. Ed. by Copper, J.P. Cambridge University Press.

Timko, P. 1976. Sand dollars as suspension feeder: A new description of feeding in *Dendrogaster excentricus. Biol. Bull., 151: 247-259.*

Trivedi, R.K. 1999. Pollution Science College, Karad, Personal communication.

Untawala, A. G., Sayeeda Wafar and T.G. Jagtap 1982. Application of remote sensing technique to study the distribution of mangroves along the estuaries of Goa. *In Wetland ecology and management Proc. I Int Wetland conference.* 10-17 September, 1980. 51-57.

Victor, A.C.C. 1993. Sea ranching of pearl oyster. *Fisheries Information Service,* No. 124. pp. 8-13.

Victor, A.C.C. and Velayudhan, T.S. 1996. Pearl Oyster farming and peral production. *Bull. Cent. Mar. Fish. Res. Inst., vol. 48.* pp. 56-63.

Wada, K. 1973. Modern and traditional methods of pearl culture. *Underwater Journal, vol. 5*(1). pp. 28-33.

Weich, P.S. 1938. *Limonology.* Mc Graw Hill Book Co. N.Y. USA.

Weidemann, V.E. and E.C. Bold, 1985. Heterotrophic growth of selected stabilization pond algae. *J. Phycol., 1:* 66p.

Welch, P.S. 1952. Limnology *II Edition. McGraw Hill Book Co., N.Y.*

White, A.W. 1977. Dinoflagellate toxins as probable cause of an atlantic herring, *Clupea herengus harengus* kill, and pteropods as apparent vector. *J. Fish. Res. Board Can., 34:* 2421-2424.

Yang, C.T. and Mitra, A., Ghosh, S.N., and Bose, A.N. 1991. *Coastal Aquaculture Engineering.* Oxford and IBH Publishing Co. Pvt. Ltd., New Delhi.

Yasumoto, T. 1980. Ciguatera In *Igaku no ayumi, 112:* 886 (In Japanese)

Yentsch, C.M. and Incze, L.S. 1980. Accumulation of algal biotoxins in Mussels. In *Mussel culture in North America,* edited by Luts, R and Incze, L.S.N.Y. Elseiver, 223-246.

www.ingramcontent.com/pod-product-compliance
Ingram Content Group UK Ltd.
Pitfield, Milton Keynes, MK11 3LW, UK
UKHW020204290926
15470UKWH00001B/7

9 789351 309888